INTERNATIONAL UNION OF PURE AND APPLIED CHEMISTRY

ANALYTICAL CHEMISTRY DIVISION
COMMISSION ON SOLUBILITY DATA

SOLUBILITY DATA SERIES

Volume 14

ALKALINE EARTH METAL HALATES

SOLUBILITY DATA SERIES

Selected Volumes in Preparation

SOLUBILITY DATA SERIES

Volume 14

ALKALINE EARTH METAL HALATES

Volume Editors

HIROSHI MIYAMOTO
Niigata University
Niigata, Japan

MARK SALOMON
US Army ERADCOM
Fort Monmouth, NJ, USA

H. LAWRENCE CLEVER
Emory University
Atlanta, GA, USA

Evaluator

HIROSHI MIYAMOTO
Niigata University
Niigata, Japan

Compilers

HIROSHI MIYAMOTO
Niigata University
Niigata, Japan

MARK SALOMON
US Army ERADCOM
Fort Monmouth, NJ, USA

BRUNO SCROSATI
University of Rome
Rome, Italy

JOHN W. LORIMER
University of Western Ontario
London, Ontario, Canada

PERGAMON PRESS

OXFORD · NEW YORK · TORONTO · SYDNEY · PARIS · FRANKFURT

U.K.	Pergamon Press Ltd., Headington Hill Hall, Oxford OX3 0BW, England
U.S.A.	Pergamon Press Inc., Maxwell House, Fairview Park, Elmsford, New York 10523, U.S.A.
CANADA	Pergamon Press Canada Ltd., Suite 104, 150 Consumers Road, Willowdale, Ontario M2J 1P9, Canada
AUSTRALIA	Pergamon Press (Aust.) Pty. Ltd., P.O. Box 544, Potts Point, N.S.W. 2011, Australia
FRANCE	Pergamon Press SARL, 24 rue des Ecoles, 75240 Paris, Cedex 05, France
FEDERAL REPUBLIC OF GERMANY	Pergamon Press GmbH, Hammerweg 6, D-6242 Kronberg-Taunus, Federal Republic of Germany

First edition 1983

Library of Congress Cataloging in Publication Data

Main entry under title:
Alkaline earth metal halates.
(Solubility data series ; v. 14)
Includes bibliographical references and indexes.
1. Alkaline earth halates—Solubility. I. Miyamoto, Hiroshi, 1917- . II. Salomon, Mark. III. Clever, H. Lawrence. IV. Series.
QD172.A42A37 1983 546'.39 83-17224

British Library Cataloguing in Publication Data

Alkaline earth metal halates.—(Solubility data series; v. 14)
1. Halides—Solubility—Tables
I. Miyamoto, Hiroshi II. Salomon, Mark
III. Clever, H. Lawrence IV. Series
541.3'42 QD165

ISBN 0 08 029212 7
ISSN 0191-5622

In order to make this volume available as economically and as rapidly as possible the authors' typescripts have been reproduced in their original forms. This method unfortunately has its typographical limitations but it is hoped that they in no way distract the reader.

Printed in Great Britain by A. Wheaton & Co. Ltd., Exeter

CONTENTS

SOLUBILITY DATA SERIES

Editor-in-Chief

INTERNATIONAL UNION OF PURE AND APPLIED CHEMISTRY

IUPAC Secretariat: Bank Court Chambers, 2-3 Pound Way,
Cowley Centre, Oxford OX4 3YF, UK

FOREWORD

If the knowledge is undigested or simply wrong, more is not better.

How to communicate and disseminate numerical data effectively in chemical science and technology has been a problem of serious and growing concern to IUPAC, the International Union of Pure and Applied Chemistry, for the last two decades. The steadily expanding volume of numerical information, the formulation of new interdisciplinary areas in which chemistry is a partner, and the links between these and existing traditional subdisciplines in chemistry, along with an increasing number of users, have been considered as urgent aspects of the information problem in general, and of the numerical data problem in particular.

Among the several numerical data projects initiated and operated by various IUPAC commissions, the *Solubility Data Project* is probably one of the most ambitious ones. It is concerned with preparing a comprehensive critical compilation of data on solubilities in all physical systems, of gases, liquids and solids. Both the basic and applied branches of almost all scientific disciplines require a knowledge of solubilities as a function of solvent, temperature and pressure. Solubility data are basic to the fundamental understanding of processes relevant to agronomy, biology, chemistry, geology and oceanography, medicine and pharmacology, and metallurgy and materials science. Knowledge of solubility is very frequently of great importance to such diverse practical applications as drug dosage and drug solubility in biological fluids, anesthesiology, corrosion by dissolution of metals, properties of glasses, ceramics, concretes and coatings, phase relations in the formation of minerals and alloys, the deposits of minerals and radioactive fission products from ocean waters, the composition of ground waters, and the requirements of oxygen and other gases in life support systems.

The widespread relevance of solubility data to many branches and disciplines of science, medicine, technology and engineering, and the difficulty of recovering solubility data from the literature, lead to the proliferation of published data in an ever increasing number of scientific and technical primary sources. The sheer volume of data has overcome the capacity of the classical secondary and tertiary services to respond effectively.

While the proportion of secondary services of the review article type is generally increasing due to the rapid growth of all forms of primary literature, the review articles become more limited in scope, more specialized. The disturbing phenomenon is that in some disciplines, certainly in chemistry, authors are reluctant to treat even those limited-in-scope reviews exhaustively. There is a trend to preselect the literature, sometimes under the pretext of reducing it to manageable size. The crucial problem with such preselection - as far as numerical data are concerned - is that there is no indication as to whether the material was excluded by design or by a less than thorough literature search. We are equally concerned that most current secondary sources, critical in character as they may be, give scant attention to numerical data.

On the other hand, tertiary sources - handbooks, reference books, and other tabulated and graphical compilations - as they exist today, are comprehensive but, as a rule, uncritical. They usually attempt to cover whole disciplines, thus obviously are superficial in treatment. Since they command a wide market, we believe that their service to advancement of science is at least questionable. Additionally, the change which is taking place in the generation of new and diversified numerical data, and the rate at which this is done, is not reflected in an increased third-level service. The emergence of new tertiary literature sources does not parallel the shift that has occurred in the primary literature.

With the status of current secondary and tertiary services being as briefly stated above, the innovative approach of the *Solubility Data Project* is that its compilation and critical evaluation work involve consolidation and reprocessing services when both activities are based on intellectual and scholarly reworking of information from primary sources. It comprises compact compilation, rationalization and simplification, and the fitting of isolated numerical data into a critically evaluated general framework.

The *Solubility Data Project* has developed a mechanism which involves a number of innovations in exploiting the literature fully, and which contains new elements of a more imaginative approach for transfer of reliable information from primary to secondary/tertiary sources. *The fundamental trend of the Solubility Data Project is toward integration of secondary and tertiary services with the objective of producing in-depth critical analysis and evaluation which are characteristic to secondary services, in a scope as broad as conventional tertiary services.*

Fundamental to the philosophy of the project is the recognition that the basic element of strength is the active participation of career scientists in it. Consolidating primary data, producing a truly critically-evaluated set of numerical data, and synthesizing data in a meaningful relationship are demands considered worthy of the efforts of top scientists. Career scientists, who themselves contribute to science by their involvement in active scientific research, are the backbone of the project. The scholarly work is commissioned to recognized authorities, involving a process of careful selection in the best tradition of IUPAC. This selection in turn is the key to the quality of the output. These top experts are expected to view their specific topics dispassionately, paying equal attention to their own contributions and to those of their peers. They digest literature data into a coherent story by weeding out what is wrong from what is believed to be right. To fulfill this task, the evaluator must cover *all* relevant open literature. No reference is excluded by design and every effort is made to detect every bit of relevant primary source. Poor quality or wrong data are mentioned and explicitly disqualified as such. In fact, it is only when the reliable data are presented alongside the unreliable data that proper justice can be done. The user is bound to have incomparably more confidence in a succinct evaluative commentary and a comprehensive review with a complete bibliography to both good and poor data.

It is the standard practice that any given solute-solvent system consists of two essential parts: I. Critical Evaluation and Recommended Values, and II. Compiled Data Sheets.

The Critical Evaluation part gives the following information:

(i) a verbal text of evaluation which discusses the numerical solubility information appearing in the primary sources located in the literature. The evaluation text concerns primarily the quality of data after consideration of the purity of the materials and their characterization, the experimental method employed and the uncertainties in control of physical parameters, the reproducibility of the data, the agreement of the worker's results on accepted test systems with standard values, and finally, the fitting of data, with suitable statistical tests, to mathematical functions;

(ii) a set of recommended numerical data. Whenever possible, the set of recommended data includes weighted average and standard deviations, and a set of smoothing equations derived from the experimental data endorsed by the evaluator;

(iii) a graphical plot of recommended data.

The compilation part consists of data sheets of the best experimental data in the primary literature. Generally speaking, such independent data sheets are given only to the best and endorsed data covering the known range of experimental parameters. Data sheets based on primary sources where the data are of a lower precision are given only when no better data are available. Experimental data with a precision poorer than considered acceptable are reproduced in the form of data sheets when they are the only known data for a particular system. Such data are considered to be still suitable for some applications, and their presence in the compilation should alert researchers to areas that need more work.

The typical data sheet carries the following information:

(i) components - definition of the system - their names, formulas and Chemical Abstracts registry numbers;

(ii) reference to the primary source where the numerical information is reported. In cases when the primary source is a less common periodical or a report document, published though of limited availability, abstract references are also given;

(iii) experimental variables;

(iv) identification of the compiler;

(v) experimental values as they appear in the primary source. Whenever available, the data may be given both in tabular and graphical form. If auxiliary information is available, the experimental data are converted also to SI units by the compiler.

Under the general heading of Auxiliary Information, the essential experimental details are summarized:

(vi) experimental method used for the generation of data;
(vii) type of apparatus and procedure employed;
(viii) source and purity of materials;
(ix) estimated error;
(x) references relevant to the generation of experimental data as cited in the primary source.

This new approach to numerical data presentation, developed during our four years of existence, has been strongly influenced by the diversity of background of those whom we are supposed to serve. We thus deemed it right to preface the evaluation/compilation sheets in each volume with a detailed discussion of the principles of the accurate determination of relevant solubility data and related thermodynamic information.

Finally, the role of education is more than corollary to the efforts we are seeking. The scientific standards advocated here are necessary to strengthen science and technology, and should be regarded as a major effort in the training and formation of the next generation of scientists and engineers. Specifically, we believe that there is going to be an impact of our project on scientific-communication practices. The quality of consolidation adopted by this program offers down-to-earth guidelines, concrete examples which are bound to make primary publication services more responsive than ever before to the needs of users. The self-regulatory message to scientists of 15 years ago to refrain from unnecessary publication has not achieved much. The literature is still, in 1983, cluttered with poor-quality articles. The Weinberg report (in "Reader in Science Information", Eds. J. Sherrod and A. Hodina, Microcard Editions Books, Indian Head, Inc., 1973, p.292) states that "admonition to authors to restrain themselves from premature, unnecessary publication can have little effect unless the climate of the entire technical and scholarly community encourages restraint..." We think that projects of this kind translate the climate into operational terms by exerting pressure on authors to avoid submitting low-grade material. The type of our output, we hope, will encourage attention to quality as authors will increasingly realize that their work will not be suited for permanent retrievability unless it meets the standards adopted in this project. It should help to dispel confusion in the minds of many authors of what represents a permanently useful bit of information of an archival value, and what does not.

If we succeed in that aim, even partially, we have then done our share in protecting the scientific community from unwanted and irrelevant, wrong numerical information.

A. S. Kertes

PREFACE

The present volume is one of four on the inorganic halates. This volume deals with the chlorates, bromates, and iodates of the alkaline earth metals (Mg, Ca, Sr, Ba). The other volumes deal with the copper and silver halates, alkali metal halates, and transition metal halates. The alkali and alkaline earth metal halates have been important in the history of both theoretical and practical analytical chemistry. In 1848 Berthet in France described the use of potassium iodate as a standard titrant for the determination of iodide. The well established method for determining phenol with excess bromate-bromine reagent in acid solution was first described by Knop in 1854 and further developed by Koppeschaar in 1875. Important practical applications of halate chemistry presently include their use in pyrotecnics, and in the paper pulp industry for the generation of chloric dioxide blanching agent.

Solubility studies involving the alkali metal halates can be classified according to the following types of studies.

1. The solubilities in pure water over a wide temperature range have been determined by the synthetic and isothermal methods. Solid phases have often been studied, and the existence of a number of hydrates has been established. Transition temperatures between the various hydrates have been determined graphically from the phase diagrams by determining the points of intersection of the various solubility branches.

2. The solubilities of the halates as a function of ionic strength were measured by the isothermal method with the objective of determining the thermodynamic solubility product, K°_{s0}. Both aqueous and mixed aqueous-organic solvents have been employed in these studies. In almost every case the solubility of the halate in aqueous solutions decreases with increasing ionic strength.

3. A number of studies on halate solubilities in mixed aqueous-organic solvents were undertaken with the major objective of verifying the Born equation: i.e. the logarithm of the solubility should be inversely proportional to the static dielectric constant. This simple relation appears to be adequate when dealing with solvents of high dielectric constant, but breaks down as the dielectric constant of the solvent decreases. Most of the earlier studies employed hydrogen-bonding solvents such as the alcohols, but recent work has been extended to include aprotic solvents such as dimethylformamide.

4. A major objective in a number of solubility studies has been the determination of formation constants of various complexes involving an alkaline earth cation with NH_3, organic compounds, and an anion (e.g. as in the formation of the ion pair $BaIO_3^+$).

5. The solubility in ternary systems containing two saturating solutes has been reported in several studies. The isothermal method is generally employed, and the nature of the solid phases have been determined.

Some general comments on the preparation of this volume are presented below.

The literature on the solubility of halates was covered through 1981. An attempt was made to survey *Chemical Abstracts* through the first half of 1982. So far as we are aware, the entire literature has been covered in this survey. In a few cases it was not possible to obtain copies of original papers published in the USSR. Thus the data from these sources have not been compiled or evaluated. For example a recently published paper on the solubility of alkaline earth halates, *Izv. Sev-Nauk. Nauch. Tsentra Vyesh. Shk. Estestv. Nauki* 1981, 55-7, could not be compiled.

Chemical Abstracts recommended names and Registry Numbers were used throughout. Common names are cross referenced to Chemical Abstracts recommended names in the INDEX. There is also a Registry Number and an Author index.

The solubilities of sparingly soluble halates as a function of ionic strength have been measured to obtain the thermodynamic solubility product by extrapolation of the solubilities to infinite dilution. Frequently the authors have not reported the actual solubilities, but only the thermodynamic solubility product. In such cases the solubilities may be estimated if the activity coefficients are known.

In the descriptions of experimental methods and purities of materials in the compilations, the compilers have, for the most part, retained the units employed in the original investigations: e.g. m for mol kg^{-1} and c for mol dm^{-3}. When possible the solubility data tabulated in the compilations have been converted to S.I. units. Temperatures have been converted to Kelvin in the critical evaluations.

The lack of reliable density data has prevented the evaluators from making conversions between mol kg^{-1} and mol dm^{-3} units. This has had the effect of reducing the possibilities of intercomparison for the evaluation and recommendation of solubility data.

All of the above considerations were taken into account in the selection of the most reliable data for the recommended solubility values. In a number of systems, no data could be recommended with sufficient confidence, and only tentative values or no values are given.

Although an attempt has been made to locate all publications on the solubilities of alkaline earth metal halates through the first half of 1982, some omissions may have occurred. The Editors will therefore be grateful to readers who will bring these omissions to their attention so that they may be included in future updates to this volume.

The editors gratefully acknowledge advice and comments from members of the IUPAC Commission on Solubility Data. In particular we would like to thank Professor A. S. Kertes, Chairman, IUPAC Commission on Solubility Data, The Hebrew University, Professor J. W. Lorimer, Chemistry Department, University of Western Ontario, and Dr. K. Loening of Chemical Abstracts Service for providing Registry Numbers for numerous compounds.

The editors thank Ms Beth Boozer of the Solubility Research and Information Project, Emory University, for technical assistance in preparing the manuscript for publication, and Ms Carolyn Dowie for typing the final manuscript.

The editors would also like to acknowledge the cooperation of the American Chemical Society and VAAP, the copyright agency of the USSR, for their permission to reproduce figures of phase diagrams from their publications.

Finally one of us (H.M.) would like to thank Professor Michihiro Fujii of Niigata University for his advice throughout the work.

Hiroshi Miyamoto
Emory University
Atlanta, GA, USA
(Permanent address: Niigata University, Niigata, Japan)

Mark Salomon
U. S. Army ERADCOM
Fort Monmouth, NJ, USA

H. Lawrence Clever
Emory University
Atlanta, GA, USA

INTRODUCTION TO THE SOLUBILITY OF SOLIDS IN LIQUIDS

Nature of the Project

The Solubility Data Project (SDP) has as its aim a comprehensive search of the literature for solubilities of gases, liquids, and solids in liquids or solids. Data of suitable precision are compiled on data sheets in a uniform format. The data for each system are evaluated, and where data from different sources agree sufficiently, recommended values are proposed. The evaluation sheets, recommended values, and compiled data sheets are published on consecutive pages.

This series of volumes includes solubilities of solids of all types in liquids of all types.

Definitions

A *mixture* (1,2) describes a gaseous, liquid, or solid phase containing more than one substance, when the substances are all treated in the same way.

A *solution* (1,2) describes a liquid or solid phase containing more than one substance, when for convenience one of the substances, which is called the *solvent* and may itself be a mixture, is treated differently than the other substances, which are called *solutes*. If the sum of the mole fractions of the solutes is small compared to unity, the solution is called a *dilute solution.*

The *solubility* of a substance B is the relative proportion of B (or a substance related chemically to B) in a mixture which is saturated with respect to solid B at a specified temperature and pressure. *Saturated* implies the existence of equilibrium with respect to the processes of dissolution and precipitation; the equilibrium may be stable or metastable. The solubility of a metastable substance is usually greater than that of the corresponding stable substance. (Strictly speaking, it is the activity of the metastable substance that is greater.) Care must be taken to distinguish true metastability from supersaturation, where equilibrium does not exist.

Either point of view, mixture or solution, may be taken in describing solubility. The two points of view find their expression in the quantities used as measures of solubility and in the reference states used for definition of activities and activity coefficients.

The qualifying phrase "substance related chemically to B" requires comment. The composition of the saturated mixture (or solution) can be described in terms of any suitable set of thermodynamic components. Thus, the solubility of a salt hydrate in water is usually given as the relative proportion of anhydrous salt in solution, rather than the relative proportions of hydrated salt and water.

Quantities Used as Measures of Solubility

1. *Mole fraction* of substance B, x_B:

$$x_B = n_B / \sum_{i=1}^{c} n_i \qquad (1)$$

where n_i is the amount of substance of substance i, and c is the number of distinct substances present (often the number of thermodynamic components in the system). *Mole per cent* of B is 100 x_B.

2. *Mass fraction* of substance B, w_B:

$$w_B = m'_B / \sum_{i=1}^{c} m'_i \qquad (2)$$

where m'_i is the mass of substance i. *Mass per cent* of B is 100 w_B. The equivalent terms weight fraction and weight per cent are not used.

3. *Solute mole (mass) fraction* of solute B (3,4):

$$x_{S,B} = n_B / \sum_{i=1}^{c'} n_i = x_B / \sum_{i=1}^{c'} x_i \qquad (3)$$

where the summation is over the solutes only. For the solvent A, $x_{S,A} = x_A$. These quantities are called *Jänecke mole (mass) fractions* in many papers.

4. *Molality* of solute B (1,2) in a solvent A:

$$m_B = n_B/n_A M_A \qquad \text{SI base units: mol kg}^{-1} \qquad (4)$$

where M_A is the molar mass of the solvent.

5. *Concentration* of solute B (1,2) in a solution of volume V:

$$c_B = [B] = n_B/V \qquad \text{SI base units: mol m}^{-3} \qquad (5)$$

The terms molarity and molar are not used.

Mole and mass fractions are appropriate to either the mixture or the solution points of view. The other quantities are appropriate to the solution point of view only. In addition of these quantities, the following are useful in conversions between concentrations and other quantities.

6. *Density*: $\rho = m/V$ SI base units: kg m^{-3} (6)

7. *Relative density*: d; the ratio of the density of a mixture to the density of a reference substance under conditions which must be specified for both (1). The symbol $d_{t'}^{t}$ will be used for the density of a mixture at t°C, 1 atm divided by the density of water at t'°C, 1 atm.

Other quantities will be defined in the prefaces to individual volumes or on specific data sheets.

Thermodynamics of Solubility

The principal aims of the Solubility Data Project are the tabulation and evaluation of: (a) solubilities as defined above; (b) the nature of the saturating solid phase. Thermodynamic analysis of solubility phenomena has two aims: (a) to provide a rational basis for the construction of functions to represent solubility data; (b) to enable thermodynamic quantities to be extracted from solubility data. Both these aims are difficult to achieve in many cases because of a lack of experimental or theoretical information concerning activity coefficients. Where thermodynamic quantities can be found, they are not evaluated critically, since this task would involve critical evaluation of a large body of data that is not directly relevant to solubility. The following discussion is an outline of the principal thermodynamic relations encountered in discussions of solubility. For more extensive discussions and references, see books on thermodynamics, e.g., (5-10).

Activity Coefficients (1)

(a) *Mixtures.* The activity coefficient f_B of a substance B is given by

$$RT\ \ell n(f_B x_B) = \mu_B - \mu_B^* \qquad (7)$$

where μ_B is the chemical potential, and μ_B^* is the chemical potential of pure B at the same temperature and pressure. For any substance B in the mixture,

$$\lim_{x_B \to 1} f_B = 1 \qquad (8)$$

(b) *Solutions.*

(i) *Solute substance, B.* The molal activity coefficient γ_B is given by

$$RT\ \ell n(\gamma_B m_B) = \mu_B - (\mu_B - RT\ \ell n\ m_B)^{\infty} \qquad (9)$$

where the superscript $^\infty$ indicates an infinitely dilute solution. For any solute B,

$$\gamma_B^{\infty} = 1 \qquad (10)$$

Activity coefficients y_B connected with concentration c_B, and $f_{x,B}$ (called the *rational activity coefficient*) connected with mole fraction x_B are defined in analogous ways. The relations among them are (1,9):

$$\gamma_B = x_A f_{x,B} = V_A^*(1 - \sum_s c_s) y_B \qquad (11)$$

or

$$f_{x,B} = (1 + M_A \sum_s m_s)\gamma_B = V_A^* y_B/V_m \quad (12)$$

or

$$y_B = (V_A + M_A \sum_s m_s V_s)\gamma_B/V_A^* = V_m f_{x,B}/V_A^* \quad (13)$$

where the summations are over all solutes, V_A^* is the molar volume of the pure solvent, V_i is the partial molar volume of substance i, and V_m is the molar volume of the solution.

For an electrolyte solute $B \equiv C_{\nu+}A_{\nu-}$, the molal activity is replaced by (9)

$$\gamma_B m_B = \gamma_\pm^{\nu} m_B^{\nu} Q^{\nu} \quad (14)$$

where $\nu = \nu_+ + \nu_-$, $Q = (\nu_+^{\nu+}\nu_-^{\nu-})^{1/\nu}$, and $\gamma_\pm$ is the mean ionic molal activity coefficient. A similar relation holds for the concentration activity $y_B c_B$. For the mol fractional activity,

$$f_{x,B}\, x_B = \nu_+^{\nu_+} \nu_-^{\nu_-} f_\pm^{\nu} x_\pm^{\nu} \quad (15)$$

The quantities x_+ and x_- are the ionic mole fractions (9), which for a single solute are

$$x_+ = \nu_+ x_B/[1+(\nu-1)x_B]; \qquad x_- = \nu_- x_B/[1+(\nu-1)x_B] \quad (16)$$

(ii) *Solvent, A:*

The *osmotic coefficient*, ϕ , of a solvent substance A is defined as (1):

$$\phi = (\mu_A^* - \mu_A)/RT\, M_A \sum_s m_s \quad (17)$$

where μ_A^* is the chemical potential of the pure solvent.

The *rational osmotic coefficient*, ϕ_x, is defined as (1):

$$\phi_x = (\mu_A - \mu_A^*)/RT \ln x_A = \phi M_A \sum_s m_s/\ln(1 + M_A \sum_s m_s) \quad (18)$$

The activity, a_A, or the activity coefficient f_A is often used for the solvent rather than the osmotic coefficient. The activity coefficient is defined relative to pure A, just as for a mixture.

The Liquid Phase

A general thermodynamic differential equation which gives solubility as a function of temperature, pressure and composition can be derived. The approach is that of Kirkwood and Oppenheim (7). Consider a solid mixture containing c' thermodynamic components i. The Gibbs-Duhem equation for this mixture is:

$$\sum_{i=1}^{c'} x_i'(S_i'dT - V_i'dp + d\mu_i) = 0 \quad (19)$$

A liquid mixture in equilibrium with this solid phase contains c thermodynamic components i, where, usually, c > c'. The Gibbs-Duhem equation for the liquid mixture is:

$$\sum_{i=1}^{c'} x_i(S_i dT - V_i dp + d\mu_i) + \sum_{i=c'+1}^{c} x_i(S_i dT - V_i dp + d\mu_i) = 0 \quad (20)$$

Eliminate $d\mu_1$ by multiplying (19) by x_1 and (20) x_1'. After some algebra, and use of:

$$d\mu_i = \sum_{j=2}^{c} G_{ij}dx_j - S_i dT + V_i dp \quad (21)$$

where (7)

$$G_{ij} = (\partial\mu_i/\partial x_j)_{T,P,x_i \neq x_j} \quad (22)$$

it is found that

$$\sum_{i=2}^{c'} \sum_{j=2}^{c} (x_i' - x_i x_1'/x_1)G_{ij}dx_j - (x_1'/x_1) \sum_{i=c'+1}^{c} \sum_{j=2}^{c} x_i G_{ij} dx_j$$

$$= \sum_{i=1}^{c'} x_i'(H_i - H_i')dT/T - \sum_{i=1}^{c'} x_i'(V_i - V_i')dp \quad (23)$$

where

$$H_i - H_i' = T(S_i - S_i') \tag{24}$$

is the enthalpy of transfer of component i from the solid to the liquid phase, at a given temperature, pressure and composition, and H_i, S_i, V_i are the partial molar enthalpy, entropy, and volume of component i. Several special cases (all with pressure held constant) will be considered. Other cases will appear in individual evaluations.

(a) *Solubility as a function of temperature.*
Consider a binary solid compound A_nB in a single solvent A. There is no fundamental thermodynamic distinction between a binary compound of A and B which dissociates completely or partially on melting and a solid mixture of A and B; the binary compound can be regarded as a solid mixture of constant composition. Thus, with c = 2, c' = 1, $x_A' = n/(n+1)$, $x_B' = 1/(n+1)$, eqn (23) becomes

$$(1/x_B - n/x_A)\{1 + \left(\frac{\partial \ln f_B}{\partial \ln x_B}\right)_{T,P}\}dx_B = (nH_A + H_B - H^*_{AB})dT/RT^2 \tag{25}$$

where the mole fractional activity coefficient has been introduced. If the mixture is a non-electrolyte, and the activity coefficients are given by the expression for a simple mixture (6):

$$RT \ln f_B = wx_A^2 \tag{26}$$

then it can be shown that, if w is independent of temperature, eqn (25) can be integrated (cf. (5), Chap. XXIII, sect. 5). The enthalpy term becomes

$$nH_A + H_B - H^*_{AB} = \Delta H_{AB} + n(H_A - H_A^*) + (H_B - H_B^*)$$

$$= \Delta H_{AB} + w(nx_B^2 + x_A^2) \tag{27}$$

where ΔH_{AB} is the enthalpy of melting and dissociation of one mole of pure solid A_nB, and H_A^*, H_B^* are the molar enthalpies of pure liquid A and B. The differential equation becomes

$$R\, d \ln\{x_B(1-x_B)^n\} = -\Delta H_{AB}\, d\left(\frac{1}{T}\right) - w\, d\left(\frac{x_A^2 + nx_B^2}{T}\right) \tag{28}$$

Integration from x_B,T to $x_B = 1/(1+n)$, T = T*, the melting point of the pure binary compound, gives:

$$\ln\{x_B(1-x_B)^n\} \simeq \ln\left\{\frac{n^n}{(1+n)^{n+1}}\right\} - \left\{\frac{\Delta H^*_{AB} - T^*\Delta C^*_p}{R}\right\}\left(\frac{1}{T} - \frac{1}{T^*}\right)$$
$$+ \frac{\Delta C_p^*}{R}\ln\left(\frac{T}{T^*}\right) - \frac{w}{R}\left\{\frac{x_A + nx_B}{T} - \frac{n}{(n+1)T^*}\right\} \tag{29}$$

where ΔC_p^* is the change in molar heat capacity accompanying fusion plus decomposition of the compound at temperature T*, (assumed here to be independent of temperature and composition), and ΔH^*_{AB} is the corresponding change in enthalpy at T = T*. Equation (29) has the general form

$$\ln\{x_B(1-x_B)^n\} = A_1 + A_2/T + A_3 \ln T + A_4(x_A^2 + nx_B^2)/T \tag{30}$$

If the solid contains only component B, n = 0 in eqn (29) and (30).

If the infinite dilution standard state is used in eqn (25), eqn (26) becomes

$$RT \ln f_{x,B} = w(x_A^2 - 1) \tag{31}$$

and (27) becomes

$$nH_A + H_B - H_{AB} = (nH_A^* + H_B^\infty - H^*_{AB}) + n(H_A - H_A^*) + (H_B - H_B^\infty) = \Delta H^\infty_{AB} + w(nx_B^2 + x_A^2 - 1) \tag{32}$$

where the first term, ΔH^∞_{AB}, is the enthalpy of melting and dissociation of solid compound A_nB to the infinitely dilute state of solute B in solvent A; H_B^∞ is the partial molar enthalpy of the solute at infinite dilution. Clearly, the integral of eqn (25) will have the same form as eqn (29), with $\Delta H^\infty_{AB}(T^*)$, $\Delta C_p^\infty(T^*)$ replacing ΔH^*_{AB} and ΔC_p^* and x_A^2-1 replacing x_A^2 in the last term.

If the liquid phase is an aqueous electrolyte solution, and the solid is a salt hydrate, the above treatment needs slight modification. Using rational mean activity coefficients, eqn (25) becomes

$$R\nu(1/x_B-n/x_A)\{1+(\partial \ln f_\pm/\partial \ln x_\pm)_{T,P}\}dx_B/\{1+(\nu-1)x_B\}$$

$$= \{\Delta H^\infty_{AB} + n(H_A-H_A^*) + (H_B-H_B^\infty)\}d(1/T) \quad (33)$$

If the terms involving activity coefficients and partial molar enthalpies are negligible, then integration gives (cf. (11)):

$$\ln\{\frac{x_B^\nu(1-x_B)^n}{1+(\nu-1)x_B^{n+\nu}}\} = \ln\{\frac{n^n}{(n+\nu)^{n+\nu}}\} - \{\frac{\Delta H^\infty_{AB}(T^*)-T^*\Delta C_p^*}{R}\}(\frac{1}{T}-\frac{1}{T^*}) + \frac{\Delta C_p^*}{R}\ln(T/T^*) \quad (34)$$

A similar equation (with ν=2 and without the heat capacity terms) has been used to fit solubility data for some $MOH=H_2O$ systems, where M is an alkali metal; the enthalpy values obtained agreed well with known values (11). In many cases, data on activity coefficients (9) and partial molal enthalpies (8,10) in concentrated solution indicate that the terms involving these quantities are not negligible, although they may remain roughly constant along the solubility temperature curve.

The above analysis shows clearly that a rational thermodynamic basis exists for functional representation of solubility-temperature curves in two-component systems, but may be difficult to apply because of lack of experimental or theoretical knowledge of activity coefficients and partial molar enthalpies. Other phenomena which are related ultimately to the stoichiometric activity coefficients and which complicate interpretation include ion pairing, formation of complex ions, and hydrolysis. Similar considerations hold for the variation of solubility with pressure, except that the effects are relatively smaller at the pressures used in many investigations of solubility (5).

(b) *Solubility as a function of composition.*

At constant temperature and pressure, the chemical potential of a saturating solid phase is constant:

$$\mu^*_{A_nB} = \mu_{A_nB}(\text{sln}) = n\mu_A + \mu_B \quad (35)$$

$$= (n\mu_A^* + \nu_+\mu_+^\infty + \nu_-\mu_-^\infty) + nRT\ln f_A x_A$$

$$+ \nu RT\ln \gamma_\pm m_\pm Q_\pm \quad (36)$$

for a salt hydrate A_nB which dissociates to water, (A), and a salt, B, one mole of which ionizes to give ν_+ cations and ν_- anions in a solution in which other substances (ionized or not) may be present. If the saturated solution is sufficiently dilute, $f_A = x_A = 1$, and the quantity K^0_{s0} in

$$\Delta G^\infty \equiv (\nu_+\mu_+^\infty + \nu_-\mu_-^\infty + n\mu_A^* - \mu_{AB}^*)$$

$$= -RT\ln K^0_{s0}$$

$$= -RT\ln Q^\nu\gamma_\pm^\nu m_+^{\nu_+} m_-^{\nu_-} \quad (37)$$

is called the *solubility product* of the salt. (It should be noted that it is not customary to extend this definition to hydrated salts, but there is no reason why they should be excluded.) Values of the solubility product are often given on mole fraction or concentration scales. In dilute solutions, the theoretical behaviour of the activity coefficients as a function of ionic strength is often sufficiently well known that reliable extrapolations to infinite dilution can be made, and values of K^0_{s0} can be determined. In more concentrated solutions, the same problems with activity coefficients that were outlined in the section on variation of solubility with temperature still occur. If these complications do not arise, the solubility of a hydrate salt $C_{\nu_+}A_{\nu_-}\cdot nH_2O$ in the presence of other solutes is given by eqn (36) as

$$\nu\ln\{m_B/m_B(0)\} = -\nu\ln\{\gamma_\pm/\gamma_\pm(0)\} - n\ln(a_{H_2O}/a_{H_2O}(0)) \quad (38)$$

where a_{H_2O} is the activity of water in the saturated solution, m_B is the molality of the salt in the saturated solution, and (0) indicates absence of other solutes. Similar considerations hold for non-electrolytes.

The Solid Phase

The definition of solubility permits the occurrence of a single solid phase which may be a pure anhydrous compound, a salt hydrate, a non-stoichiometric compound, or a solid mixture (or solid solution, or "mixed crystals"), and may be stable or metastable. As well, any number of solid phases consistent with the requirements of the phase rule may be present. Metastable solid phases are of widespread occurrence, and may appear as polymorphic (or allotropic) forms or crystal solvates whose rate of transition to more stable forms is very slow. Surface heterogeneity may also give rise to metastability, either when one solid precipitates on the surface of another, or if the size of the solid particles is sufficiently small that surface effects become important. In either case, the solid is not in stable equilibrium with the solution. The stability of a solid may also be affected by the atmosphere in which the system is equilibrated.

Many of these phenomena require very careful, and often prolonged, equilibration for their investigation and elimination. A very general analytical method, the "wet residues" method of Schreinemakers (12) (see a text on physical chemistry) is usually used to investigate the composition of solid phases in equilibrium with salt solutions. In principle, the same method can be used with systems of other types. Many other techniques for examination of solids, in particular X-ray, optical, and thermal analysis methods, are used in conjunction with chemical analyses (including the wet residues method).

COMPILATIONS AND EVALUATIONS

The formats for the compilations and critical evaluations have been standardized for all volumes. A brief description of the data sheets has been given in the FOREWORD; additional explanation is given below.

Guide to the Compilations

The format used for the compilations is, for the most part, self-explanatory. The details presented below are those which are not found in the FOREWORD or which are not self-evident.

Components. Each component is listed according to IUPAC name, formula, and Chemical Abstracts (CA) Registry Number. The formula is given either in terms of the IUPAC or Hill (13) system and the choice of formula is governed by what is usual for most current users: i.e. IUPAC for inorganic compounds, and Hill system for organic compounds. Components are ordered according to:

(a) saturating components;
(b) non-saturating components in alphanumerical order;
(c) solvents in alphanumerical order.

The saturating components are arranged in order according to a 18-column, 2-row periodic table:

Columns 1,2: H, groups IA, IIA;
3,12: transition elements (groups IIIB to VIIB, group VIII, groups IB, IIB);
13-18: groups IIIA-VIIA, noble gases.

Row 1: Ce to Lu;
Row 2: Th to the end of the known elements, in order of atomic number.

Salt hydrates are generally not considered to be saturating components since most solubilities are expressed in terms of the anhydrous salt. The existence of hydrates or solvates is carefully noted in the texts, and CA Registry Numbers are given where available, usually in the critical evaluation. Mineralogical names are also quoted, along with their CA Registry Numbers, again usually in the critical evaluation.

Original Measurements. References are abbreviated in the forms given by *Chemical Abstracts Service Source Index (CASSI)*. Names originally in other than Roman alphabets are given as transliterated by *Chemical Abstracts*.

Experimental Values. Data are reported in the units used in the original publication, with the exception that modern *names* for units and quantities are used; e.g., mass per cent for weight per cent; mol dm^{-3} for molar; etc. Both mass and molar values are given. Usually, only one type of value (e.g., mass per cent) is found in the original paper, and the compiler has added the other type of value (e.g., mole per cent) from computer calculations based on 1976 atomic weights (14). Errors in calculations and fitting equations in original papers have been noted and corrected, by computer calculations where necessary.

Method. Source and Purity of Materials. Abbreviations used in *Chemical Abstracts* are often used here to save space.

Estimated Error. If these data were omitted by the original authors, and if relevant information is available, the compilers have attempted to estimate errors from the internal consistency of data and type of apparatus used. Methods used by the compilers for estimating and reporting errors are based on the papers by Ku and Eisenhart (15).

Comments and/or Additional Data. Many compilations include this section which provides short comments relevant to the general nature of the work or additional experimental and thermodynamic data which are judged by the compiler to be of value to the reader.

References. See the above description for Original Measurements.

Guide to the Evaluations

The evaluator's task is to check whether the compiled data are correct, to assess the reliability and quality of the data, to estimate errors where necessary, and to recommend "best" values. The evaluation takes the form of a summary in which all the data supplied by the compiler have been critically reviewed. A brief description of the evaluation sheets is given below.

Components. See the description for the Compilations.

Evaluator. Name and date up to which the literature was checked.

Critical Evaluation

(a) Critical text. The evaluator produces text evaluating *all* the published data for each given system. Thus, in this section the evaluator review the merits or shortcomings of the various data. Only published data are considered; even published data can be considered only if the experimental data permit an assessment of reliability.

(b) Fitting equations. If the use of a smoothing equation is justifiable, the evaluator may provide an equation representing the solubility as a function of the variables reported on all the compilation sheets.

(c) Graphical summary. In addition to (b) above, graphical summaries are often given.

(d) Recommended values. Data are *recommended* if the results of at least two independent groups are available and they are in good agreement, and if the evaluator has no doubt as to the adequacy and reliability of the applied experimental and computational procedures. Data are reported as *tentative* if only one set of measurements is available, or if the evaluator considers some aspect of the computational or experimental method as mildly undesirable but estimates that it should cause only minor errors. Data are considered as *doubtful* if the evaluator considers some aspect of the computational or experimental method as undesirable but still considers the data to have some value in those instances where the order of magnitude of the solubility is needed. Data determined by an inadequate method or under ill-defined conditions are *rejected.* However references to these data are included in the evaluation together with a comment by the evaluator as to the reason for their rejection.

(e) References. All pertinent references are given here. References to those data which, by virtue of their poor precision, have been rejected and not compiled are also listed in this section.

(f) Units. While the original data may be reported in the units used by the investigators, the final recommended values are reported in S.I. units (1,16) when the data can be accurately converted.

References

1. Whiffen, D. H., ed., *Manual of Symbols and Terminology for Physico-chemical Quantities and Units.* *Pure Applied Chem.* 1979, *51*, No. 1.
2. McGlashan, M.L. *Physicochemical Quantities and Units.* 2nd ed. Royal Institute of Chemistry. London. 1971.
3. Jänecke, E. *Z. Anorg. Chem.* 1906, *51*, 132.
4. Friedman, H.L. *J. Chem. Phys.* 1960, *32*, 1351.
5. Prigogine, I.; Defay, R. *Chemical Thermodynamics.* D.H. Everett, transl. Longmans, Green. London, New York, Toronto. 1954.
6. Guggenheim, E.A. *Thermodynamics.* North-Holland. Amsterdam. 1959. 4th ed.
7. Kirkwood, J.G.; Oppenheim, I. *Chemical Thermodynamics.* McGraw-Hill, New York, Toronto, London. 1961.
8. Lewis, G.N.; Randall, M. (rev. Pitzer, K.S.; Brewer, L.). *Thermodynamics.* McGraw Hill. New York, Toronto, London. 1961. 2nd ed.
9. Robinson, R.A.; Stokes, R.H. *Electrolyte Solutions.* Butterworths. London. 1959, 2nd ed.
10. Harned, H.S.; Owen, B.B. *The Physical Chemistry of Electrolytic Solutions.* Reinhold. New York. 1958. 3rd ed.
11. Cohen-Adad, R.; Saugier, M.T.; Said, J. *Rev. Chim. Miner.* 1973, *10*, 631.
12. Schreinemakers, F.A.H. *Z. Phys. Chem., stoechiom. Verwandschaftsl.* 1893, *11*, 75.
13. Hill, E.A. *J. Am. Chem. Soc.* 1900, *22*, 478.
14. IUPAC Commission on Atomic Weights. *Pure Appl. Chem.*, 1976, *47*, 75.

15. Ku, H.H., p. 73; Eisenhart, C., p. 69; in Ku, H.H., ed. *Precision Measurement and Calibration*. NBS Special Publication 300. Vol. 1. Washington. 1969.
16. *The International System of Units*. Engl. transl. approved by the BIPM of *Le Système International d'Unités*. H.M.S.O. London. 1970.

R. Cohen-Adad, Villeurbanne, France
J.W. Lorimer, London, Canada
M. Salomon, Fair Haven, New Jersey, U.S.A.

COMPONENTS:

(1) Magnesium chlorate; $Mg(ClO_3)_2$; [10326-21-3]

(2) Water; H_2O; [7732-18-5]

EVALUATOR:

Hiroshi Miyamoto
Department of Chemistry
Niigata University
Niigata, Japan

May, 1982

CRITICAL EVALUATION:

Solubility in the binary $Mg(ClO_3)_2$ - H_2O system

Solubilities in the binary $Mg(ClO_3)_2$-H_2O system have been reported in 2 publications (1,2).

Mylius and Funk (1) determined the solubility of magnesium chlorate in pure water at 291K gravimetrically. The degree of hydration for the salt used was not reported, but the evaluator assumes that the hexahydrate was used for the determination of solubility because the hexahydrate is the stable solid phase at 291K.

Meusser (2) measured solubilities of magnesium chlorate in water over the wide temperature range of 255 to 366K. The magnesium content of the saturated solutions was weighed as sulfate.

Depending upon temperature and composition, equilibrated solid phases of varying the degrees of hydration have been reported. The following solid phases have been identified:

$Mg(ClO_3)_2 \cdot 6H_2O$	[7791-19-7]
$Mg(ClO_3)_2 \cdot 4H_2O$	[82150-38-7]
$Mg(ClO_3)_2 \cdot 2H_2O$	[36355-97-2]
$Mg(ClO_3)_2$	[10326-21-3]

Meusser did not report the presence of the anhydrous salt in his solubility study.

The relation between the solubility and the temperature is shown in Fig. 1.

Solubility at 291.2K. This value has been reported in 2 publications (1,2). The result calculated by the compiler from the original data reported in (1) is 6.73 mol kg^{-1}. The agreement between the result of Meusser (2) and that of Mylius and Funk is within the limit of the estimated error. The arithmetic mean of two results is 6.76 mol kg^{-1} and the standard deviation is 0.04 mol kg^{-1}. The mean is designated as a recommended value.

Solubility at other temperatures. Only one publication (2) is available for solubilities of $Mg(ClO_3)_2$ at temperatures other than 298.2K. The results of Meusser are designated as tentative values.

The recommended and tentative values are given in Table 1.

COMPONENTS:	EVALUATOR:
(1) Magnesium chlorate; $Mg(ClO_3)_2$; [10326-21-3] (2) Water; H_2O; [7732-18-5]	Hiroshi Miyamoto Department of Chemistry Niigata University Niigata, JAPAN May, 1982

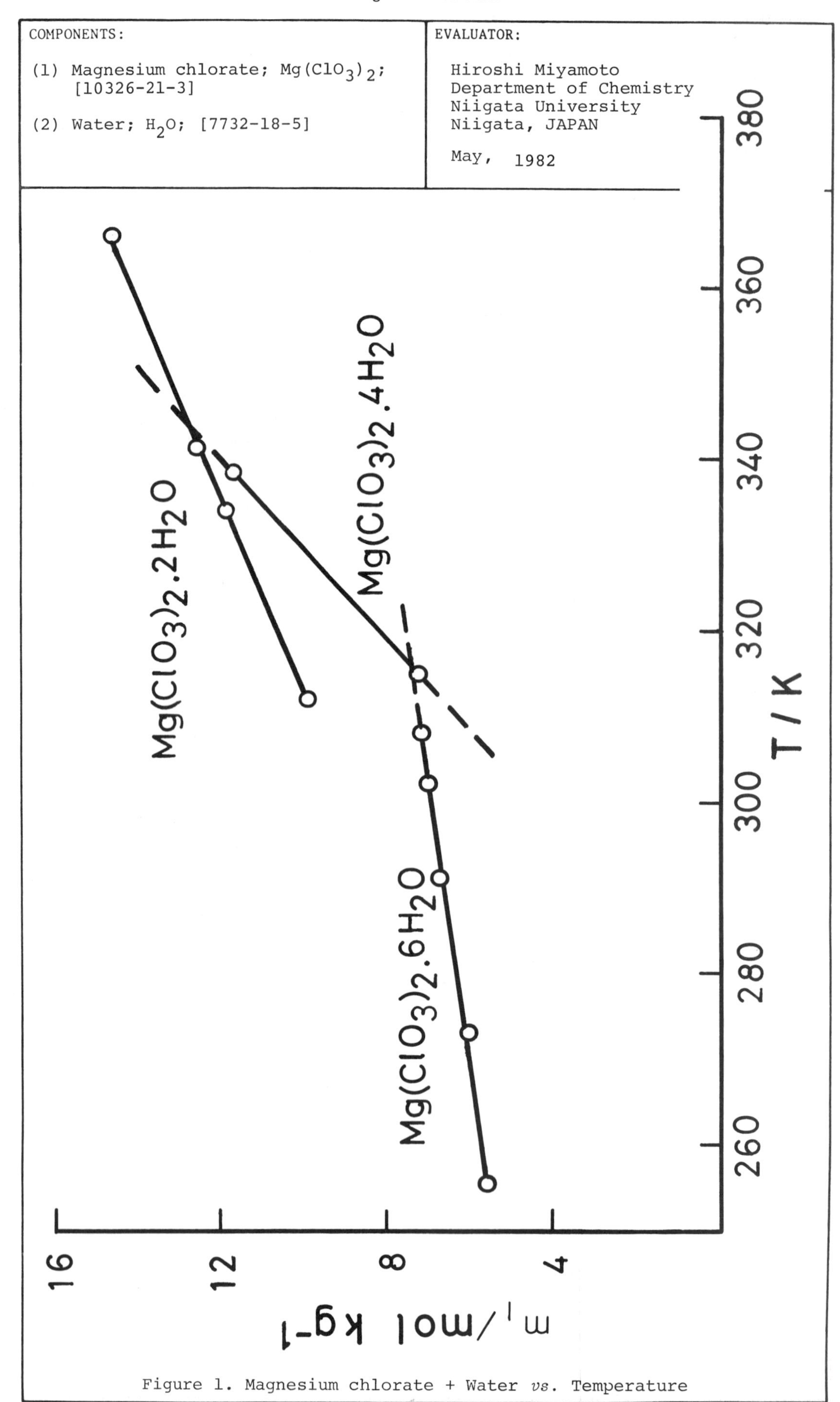

Figure 1. Magnesium chlorate + Water *vs.* Temperature

COMPONENTS:

(1) Magnesium chlorate; $Mg(ClO_3)_2$; [10326-21-3]

(2) Water; H_2O; [7732-18-5]

EVALUATOR:
Hiroshi Miyamoto
Department of Chemistry
Niigata University
Niigata, Japan

May, 1982

CRITICAL EVALUATION:

Table 1 Recommended and tentative values for the solubility of magnesium chlorate in water

T/K	m_1/mol kg^{-1}	m_1'/mol kg^{-1}	Nature of the Solid Phase
261.2	1.871		Ice
265.2	1.496		"
255.2	5.585	5.632	$Mg(ClO_3)_2 \cdot 6H_2O$
273.2	5.962	5.810	"
291.2	6.76[a]	6.90	"
302.2	7.920	8.085	"
308.2	9.158	8.949	"
315.2	9.225		$Mg(ClO_3)_2 \cdot 4H_2O$
338.7	11.71		"
312.7	9.872	9.868	$Mg(ClO_3)_2 \cdot 2H_2O$
334.2	11.89	11.93	"
341.2	12.61	12.57	"
366.2	14.66	14.67	"

m_1: experimental value

m_1': calculated value

a: recommended value

The data in Table 1 were fitted to the following smoothing equations:

$$\ln(S_6/\text{mol kg}^{-1}) = -67.42808 + 92.03835/(T/100\text{K}) + 35.32089 \ln(T/100\text{K}) \quad : \quad \sigma = 0.24$$

$$\ln(S_2/\text{mol kg}^{-1}) = 16.10542 - 25.38126/(T/100\text{K}) - 4.999113 \ln(T/100\text{K}) \quad : \quad \sigma = 0.058$$

where S_6 and S_2 are the solubilities of the hexahydrate and dihydrate, respectively.

REFERENCES:

1. Mylius, F.; Funk, R. *Ber. Dtsch. Chem. Ges.* 1897, *30*, 1716.

2. Meusser, A. *Ber. Dtsch. Chem. Ges.* 1902, *35*, 1414.

COMPONENTS:

(1) Magnesium chlorate; $Mg(ClO_3)_2$; [10326-21-3]

(2) Water; H_2O; [7732-18-5]

ORIGINAL MEASUREMENTS:

Mylius, F.; Funk, R.

Ber. Dtsch. Chem. Ges. 1897, *30*, 1716-25.

VARIABLES:

T/K = 291

PREPARED BY:

Hiroshi Miyamoto

EXPERIMENTAL VALUES:

The solubility of $Mg(ClO_3)_2 \cdot 6H_2O$ in water at 18°C is given below:

56.3 mass % (authors)
128.6 g/100g[a] H_2O (authors)

6.726 mol kg^{-1} (compiler)

The density of the saturated solution is given:

1.594 g cm^{-3}

Based on this density, the compiler calculated the solubility in volume units as

4.690 mol dm^{-3}

[a] The compiler presumes that the first word in the fifth line from the end of page 1717 should read 100g.

AUXILIARY INFORMATION

METHOD/APPARATUS/PROCEDURE:

The salt and water were placed in a bottle and the bottle was shaken in a thermostat for a long time. After the saturated solution settled, an aliquot of solution was removed with a pipet. Magnesium chlorate was determined by evaporation of the solution to dryness. The density of the saturated solution was also determined.

SOURCE AND PURITY OF MATERIALS:

The salt used was purchased as a "pure" chemical, and traces of impurities were not present. The purity sufficed for the solubility determination.

ESTIMATED ERROR:

Soly: precision within 1 %
Temp: nothing specified

REFERENCES:

COMPONENTS:

(1) Magnesium chlorate; $Mg(ClO_3)_2$; [10326-21-3]

(2) Water; H_2O; [7732-18-5]

ORIGINAL MEASUREMENTS:

Meusser, A.

Ber. Dtsch. Chem. Ges. 1902, *35*, 1414-24.

VARIABLES:

T/K = 255 to 366

PREPARED BY:

Hiroshi Miyamoto

EXPERIMENTAL VALUES:

t/°C	Magnesium Chlorate[a]			Nature of the solid phase
	mass %	mol/100 mol H_2O (compiler)	m_1/mol kg^{-1} (compiler)	
-12	26.35	3.371	1.871	Ice
- 8	22.24	2.695	1.496	"
-18	51.64	10.061	5.585	$Mg(ClO_3)_2\cdot 6H_2O$
0	53.27	10.740	5.962	"
18	56.50	12.238	6.793	"
29	60.23	16.269	7.920	"
35	63.65	16.498	9.158	"
42	63.82	16.620	9.225	$Mg(ClO_3)_2\cdot 4H_2O$
65.5	69.12	21.089	11.71	"
39.5	65.37	17.785	9.872	$Mg(ClO_3)_2\cdot 2H_2O$
61	69.46	21.429	11.89	"
68	70.69	22.724	12.61	"
93	(73.71)[b]	26.416	14.66	"

[a] Molalities and mol/100 mol H_2O calculated by compiler using 1977 IUPAC recommended atomic weights.

[b] No explanation for parenthesis is given.

AUXILIARY INFORMATION

METHOD/APPARATUS/PROCEDURE:

The salt and water were placed into a test tube and agitated for one hour. The saturated solutions were withdrawn with a pipet, and Mg content of the solution was weighed as sulfate.

SOURCE AND PURITY OF MATERIALS:

Pure $Mg(ClO_3)_2\cdot 6H_2O$ was recrystallized. The product was used in solubility determination. No other information was given.

ESTIMATED ERROR:

Nothing specified.

REFERENCES:

COMPONENTS:

(1) Magnesium bromate; $Mg(BrO_3)_2$; [14519-17-6]

(2) Water; H_2O; [7732-18-5]

EVALUATOR:

Hiroshi Miyamoto
Department of Chemistry
Niigata University
Niigata, Japan

May, 1982

CRITICAL EVALUATION:

The solubility in the binary $Mg(BrO_3)_2$-H_2O system

Solubilities in the binary $Mg(BrO_3)_2$-H_2O system have been reported in (1) obtained by Linke. Linke states that a single determination by Kohlrausch (2) is available, but the evaluator was unable to obtain Kohlrausch's paper. Only the results of Linke are considered in this critical evaluation.

Linke measured solubilities over the temperature range of 260 to 407K. The bromate content was determined iodometrically. The magnesium content was determined gravimetrically.

Depending upon temperature and composition, equilibrated solid phases of varying degrees of hydration have been reported by Linke. The following solid phases have been identified:

$Mg(BrO_3)_2 \cdot 6H_2O$	[7789-36-8]
$Mg(BrO_3)_2 \cdot 2H_2O$	[82150-36-5]
$Mg(BrO_3)_2$	[14519-17-6]

The relation between temperature and solubility in the binary $Mg(BrO_3)_2$-H_2O system at 1 atm is shown in Fig. 1 as reported in (1).

The eutectic of the system $Mg(BrO_3)_2$-H_2O lies at 260.2K and 38.5 mass % $Mg(BrO_3)_2$ with ice and $Mg(BrO_3)_2 \cdot 6H_2O$ as saturating solids. The transition from the hexahydrate to the dihydrate occurs at 353.2K. A saturated solution boils at 407K and contains 74.6 mass % $Mg(BrO_3)_2$.

The data reported in (1) obtained by Linke are designated as tentative values, and were fitted to the following smoothing equations:

$$\ln(S_6/\text{mol kg}^{-1}) = -33.93494 + 44.44325/(T/100\text{K}) + 18.56304 \ln(T/100\text{K}) \quad : \quad \sigma = 0.19$$

$$\ln(S_2/\text{mol kg}^{-1}) = 15.23113 - 23.79862/(T/100\text{K}) - 5.051857 \ln(T/100\text{K}) \quad : \quad \sigma = 0.066$$

where S_6 and S_2 are the solubilities of the hexahydrate and the dihydrate, respectively.

The tentative values and the values calculated from the smoothing equation are given in Table 1.

COMPONENTS:

(1) Magnesium bromate; $Mg(BrO_3)_2$; [14519-17-6]

(2) Water; H_2O; [7732-18-5]

EVALUATOR:

Hiroshi Miyamoto
Department of Chemistry
Niigata University
Niigata, Japan

May, 1982

CRITICAL EVALUATION:

Table 1 Tentative values of the solubility of magnesium bromate in water

T/K	m_1/mol kg^{-1}	m_1'/mol kg^{-1}	Solid Phase
271.6	0.368		Ice
269.0	0.9456		"
263.1	1.894		"
260.2	2.23		Ice + $Mg(BrO_3)_2 \cdot 6H_2O$
273.2	2.62	2.69	$Mg(BrO_3)_2 \cdot 6H_2O$
283.2	2.99	2.95	"
293.2	3.38	3.29	"
303.2	3.78	3.72	"
313.2	4.28	4.25	"
323.2	4.79	4.92	"
338.2	5.95	6.20	"
348.2	7.18	7.30	"
351.2	7.69	7.68	"
353.7	8.39	8.01	"
353.2	8.37		$Mg(BrO_3)_2 \cdot 6H_2O$ + $Mg(BrO_3)_2 \cdot 2H_2O$
354.2	8.37	8.36	$Mg(BrO_3)_2 \cdot 2H_2O$
358.2	8.49	8.51	"
363.2	8.66	8.70	"
373.2	9.13	9.04	"
390.2	9.46	9.53	"
403.2	9.80	9.83	"
407.2	9.95	9.91	"

m_1: experimental value

m_1': calculated value

REFERENCES:

1. Linke, W. F. *J. Am. Chem. Soc.* 1955, *77*, 866.

2. Kohlrausch, F. *Sitzb. K. Akad. Wiss. (Berlin)* 1897, *1*, 90.

COMPONENTS:

(1) Magnesium bromate; $Mg(BrO_3)_2$; [14519-17-6]

(2) Water; H_2O; [7732-18-5]

ORIGINAL MEASUREMENTS:

Linke, W. F.

J. Am. Chem. Soc. 1955, *77*, 866-7.

EXPERIMENTAL VALUES:

t/°C	Magnesium Bromate			Density	Nature of the Solid Phase[a]
	mass %	mol % (compiler)	m_1/mol kg^{-1} (compiler)	ρ/g cm^{-3}	
- 1.6	9.34	0.658	0.368	--	I
- 4.2	20.94	1.675	0.9456	--	"
-10.1	34.66	3.299	1.894	--	"
-13.0	38.5[b]	3.87	2.23	1.448	I + A
0	42.3_4	4.51_0	2.62_1	1.512	A
10	45.58	5.11_1	2.99_0	1.562	"
20	48.6_6	5.74_6	3.38_4	1.609	"
30	51.4	6.37	3.78	1.662	"
40	54.5	7.15	4.28	1.722	"
50	57.3	7.94	4.79	1.787	"
65	62.5	9.68	5.95	1.900	"
75	66.8	11.5	7.18	2.013	"
78	68.3	12.2	7.69	2.070	"
80.5[b]	70.15	13.13	8.390	--	"
80.0[b]	70.1[b]	13.1	8.37	--	A + B
81	70.1	13.1	8.37	--	B
90	70.8	13.5	8.66	--	"
100	71.9	14.1	9.13	--	B
117	72.6	14.6	9.46	--	"
130	73.3	15.0	9.80	--	"
134[b,c]	73.6[c]	15.2	9.95	--	"

[a] I = Ice; A = $Mg(BrO_3)_2\cdot 6H_2O$; B = $Mg(BrO_3)_2\cdot 2H_2O$

[b] Estimated graphically;   c: Boiling point.

The solubility (S) of $Mg(BrO_3)_2\cdot 6H_2O$ increases linearly from the eutectic (-13.0°C) to 65°C, and the relation was given as follows:

$$S = 4.24 + 0.300t$$

over this range with an average deviation 0.1 from the experimental values. The relation between the solubility (S') of $Mg(BrO_3)_2\cdot 2H_2O$ and the temperature (from 80°C to boiling point) is given as follows:

$$S' = 65.1 + 0.064t$$

with an accuracy of ± 0.1.

COMPONENTS:

(1) Magnesium bromate; $Mg(BrO_3)_2$; [14519-17-6]

(2) Water; H_2O; [7732-18-5]

ORIGINAL MEASUREMENTS:

Linke, W. F.

J. Am. Chem. Soc. 1955, *77*, 866-7.

VARIABLES:

T/K = 260.2 to 407

PREPARED BY:

Hiroshi Miyamoto

EXPERIMENTAL VALUES:

AUXILIARY INFORMATION

METHOD/APPARATUS/PROCEDURE:

Isothermal method used. Below 80°C solutions were equilibrated in a water thermostat, and above 80°C a vapor bath (1) was used. All samples were stirred internally. Repeated analysis showed that equilibrium was attained within an hour in every case. In a few cases equilibrium was checked by approach from supersaturation. Each reported value is the average of at least two independent determinations. The bromate content was determined by iodometry. Analysis for magnesium by precipitation of the oxime was done by reducing the bromate ions prior to the addition of the oxime; KBr or KI plus HCl were added, and the solution was boiled to expel the liberated halogens.
The densities of solutions in equil. with the hexahydrate were determined in a small pyknometer.

SOURCE AND PURITY OF MATERIALS:

Magnesium bromate hexahydrate was prepared by the addition of $MgSO_4$ solution to a hot suspension of $Ba(BrO_3)_2 \cdot H_2O$. The precipitate was allowed to digest overnight and then separated by filtration. The solution was evaporated by boiling until it became rather sirupy. Upon cooling the mass solidified completely. The salt was recrystallized twice and then air-dried. Found: $Mg(BrO_3)_2$, 72.03%. Calcd. for $Mg(BrO_3)_2 \cdot 6H_2O$: $Mg(BrO_3)_2$, 72.15%. Magnesium bromate dihydrate was prepared by heating the hexahydrate to 50-60°C. Found: $Mg(BrO_3)_2$, 88.67%. Calcd. for $Mg(BrO_3)_2 \cdot 2H_2O$: $Mg(BrO_3)_2$, 88.61%.

ESTIMATED ERROR:

Soly: precision 0.2%
Temp: below 80°C, ± 0.05°C; above 80°C, ± 0.5°C

REFERENCES:

1. Linke, W. F. *J. Chem. Educ.* 1952, *29*, 492.

COMPONENTS:

(1) Magnesium bromate; $Mg(BrO_3)_2$; [14519-17-6]

(2) Magnesium nitrate; $Mg(NO_3)_2$; [10377-60-3]

(3) Water; H_2O; [7732-18-5]

ORIGINAL MEASUREMENTS:

Linke, W. F.

J. Am. Chem. Soc. 1955, *77*, 866-7.

VARIABLES:

T/K = 358
$Mg(NO_3)_2$/mass % = 0 - 9.50

PREPARED BY:

Hiroshi Miyamoto

EXPERIMENTAL VALUES:

t/°C	Composition of the Saturated Solutions				Nature of the Solid Phase
	Magnesium Nitrate		Magnesium Bromate		
	mass %	mol % (compiler)	mass %	mol % (compiler)	
85	--	--	70.4[a]	13.27	$Mg(BrO_3)_2 \cdot 2H_2O$
	4.72	1.61	64.30	11.59	"
	6.00	2.04	62.99	11.32	"
	7.35	2.46	61.12	10.81	"
	8.85	2.95	59.62	10.52	"
	9.50	3.09	58.04	10.00	"

[a] For binary system the compiler computes the following

Soly of $Mg(BrO_3)_2$ = 8.49 mol kg^{-1} at 85°C

AUXILIARY INFORMATION

METHOD/APPARATUS/PROCEDURE:

Isothermal method used.
$Mg(BrO_3)_2 \cdot 2H_2O$, a solution of $Mg(NO_3)_2$ of known concentration and water were stirred in a vapor bath for from 1 to 5 hours at 85°C.
The bromate content was determined by iodometry. Analysis for total magnesium by precipitation of the oxime was done by reducing the bromate ion prior to the addition of the oxime; KBr and KI plus HCl were added, and the solution was boiled to expel the liberated halogens. The composition in solid phase was determined by algebric extrapolation of the tie line.
Each reported value is the average of at least two independent determinations.

SOURCE AND PURITY OF MATERIALS:

Magnesium bromate hexahydrate was prepared by the addition of $MgSO_4$ solution to a hot suspension of $Ba(BrO_3)_2 \cdot H_2O$. The precipitate was allowed to digest overnight and then separated by filtration. The solution was evaporated by boiling until it became rather sirupy. Upon cooling the mass solidified completely. This salt was recrystallized twice and then air-dried.
Found: $Mg(BrO_3)_2$, 72.03%. Calcd for $Mg(BrO_3)_2 \cdot 6H_2O$: $Mg(BrO_3)_2$, 72.15%.
Magnesium bromate dihydrate was prepared by heating the hexahydrate to 50-60°C. Found: $Mg(BrO_3)_2$, 88.67%. Calcd. for $Mg(BrO_3)_2 \cdot 2H_2O$: $Mg(BrO_3)_2$, 88.61%.

ESTIMATED ERROR:

Soly: precision 0.2%
Temp: ± 0.5°C (author)

COMPONENTS:

(1) Magnesium iodate; $Mg(IO_3)_2$; [7790-32-1]

(2) Water; H_2O; [7732-18-5]

EVALUATOR:

Hiroshi Miyamoto
Department of Chemistry
Niigata University
Niigata, Japan

April, 1982

CRITICAL EVALUATION:

1. The binary system; $Mg(IO_3)_2$-H_2O

Solubilities in the binary $Mg(IO_3)_2$-H_2O system have been reported in 10 publications (1-9, 11).

Table 1 Solubility studies of magnesium iodate in water

Reference	T/K	Solid Phase	Method of Analysis
Mylius; Funk(1)	273-325	$Mg(IO_3)_2\cdot 10H_2O$	gravimetric($Mg(IO_3)_2$)
	273-373	$Mg(IO_3)_2\cdot 4H_2O$	"
Hill; Moskowitz(2)	273-288	$Mg(IO_3)_2\cdot 10H_2O$	gravimetric(Mg^{2+}) iodometric(IO_3^-)
	278-363	$Mg(IO_3)_2\cdot 4H_2O$	"
	233-363	$Mg(IO_3)_2$	"
Hill;Ricci(3)	278	$Mg(IO_3)_2\cdot 10H_2O$	acidimetric(Mg^{2+}) iodometric(IO_3^-)
	298,323	$Mg(IO_3)_2\cdot 4H_2O$	"
Ricci; Freedman(4)	298	$Mg(IO_3)_2\cdot 4H_2O$	iodometric(IO_3^-)
Vinogradov; Karataeva(5)	323	$Mg(IO_3)_2\cdot 4H_2O$	complexometric(Mg^{2+}) iodometric(IO_3^-)
Azarova; Vinogradov; Pakhomov(6)	323	$Mg(IO_3)_2\cdot 4H_2O$	complexometric(Mg^{2+}) iodometric(IO_3^-)
Tarasova; Vinogradov; Lepeshkov(7)	298	$Mg(IO_3)_2\cdot 4H_2O$	complexometric(Mg^{2+}) iodometric(IO_3^-)
Shklovskaya; Arkhipov; Kidyarov(8)	298	$Mg(IO_3)_2\cdot 4H_2O$	complexometric(Mg^{2+}) iodometric(IO_3^-)
Shklovskaya; Arkipov; Kidyarov; Poleva(9)	298	$Mg(IO_3)_2\cdot 4H_2O$	complexometric(Mg^{2+})
Tarasova; Vinogradov(11)	298	$Mg(IO_3)_2\cdot 4H_2O$	complexometric(Mg^{3+}) iodometric(IO_3^-)

Mylius and Funk (1), and Hill and Moskowitz (2) have measured solubilities in the binary $Mg(IO_3)_2$-H_2O only over a wide temperature range from 273 to 373K. All other investigations deal with ternary systems, and the solubility in the binary system is given as one point in a phase diagram, and they employed the isothermal method. With few exceptions, most papers do not report experimental errors.

Vinogradov and Azarova (10) studied the solubility in the ternary system HIO_3-$Mg(IO_3)_2$-H_2O. However they did not report the solubility in binary $Mg(IO_3)_2$-H_2O system.

COMPONENTS:

(1) Magnesium iodate; $Mg(IO_3)_2$; [7790-32-1]

(2) Water; H_2O; [7732-18-5]

EVALUATOR:

Hiroshi Miyamoto
Department of Chemistry
Niigata University
Niigata, Japan

April, 1982

CRITICAL EVALUATION:

Depending upon temperature and composition, equilibrated solid phases of varying degrees of hydration have been reported. The following solid phases have been identified

$Mg(IO_3)_2 \cdot 10H_2O$	[82150-39-8]
$Mg(IO_3)_2 \cdot 4H_2O$	[13446-17-8]
$Mg(IO_3)_2 \cdot 2H_2O$	[76629-97-5]
$Mg(IO_3)_2$	[7790-32-1]

The relation between temperature and the solubility of magnesium iodate in pure water studied by Hill and Moskowitz (2) is shown in Fig. 1.

The system is found to show the following invariant points: a eutectic at 272.79K found by usual thermal means, a transition of the decahydrate to the tetrahydrate at 286.5K (by interpolation), and a transition of the tetrahydrate to the anhydrous form at 330.7K. The transition points could not be determined by direct measurements of the melting points due to the marked metastability of the various hydrated forms.

Mylius and Funk (1) also studied the relation between the solubility of magnesium iodate in pure water and temperature. The tetra- and decahydrate of magnesium iodate are the reported solid phases, and the transition temperature of 286K was determined graphically. This value is in agreement with that reported by Hill and Moskowitz (2) who did not study the solubility of the anhydrous salt. Mylius and Funk (1) did not study the solubility of the anhydrous salt, and did not report the existence of the metastable tetrahydrate.

The evaluator found using the recommended and tentative values that the monohydrate→hexahydrate transition temperature is 328.5K and the hexahydrate→decahydrate transition temperature is 286.4K.

Hill and Ricci (3) measured the solubility of magnesium iodate in water at 278, 298 and 323K; the solid phase is the decahydrate at 278K and the tetrahydrate at 298 and 323K.

Other investigators (5-9) used the tetrahydrate salt in solubility determinations.

The data to be considered in the critical evaluation are summarized in Table 2.

COMPONENTS:	EVALUATOR:
(1) Magnesium iodate; $Mg(IO_3)_2$; [7790-32-1] (2) Water; H_2O; [7732-18-5]	Hiroshi Miyamoto Department of Chemistry Niigata University Niigata, JAPAN April, 1982

CRITICAL EVALUATION:

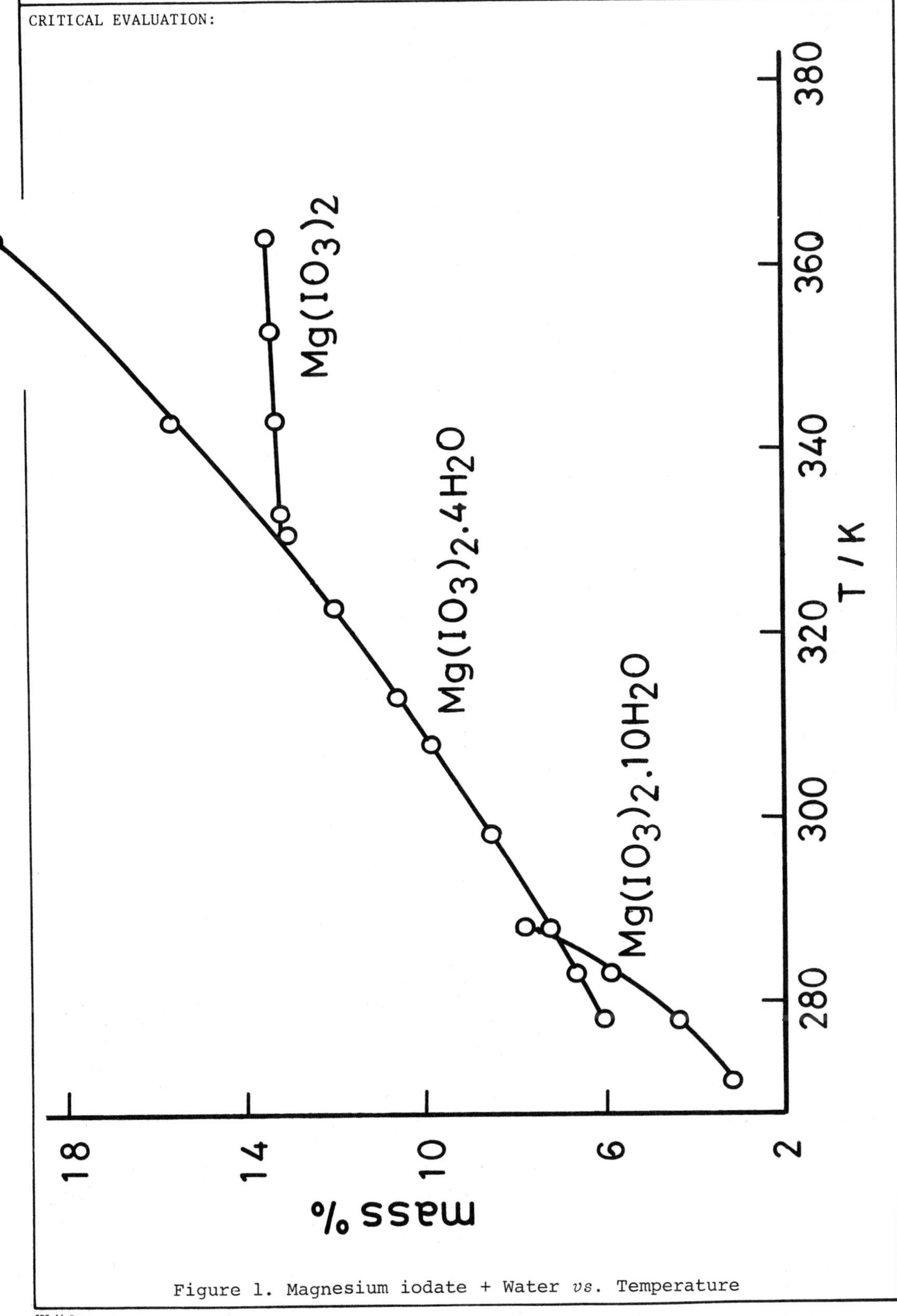

Figure 1. Magnesium iodate + Water *vs.* Temperature

COMPONENTS:

(1) Magnesium iodate; $Mg(IO_3)_2$; [7790-32-1]

(2) Water; H_2O; [7732-18-5]

EVALUATOR:
Hiroshi Miyamoto
Department of Chemistry
Niigata University
Niigata, Japan

April, 1982

CRITICAL EVALUATION:

Table 2 Summary of solubility data in the binary $Mg(IO_3)_2$-H_2O system

T/K	m_1/mol kg^{-1}	ref
$Mg(IO_3)_2 \cdot 10H_2O$		
272.79	0.0878	(2)
273.2	0.086	(1)
278.2	0.122	(3)
"	0.1227	(2)
283.2	0.1667	(2)
288.2	0.2258	(2)
293.2	0.304	(1)
303.2	0.563	(1)
308.2	0.750	(1)
323.2	5.55	(1)
$Mg(IO_3)_2$		
333.2	0.406	(2)
343.2	0.410	(2)
353.2	0.414	(2)
363.2	0.417	(2)

T/K	m_1/mol kg^{-1}	ref
$Mg(IO_3)_2 \cdot 4H_2O$		
278.2	0.1733	(2)
"	0.20	(1)
283.2	0.18	(1)
"	0.1913	(2)
288.2	0.2102	(2)
291.2	0.184	(1)
293.2	0.22	(1)
298.2	0.248	(8)
"	0.248	(9)
"	0.2499	(2)
"	0.250	(3)
"	0.250	(4)
"	0.250	(7)
"	0.250	(11)
308.2	0.26	(1)
"	0.2914	(2)
313.2	0.3139	(2)
323.2	0.3593	(5)
"	0.3635	(3)
"	0.3662	(2)
"	0.3669	(6)
330.7	0.403	(2)
336.2	0.385	(1)
343.2	0.498	(2)
363.2	0.652	(2)
373.2	0.639	(1)

The smoothing equations for the results of Hill and Moskowitz are given as follows:

$$\ln(S_{10}/\text{mol kg}^{-1}) = -27.05375 + 10.09645/(T/100\text{K}) + 20.84322 \ln(T/100\text{K}) \quad : \sigma = 0.0001$$

$$\ln(S_4/\text{mol kg}^{-1}) = -6.196006 - 0.8553821/(T/100\text{K}) + 4.653274 \ln(T/100\text{K}) \quad : \sigma = 0.0039$$

$$\ln(S_0/\text{mol kg}^{-1}) = 1.496293 - 4.293625/(T/100\text{K}) - 0.9215992 \ln(T/100\text{K}) \quad : \sigma = 0.0002$$

where S_{10}, S_4 and S_0 are the solubilities for the decahydrate, the tetrahydrate and the anhydrous salt, respectively.

COMPONENTS:

(1) Magnesium iodate; $Mg(IO_3)_2$; [7790-32-1]

(2) Water; H_2O; [7732-18-5]

EVALUATOR:

Hiroshi Miyamoto
Department of Chemistry
Niigata University
Niigata, Japan

April, 1982

CRITICAL EVALUATION:

EVALUATION OF THE DATA

Solubility at 273.2K (solid phase: decahydrate). Only one result has been reported by Mylius and Funk (1), but the result at this temperature is distinctly lower than that of Hill and Moskowitz (2) at 272.79K, and this value is therefore rejected.

Solubility at 278.2K (solid phase: decahydrate). The result has been reported in 2 publications (2,3). The arithmetic mean of the two results is 0.122 mol kg^{-1} and the standard deviation is 0.0005 mol kg^{-1}. The mean is a recommended value.

Solubility at 278.2K (solid phase: tetrahydrate). The result reported in (1) obtained by Mylius and Funk is higher than that of other investigators (2). It is felt the data in (2) are more accurate, therefore, the result of Mylius and Funk is rejected. The tentative value based on the result reported by Hill and Moskowitz (2) is 0.1733 mol kg^{-1}.

Solubility at 283.2K (solid phase: tetrahydrate). This result has been reported in 2 publications (1,2). The result reported in (1) obtained by Mylius and Funk is considerably lower than that of Hill and Moskowitz (2). It is felt that the data in (2) are more accurate because the result of Mylius and Funk is also lower than the data calculated from the smoothing equation using the result of Hill and Moskowitz. The tentative value based on the work reported by Hill and Moskowitz is 0.1913 mol kg^{-1}.

Solubility at 291.2K (solid phase: tetrahydrate). Only one result has been reported by Mylius and Funk (1), however, the result at this temperature is distinctly lower than that of Hill and Moskowitz at 283 and 288K. This value is therefore rejected.

Solubility at 298.2K (solid phase: tetrahydrate). The result has been reported in 7 publications (2,3,4,7,8,9,11). The arithmetic mean of all results is 0.249 mol kg^{-1}, and the standard deviation is 0.001 mol kg^{-1}. The mean is designated as a recommended value.

Solubility at 308.2K (solid phase: tetrahydrate). This result has been reported in 2 publications (1,2). The result reported in (1) obtained by Mylius and Funk is considerably lower than that of Hill and Moskowitz (2), and as in the above analysis it is felt that the data in (2) are more accurate. The tentative value based on the result of Hill and Moskowitz is 0.291^4 mol kg^{-1}.

Solubility at 323.2K (solid phase: tetrahydrate). This value has been reported in 4 publications (2,3,5,6). The arithmetic mean of all results is 0.364 mol kg^{-1} and the standard deviation is 0.003 mol kg^{-1}. The mean is designated as a recommended value.

Solubility at 336.2K (solid phase: tetrahydrate). Only one result has been reported by Mylius and Funk (1), but the result at this temperature is considerably lower than that of Hill and Moskowitz at 330.7K. The value is therefore rejected.

COMPONENTS:

(1) Magnesium iodate; $Mg(IO_3)_2$; [7790-32-1]

(2) Water; H_2O; [7732-18-5]

EVALUATOR:

Hiroshi Miyamoto
Department of Chemistry
Niigata University
Niigata, Japan

April, 1982

CRITICAL EVALUATION:

Solubility at other temperatures. Only one result at each temperature has been reported by Mylius and Funk (1) and by Hill and Moskowitz (2). As in the above analysis, the results of Hill and Moskowitz are more accurate and reliable. The results of Hill and Moskowitz are designated as tentative values.

The recommended and tentative values with the calculated values from the smoothing equations are given in Table 3.

Table 3 Recommended and tentative values for solubility of magnesium iodate in water

T/K	m_1/mol kg^{-1}	σ/mol kg^{-1}	m_1'(calcd)/mol kg^{-1}
Solid phase: $Mg(IO_3)_2 \cdot 10H_2O$			
272.79	0.0878	--	0.0877
278.2	0.122[a]	0.0005	0.122
283.2	0.167	--	0.166
288.2	0.226	--	0.226
Solid phase: $Mg(IO_3)_2 \cdot 4H_2O$			
278.2	0.1733	--	0.1751
283.2	0.1913	--	0.1912
288.2	0.2102	--	0.2084
298.2	0.249[a]	0.001	0.2466
308.2	0.2914	--	0.2900
313.2	0.3139	--	0.3139
323.2	0.364[a]	0.003	0.3665
330.7	0.403	--	0.410
343.2	0.498	--	0.492
363.2	0.652	--	0.651
Solid phase: $Mg(IO_3)_2$			
333.2	0.406	--	0.406
343.2	0.410	--	0.410
353.2	0.414	--	0.414
363.2	0.417	--	0.417

m_1: experimental value

σ: standard deviation

m_1': calculated value

a: recommended value

The fitting equation used was:

$$\ln S = A + B/(T/100K) + C \ln (T/100K)$$

By using T/100K as the variable rather than T/K the coefficients in the smoothing equation are of roughly equal magnitude.

COMPONENTS:

(1) Magnesium iodate; $Mg(IO_3)_2$; [7790-32-1]

(2) Water; H_2O; [7732-18-5]

EVALUATOR:

Hiroshi Miyamoto
Department of Chemistry
Niigata University
Niigata, Japan

April, 1982

CRITICAL EVALUATION:

The data in Table 3 were fitted to the following equations:

$$\ln(S_{10}/\text{mol kg}^{-1}) = -52.36353 + 45.02929/(T/100\text{K}) + 33.30526 \ln(T/100\text{K}) \; : \; \sigma = 0.0007$$

$$\ln(S_4/\text{mol kg}^{-1}) = -6.714708 - 9.573980/(T/100\text{K}) + 4.893593 \ln(T/100\text{K}) \; : \; \sigma = 0.0039$$

$$\ln(S_0/\text{mol kg}^{-1}) = 1.496293 - 4.293625/(T/100\text{K}) - 0.9215992 \ln(T/100\text{K}) \; : \; \sigma = 0.0002$$

where S_{10}, S_4, and S_0 are the concentrations of the saturated solutions in equilibrium with the decahydrate, the tetrahydrate and the anhydrous salt, respectively.

The values calculated from the smoothing equations also are given in Table 3.

2. Ternary systems

Systems with alkali metal iodates. The ternary systems $Mg(IO_3)_2$-$LiIO_3$-H_2O, $Mg(IO_3)_2$-$NaIO_3$-H_2O, $Mg(IO_3)_2$-KIO_3-H_2O, $Mg(IO_3)_2$-$RbIO_3$-H_2O and $Mg(IO_3)_2$-$CsIO_3$-H_2O have been studied. The existence of double salts in the ternary $Mg(IO_3)_2$-$LiIO_3$-H_2O and $Mg(IO_3)_2$-KIO_3-H_2O is reported in (6) obtained by Azarova, Vinogradov, and Pakhomov, and in (5) obtained by Vinogradov and Karataeva, respectively. The double salts found experimentally are:

$mLiIO_3 \cdot nMg(IO_3)_2 \cdot 4H_2O$ (6)

$2KIO_3 \cdot Mg(IO_3)_2 \cdot 4H_2O$ (5)

Double salts in the other systems were not formed. The dominant feature in these systems are simple eutonic type phase diagrams.

Systems with alkaline earth iodates. Solubilities in the ternary $Mg(IO_3)_2$-$Ca(IO_3)_2$-H_2O and $Mg(IO_3)_2$-$Ba(IO_3)_2$-H_2O systems have been reported in (9) obtained by Shklovskaya's group, and in (4) obtained by Ricci and Freedman, respectively. The system with $Ca(IO_3)_2$ is of the simple eutonic type. The solubility of $Ba(IO_3)_2$ in the ternary $Mg(IO_3)_2$-$Ba(IO_3)_2$-H_2O system is negligible, and only the solubility of $Mg(IO_3)_2$ was reported.

Systems with other compounds. The ternary $Mg(IO_3)_2$-HIO_3-H_2O, $Mg(IO_3)_2$-NH_4IO_3-H_2O, and $Mg(IO_3)_2$-$Mg(NO_3)_2$-H_2O systems have been studied. The dominant feature in these systems are phase diagrams of the eutonic type: double salts were not reported.

COMPONENTS:	EVALUATOR:
(1) Magnesium iodate; $Mg(IO_3)_2$; [7790-32-1] (2) Water; H_2O; [7732-18-5]	Hiroshi Miyamoto Department of Chemistry Niigata University Niigata, Japan April, 1982

CRITICAL EVALUATION:

REFERENCES:

1. Mylius, F.; Funk, R. *Ber. Dtsch. Chem. Ges.* 1897, *30*, 1716.
2. Hill, A. E.; Moskowitz, S. *J. Am. Chem. Soc.* 1931, *53*, 941.
3. Hill, A. E.; Ricci, J. E. *J. Am. Chem. Soc.* 1931, *53*, 4305.
4. Ricci, J. E.; Freedman, A. J. *J. Am. Chem. Soc.* 1952, *74*, 1769.
5. Vinogradov, E. E.; Karataeva, I. M. *Zh. Neorg. Khim.* 1976, *21*, 1666; *Russ. J. Inorg. Chem. (Engl. Transl.)* 1976, *21*, 912.
6. Azarova, L. A.; Vinogradov, E. E.; Pakhomov, V. I. *Zh. Neorg. Khim.* 1976, *21*, 2801; *Russ. J. Inorg. Chem. (Engl. Transl.)* 1976, *21*, 1545.
7. Tarasova, G. N.; Vinogradov, E. E.; Lepeshkov, I. N. *Zh. Neorg. Khim.* 1977, *22*, 809; *Russ. J. Inorg. Chem. (Engl. Transl.)* 1977, *22*, 448.
8. Shklovskaya, R. M.; Arkhipov, S. M.; Kidyarov, B. I. *Izv. Sib. Otd. Akad. Nauk. SSSR Ser. Khim. Nauk* 1979, *(9)*, 75.
9. Shklovskaya, R. M.; Arkhipov, S. M.; Kidyarov, B. I.; Poleva, G. V. *Zh. Neorg. Khim.* 1979, *24*, 1416; *Russ. J. Inorg. Chem. (Engl. Transl.)* 1979, *24*, 786.
10. Vinogradov, E. E.; Azarova, L. A. *Zh. Neorg. Khim.* 1977, *22*, 1666; *Russ. J. Inorg. Chem. (Engl. Transl.)* 1977, *22*, 903.
11. Tarasova, G. N.; Vinogradov, E. E. *Zh. Neorg. Khim.* 1981, *26*, 2283. *Russ. J. Inorg. Chem. (Engl. Transl.)* 1981, *26*, 1544.

COMPONENTS:

(1) Magnesium iodate; $Mg(IO_3)_2$; [7790-32-1]

(2) Water; H_2O; [7732-18-5]

ORIGINAL MEASUREMENTS:

Mylius, F.; Funk, R.

Ber. Dtsch. Chem. Ges. 1897, *30*, 1716-25.

VARIABLES:

T/K = 273 - 373

PREPARED BY:

Hiroshi Miyamoto

EXPERIMENTAL VALUES:

t/°C	Magnesium Iodate mass %	m_1/mol kg^{-1} [a]	Nature of the solid phase
0	3.1	0.086	$Mg(IO_3)_2 \cdot 10H_2O$
20	10.2	0.304	"
30	17.4	0.563	"
35	21.9	0.750	"
50	67.5	5.55	"
0	6.8	0.20	$Mg(IO_3)_2 \cdot 4H_2O$
10	6.4	0.18	"
18[b]	6.44	0.184	"
20	7.7	0.22	"
35	8.9	0.26	"
63	12.6	0.385	"
100	19.3	0.639	"

[a] Molalities calculated by compiler

[b] The solubility, 6.88 g/100g H_2O, and the density of the saturated solution, 1.078, were also reported.

The compiler presumes that the first word in the fifth line from the end of page 1717 should read 100g.

AUXILIARY INFORMATION

METHOD/APPARATUS/PROCEDURE:

The salt and water were placed in a bottle. The bottle was shaken in a constant temperature bath for a long time.
After the saturated solution settled, an aliquot of solution was removed with a pipet. Magnesium iodate was determined by evaporation of the solution to dryness.
The density of the saturated solution was also determined.

SOURCE AND PURITY OF MATERIALS:

The salt used was purchased as "pure" chemical, and traces of impurities were not present. The purity sufficed for the solubility determination.

ESTIMATED ERROR:

Soly: precision within 1 %.
Temp: nothing specified

REFERENCES:

COMPONENTS:

(1) Magnesium iodate; $Mg(IO_3)_2$; [7790-32-1]

(2) Water; H_2O; [7732-18-5]

ORIGINAL MEASUREMENTS:

Hill, A. E.; Moskowitz, S.

J. Am. Chem. Soc. 1931, *53*, 941-6.

VARIABLES:

T/K = 272.79 - 363

PREPARED BY:

Hiroshi Miyamoto

EXPERIMENTAL VALUES:

t/°C	Magnesium Iodate mass %	m_1/mol kg^{-1} [a]	Nature of the Solid Phase[b]
- 0.36	3.18	0.0878	A + ice
+ 5	4.39	0.1227	A
10	5.87	0.1667	A
15	7.79	0.2258	A(m)
5	6.09	0.1733	B(m)
10	6.68	0.1913	B(m)
15	7.29	0.2102	B
25	8.55	0.2499	B
35	9.83	0.2914	B
40	10.51	0.3139	B
50	12.05	0.3662	B
57.5	13.1	0.403	B
70	15.7	0.498	B(m)
90	19.6	0.652	B(m)
60	13.2	0.406	C
70	13.3	0.410	C
80	13.4	0.414	C
90	13.5	0.417	C

[a] Molalities calculated by compiler.

[b] A = $Mg(IO_3)_2 \cdot 10H_2O$; B = $Mg(IO_3)_2 \cdot 4H_2O$; C = $Mg(IO_3)_2$
(m) indicates that the solid phase was metastable.

AUXILIARY INFORMATION

METHOD/APPARATUS/PROCEDURE:

Excess magnesium iodate was added to distilled water in glass-stoppered Pyrex tubes. For the lower temperatures the tubes were rotated in a water thermostat. For temperatures above 50°C the tubes were placed in an air thermostat and stirred by means of a brass paddle on a mechanical stirrer. Filtered samples were withdrawn by calibrated pipets at the lower temperatures and weighed to give approximate figures for density. At the higher temperatures the samples were forced through a tube into a weighing bottle by air pressure. At lower temperatures, one to three days were found sufficient to attain equilibrium from undersaturation. The time required for the change from the tetrahydrate to the anhydrate was 4 days at 80°C and 2 days at 90°C. At 70°C the solid phase used was anhydrous salt. Equilibrium at the higher temperatures being reached within one days. The iodate content was determined iodometrically.

SOURCE AND PURITY OF MATERIALS:

Magnesium iodate was prepared by neutralizing an aqueous solution of HIO_3 with $MgCO_3$ and evaporating the slightly acidified solution at above 40-50°C for 4 days, which gave a copious crystallization of the tetrahydrate. The crystals were washed with water, air-dried, ground and placed in a desiccator.

ESTIMATED ERROR:

Soly: nothing specified
Temp: water thermostat, constant to about 0.03°C

REFERENCES:

COMPONENTS:

(1) Magnesium iodate; $Mg(IO_3)_2$; [7790-32-1]

(2) Iodic acid; HIO_3; [7782-68-5]

(3) Water; H_2O; [7732-18-5]

ORIGINAL MEASUREMENTS:

Ricci, J. E.; Freedman, A. J.

J. Am. Chem. Soc. 1952, *74*, 1769-73.

VARIABLES:

T/K = 298

HIO_3/ mass % = 0 - 75.32

PREPARED BY:

Hiroshi Miyamoto

EXPERIMENTAL VALUES:

Composition of the Saturated Solutions

t/°C	Iodic Acid		Magnesium Iodate		Nature of the Solid Phase[a]
	mass %	mol % (compiler)	mass %	mol % (compiler)	
25	0.0	0	8.55[b]	0.448	A
	10.04	1.278	11.04	0.6605	A
	13.99	1.901	12.68	0.8101	A
	21.38	3.384	16.93	1.260	A
	29.61	5.676	21.01	1.894	A
	36.63	8.524	24.28	2.657	A
	45.17	12.945	24.92	3.358	A
	47.71	14.559	24.81	3.560	A
	66.85	23.345	11.21	1.841	B
	70.37	23.653	6.69	1.057	B
	72.71	23.511	3.22	0.490	B
	75.32	23.812	0.00	0.00	B

[a] A = $Mg(IO_3)_2 \cdot 4H_2O$; B = HIO_3.

[b] For binary systems the compiler computes the following:

Soly of $Mg(IO_3)_2$ = 0.250 mol kg^{-1}

Soly of HIO_3 =]7.35 mol kg^{-1}.

AUXILIARY INFORMATION

METHOD/APPARATUS/PROCEDURE:

The details of equilibrium procedure were not given in the paper. For the analysis of the solution, HIO_3 was determined by titration with NaOH using methyl red as an indicator. Total iodate was then determined iodometrically on the neutralized sample.

SOURCE AND PURITY OF MATERIALS:

Magnesium iodate was prepared by neutralizing an aqueous solution of HIO_3 with $MgCO_3$ and evaporating the slightly acidified solution at above 40-50°C for 4 days, which gave a copious crystallization of the tetrahydrate. The crystals were washed with water, air-dried, ground and placed in a desiccator. The product contained 83.76% $Mg(IO_3)_2$ as compared with the theoretical 83.86%.

One sample of HIO_3 used was a commercial c.p. product containing 99.82% HIO_3 by determination of iodate and of acid. Another sample was made from I_2O_5 (c.p. grade) and water. The solution was evaporated at ∿ 40°C in a steam of air. When ground and stored in vacuum, constant composition was reached after two weeks, at 99.66% HIO_3. The authors state that the solubility of the two samples was the same.

COMPONENTS:

(1) Magnesium iodate; $Mg(IO_3)_2$; [7790-32-1]

(2) Iodic acid; HIO_3; [7782-68-5]

(3) Water; H_2O; [7732-18-5]

ORIGINAL MEASUREMENTS:

Ricci, J. E.; Freedman, A. J.

J. Am. Chem. Soc. 1952, *74*, 1769-73.

COMMENTS AND/OR ADDITIONAL DATA:

The phase diagram is given below (based on mass%).

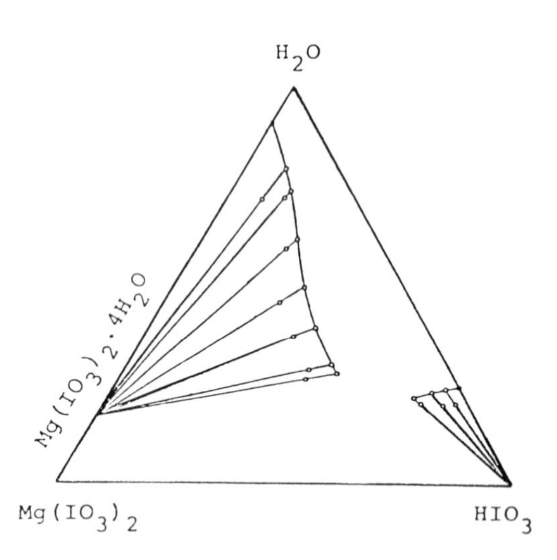

AUXILIARY INFORMATION

METHOD/APPARATUS/PROCEDURE:

SOURCE AND PURITY OF MATERIALS:

ESTIMATED ERROR:

ACKNOWLEDGEMENT:
The figure reprinted from the *J. Am. Chem. Soc.* by permission of the copyright owners, The American Chemical Society.

COMPONENTS:

(1) Magnesium iodate; $Mg(IO_3)_2$; [7790-32-1]

(2) Iodic acid; HIO_3; [7782-68-5]

(3) Water; H_2O; [7732-18-5]

ORIGINAL MEASUREMENTS:

Vinogradov, E. E.; Azarova, L. A.

Zh. Neorg. Khim. 1977, *22*, 1666-8; *Russ. J. Inorg. Chem. (Engl. Transl.)* 1977, *22*, 903-5.

VARIABLES:

T/K = 323
composition

PREPARED BY:

Hiroshi Miyamoto

EXPERIMENTAL VALUES:

t/°C	Composition of Saturated Solutions				Nature of the Solid Phase[a]
	Iodic Acid		Magnesium Iodate		
	mass %	mol % (compiler)	mass %	mol % (compiler)	
50	0.31	0.088	66.77	8.890	A
	0.31	0.099	71.10	10.68	A
	trace	--	75.14	12.71	A
	1.29	0.629	81.76	18.73	A
	2.96	1.24	76.52	15.04	A
	4.03	1.77	76.77	15.86	A
	3.77	1.64	76.78	15.71	A
	9.48	3.94	70.24	13.73	A
	61.83	39.57	29.94	9.009	A + B
	45.12	21.23	39.64	8.768	A + B
	85.91	46.83	4.31	1.10	B
	100.00	--[b]	0.37	--[b]	B
	98.13	87.09	0.40	0.17	B
	91.49	79.68	6.43	2.63	B

[a] A = $Mg(IO_3)_2 \cdot 4H_2O$; B = HIO_3.

[b] Compiler was unable to calculate mol % from given data.

AUXILIARY INFORMATION

METHOD/APPARATUS/PROCEDURE:

The solubility of $Mg(IO_3)_2$ in HIO_3-water system was studied in a water thermostat at 50°C. The equilibrium of the system was established in 10 days.
The magnesium content in liquid and solid phases were determined complexometrically, and iodate iodometrically. The composition and nature of the solid phases were found by Schreinemakers' method of "residues" crystal-optically, and by X-ray diffraction. The x-ray diffraction patterns were recorded on a Rigaku-Denki Geigerfleks diffractometer with Cu Kα radiation.

SOURCE AND PURITY OF MATERIALS:

Magnesium iodate was made from magnesium carbonate and HIO_3. The chemical analysis and mass losses in the recording of the derivatograms showed that the product corresponded to the formula $Ma(IO_3)_2 \cdot 4H_2O$.
Chemical pure grade iodic acid was used.

ESTIMATED ERROR:

Soly: nothing specified
Temp: ± 0.1 % (authors)

REFERENCES:

COMPONENTS:

(1) Magnesium iodate; $Mg(IO_3)_2$; [7790-32-1]

(2) Iodic acid; HIO_3; [7782-68-5]

(3) Water; H_2O; [7732-18-5]

ORIGINAL MEASUREMENTS:

Vinogradov, E. E.; Azarova, L. A.

Zh. Neorg. Khim. 1977, 22, 1666-8; *Russ. J. Inorg. Chem. (Engl. Transl.)* 1977, 22, 903-5.

COMMENTS AND/OR ADDITIONAL DATA: continued

The phase diagram is given below (based on mass%).

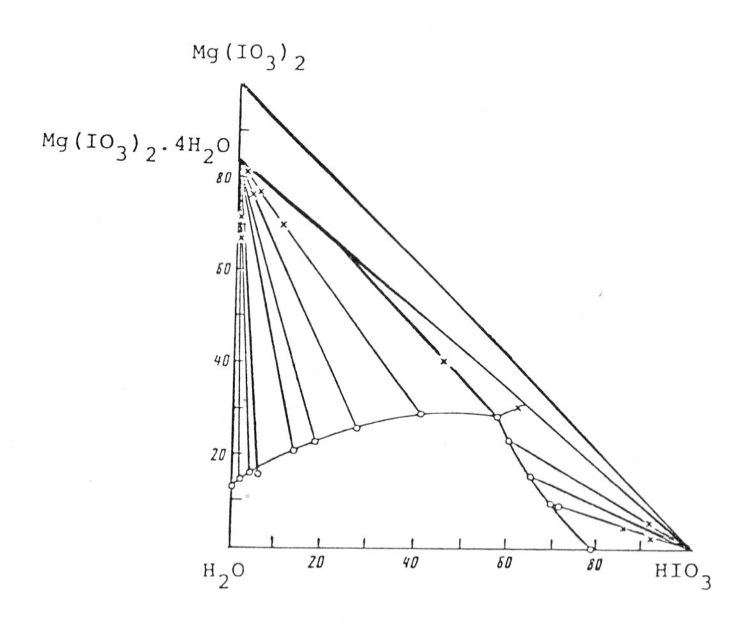

AUXILIARY INFORMATION

METHOD/APPARATUS/PROCEDURE:

SOURCE AND PURITY OF MATERIALS:

ESTIMATED ERROR:

ACKNOWLEDGEMENT:

The figure reprinted from *Zh. Neorg. Khim.* by permission of the copyright owners, VAAP, The Copyright Agency of the USSR.

COMPONENTS:

(1) Magnesium iodate; $Mg(IO_3)_2$; [7790-32-1]

(2) Ammonium iodate; NH_4IO_3; [13446-09-8]

(3) Water; H_2O; [7732-18-5]

ORIGINAL MEASUREMENTS:

Tarasova, G. N.; Vinogradov, E. E.; Lepeshkov, I. N.

Zh. Neorg. Khim. 1977, *22*, 809-11; *Russ. J. Inorg. Chem. (Engl. Transl.)* 1977, *22*, 448-9.

VARIABLES:

T/K = 298
composition

PREPARED BY:

Hiroshi Miyamoto

EXPERIMENTAL VALUES:

Composition of Saturated Solutions

t/°C	Ammonium Iodate mass %	Ammonium Iodate mol % (compiler)	Magnesium Iodate mass %	Magnesium Iodate mol % (compiler)	Nature of the Solid Phase[a]
25	3.72	0.359	--	--	A
	3.29	0.322	1.26	0.0633	A
	3.06	0.299	1.85	0.0933	A
	2.88	0.286	3.62	0.186	A
	2.68	0.268	4.41	0.227	A
	2.08	0.209	5.39	0.279	A
	1.85	0.190	7.77	0.411	A + B
	1.75	0.180	7.76	0.411	A + B
	1.98	0.204	8.00	0.425	A + B
	1.58	0.163	7.76	0.410	B
	0.49	0.050	8.37	0.440	B
	--	--	8.55[b]	0.448	B

[a] A = NH_4IO_3; B = $Mg(IO_3)_2 \cdot 4H_2O$

[b] For binary system the compiler computes the following

Soly of $Mg(IO_3)_2$ = 0.250 mol kg^{-1} at 25°C

AUXILIARY INFORMATION

METHOD/APPARATUS/PROCEDURE:

The experiment was carried out in a water thermostat with an electric heater. Equilibrium was reached in 12-14 days at 25°C with continuous stirring.
The magnesium content in liquid phase was analysed by complexometric titration. The iodate content was determined iodometrically, and the ammonium content gravimetrically with use of sodium tetraphenylborate. The solid phases were determined by Schreinmakers' method of "residues".

SOURCE AND PURITY OF MATERIALS:

Magnesium iodate was made from HIO_3 and magnesium carbonate.
NH_4IO_3 was prepared by mixing a slight excess of NH_4OH with HIO_3 in water. The precipitate was then filtered, washed to remove the excess NH_3, and redissolved in the growth solution. Crystals were then grown by evaporation at room temperature.

ESTIMATED ERROR:

Soly: nothing specified
Temp: ± 0.1°C (authors)

REFERENCES:

COMPONENTS:

(1) Magnesium iodate; $Mg(IO_3)_2$; [7790-32-1]

(2) Ammonium iodate; NH_4IO_3; [13446-09-8]

(3) Water; H_2O; [7732-18-5]

ORIGINAL MEASUREMENTS:

Tarasova, G. N.; Vinogradov, E. E.; Lepeshkov, I. N.

Zh. Neorg. Khim. 1977, *22*, 809-11; *Russ. J. Inorg. Chem. (Engl. Trans.)* 1977, *22*, 448-9.

COMMENTS AND/OR ADDITIONAL DATA: continued

The phase diagram is given below (based on mass%).

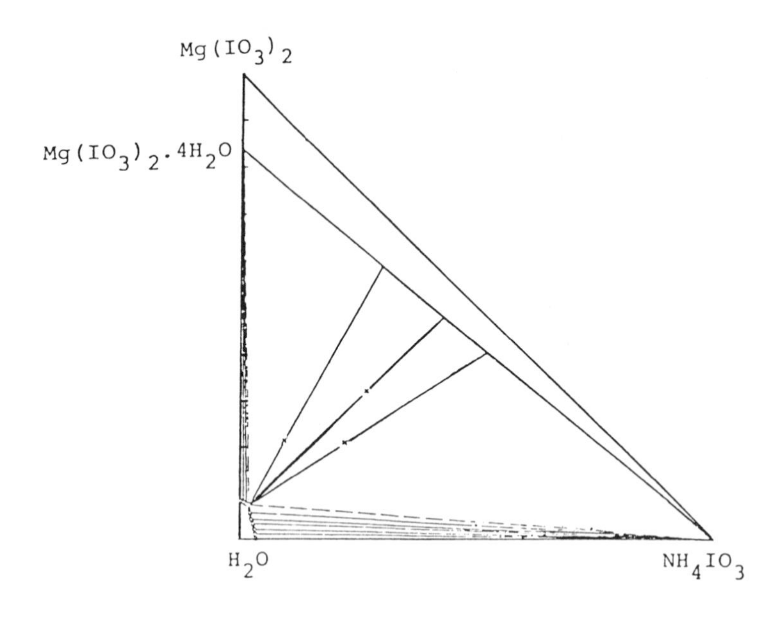

AUXILIARY INFORMATION

METHOD/APPARATUS/PROCEDURE:

SOURCE AND PURITY OF MATERIALS:

ESTIMATED ERROR:

ACKNOWLEDGEMENT:

The figure reprinted from *Zh. Neorg. Khim.* by permission of the copyright owners, VAAP, The Copyright Agency of the USSR.

COMPONENTS:

(1) Magnesium iodate; $Mg(IO_3)_2$; [7790-32-1]

(2) Lithium iodate; $LiIO_3$; [13765-03-2]

(3) Water; H_2O; [7732-18-5]

ORIGINAL MEASUREMENTS:

Azarova, L. A.; Vinogradov, E. E.; Pakhnomov, V. I.

Zh. Neorg. Khim. 1976, *21*, 2801-4; *Russ. J. Inorg. Chem. (Engl. Transl.)* 1976, *21*, 1545-7.

VARIABLES:

T/K = 323
Composition

PREPARED BY:

Hiroshi Miyamoto

EXPERIMENTAL VALUES:

t/°C	Composition of Saturated Solutions				Nature of the Solid Phase[a]
	Lithium Iodate		Magnesium Iodate		
	mass %	mol % (compiler)	mass %	mol % (compiler)	
50	--	--	12.07[b]	0.6567	A
	1.99	0.23	11.14	0.6124	B
	8.93	1.05	7.83	0.446	B
	30.34	4.386	4.35	0.306	B
	37.25	5.966	4.82	0.375	B
	43.48	7.252	1.50	0.122	B + C
	43.44	7.166	0.85	0.0868	B + C
	41.97	6.782	0.92	0.072	B + C
	42.49	6.842	0.21	0.016	C
	43.28	7.028	--	--	C
	Metastable branch				
	4.35	0.504	11.03	0.6206	A
	18.00	2.344	8.10	0.513	A
	26.99	3.770	5.00	0.339	B
	37.26	5.939	4.49	0.348	B

[a] A = $Mg(IO_3)_2 \cdot 4H_2O$; B = $mMg(IO_3)_2 \cdot 4H_2O \cdot nLiIO_3$; C = $LiIO_3$.

[b] For binary system the compiler computes the following

Soly of $Mg(IO_3)_2$ = 0.3669 mol kg^{-1} at 50°C

AUXILIARY INFORMATION

METHOD/APPARATUS/PROCEDURE:

Equilibrium was reached in the ternary system in 12 days. The magnesium content in the liquid phase was determined complexometrically, IO_3^- iodometrically. The lithium content was determined by difference. Schreinemakers' method of "residues" and X-ray diffraction were used to find the composition and nature of the solid.

SOURCE AND PURITY OF MATERIALS:

Lithium and magnesium iodate were made from the corresponding carbonate and iodic acid. The chemical analysis and mass loss on the derivatogram showed that the magnesium iodate obtained has the formula $Mg(IO_3)_2 \cdot 4H_2O$.

ESTIMATED ERROR:

Nothing specified

REFERENCES:

COMPONENTS:

(1) Magnesium iodate; $Mg(IO_3)_2$; [7790-32-1]

(2) Lithium iodate; $LiIO_3$; [13765-03-2]

(3) Water; H_2O; [7732-18-5]

ORIGINAL MEASUREMENTS:

Azarova, L. A.; Vinogradov, E. E.; Pakhnomov, V. I.

Zh. Neorg. Khim. 1976, *21*, 2801-4; *Russ. J. Inorg. Chem. (Engl. Transl.)* 1976, *21*, 1545-7.

COMMENTS AND/OR ADDITIONAL DATA: continued

The phase diagram is given below (based on mass%).

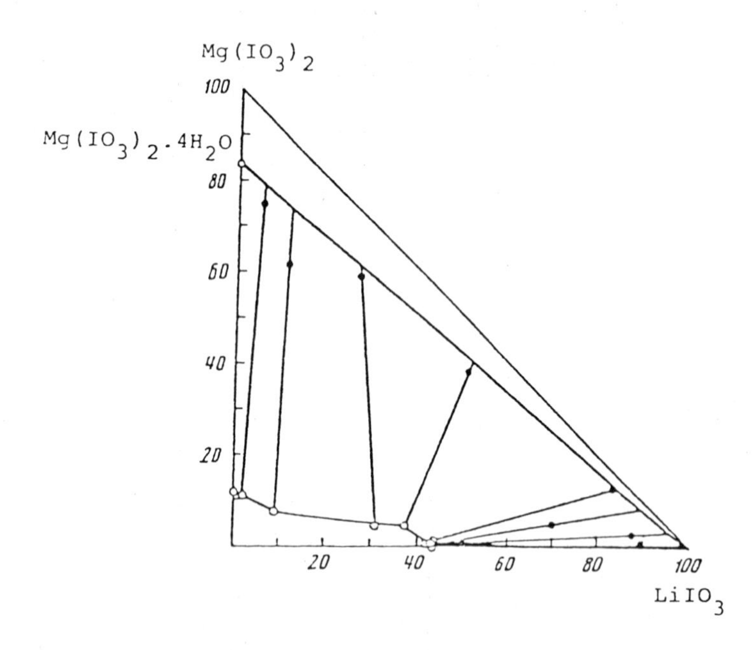

AUXILIARY INFORMATION

METHOD/APPARATUS/PROCEDURE:

SOURCE AND PURITY OF MATERIALS:

ESTIMATED ERROR:

ACKNOWLEDGEMENT:

The figure reprinted from *Zh. Neorg. Khim.* by permission of the copyright owners, VAAP, The Copyright Agency of the USSR.

COMPONENTS:

(1) Magnesium iodate; $Mg(IO_3)_2$; [7790-32-1]

(2) Lithium iodate; $LiIO_3$; [13765-03-2]

(3) Water; H_2O; [7732-18-5]

ORIGINAL MEASUREMENTS:

Shklovskaya, R. M.; Arkhipov, S. M.; Kidyarov, B. I.

Izv. Sib. Otd. Akad. Nauk. SSSR Ser. Khim. Nauk 1979, (9) 75-7.

VARIABLES:

T/K = 298
$LiIO_3$/mass % = 0 - 41.50

PREPARED BY:

Hiroshi Miyamoto

EXPERIMENTAL VALUES:

t/°C	Composition of Saturated Solutions				Nature of the Solid Phase[a]
	Lithium Iodate		Magnesium Iodate		
	mass %	mol % (compiler)	mass %	mol % (compiler)	
25	--	--	8.49[b]	0.445	A
	2.48	0.271	7.35	0.390	A
	5.89	0.657	6.13	0.332	A
	9.22	1.06	5.43	0.302	A
	13.57	1.600	3.95	0.226	A
	19.62	2.457	3.38	0.206	A
	23.66	3.101	3.24	0.206	A
	28.27	3.901	2.88	0.193	A
	33.63	4.960	2.66	0.191	A
	36.39	5.555	2.43	0.180	A
	39.45	6.286	2.40	0.186	A
	41.50	6.808	2.33	0.186	A + B
	43.82	7.173	--	--	B

[a] A = $Mg(IO_3)_2 \cdot 4H_2O$; B = α-$LiIO_3$.

[b] For binary system the compiler computes the following

Soly of $Mg(IO_3)_2$ = 0.248 mol kg^{-1} at 25°C.

AUXILIARY INFORMATION

METHOD/APPARATUS/PROCEDURE:

The isothermal method was employed. Ternary complexes, $Mg(IO_3)_2$-α-$LiIO_3$-H_2O, of known composition were made to come to equilibrium at 25°C. The mixture was stirred for 30 days. The iodate content in saturated solution was determined iodometrically, the magnesium content was determined complexometrically. The lithium content was calculated by difference. The composition of the solid phases was determined by the "residues" method and verified by X-ray diffraction.

SOURCE AND PURITY OF MATERIALS:

$Mg(IO_3)_2 \cdot 4H_2O$ was prepared from $MgCO_3$ and KIO_3.
α-$LiIO_3$ used was of guarantee reagent.

ESTIMATED ERROR:

Nothing specified.

REFERENCES:

COMPONENTS:

(1) Magnesium iodate; $Mg(IO_3)_2$; [7790-32-1]

(2) Lithium iodate; $LiIO_3$; [13765-03-2]

(3) Water; H_2O; [7732-18-5]

ORIGINAL MEASUREMENTS:

Shklovskaya, R. M.; Arkhipov, S. M. Kidyarov, B. I.

Izv. Sib. Otd. Akad. Nauk. SSSR Ser. Khim. Nauk 1979, (9) 75-7.

COMMENTS AND/OR ADDITIONAL DATA: (continued)

The phase diagram is given below (based on mass%).

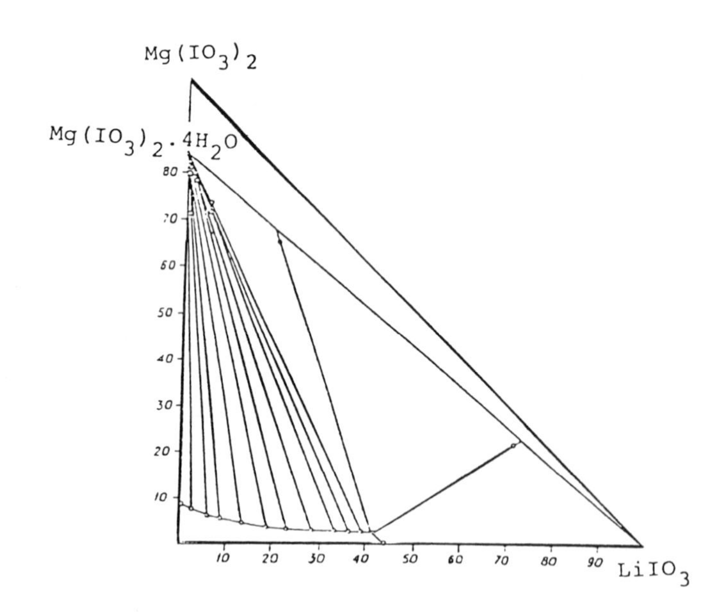

AUXILIARY INFORMATION

METHOD/APPARATUS/PROCEDURE:

SOURCE AND PURITY OF MATERIALS:

ESTIMATED ERROR:

ACKNOWLEDGEMENT:

The figure reprinted from the journal by permission of the copyright owners, VAAP, The Copyright Agency of the USSR.

COMPONENTS:
(1) Magnesium iodate; $Mg(IO_3)_2$; [7790-32-1]
(2) Sodium iodate; $NaIO_3$; [7681-55-2]
(3) Water; H_2O; [7732-18-5]

ORIGINAL MEASUREMENTS:
Hill, A. E.; Ricci, J. E.

J. Am. Chem. Soc. 1931, *53*, 4305-15.

EXPERIMENTAL VALUES:

t/°C	Composition of Saturated Solutions				Nature of the Solid Phase[a]
	Sodium Iodate		Magnesium Iodate		
	mass %	mol % (compiler)	mass %	mol % (compiler)	
5	0.00	--	4.37[b]	0.220	A
	1.45	0.139	3.74	0.189	A
	2.53	0.244	3.33	0.170	A + C
	2.68	0.256	2.19	0.110	C
	3.28	0.308	0.00	--	C
25	0.00	--	8.54[b]	0.448	B
	2.80	0.281	7.24	0.385	B
	6.16	0.632	6.05	0.353	B
	7.49	0.778	5.86	0.322	B + C
	7.66	0.796	5.73	0.315	B(m)
	8.79	0.922	5.57	0.309	B + C(m)
	8.74	0.916	5.42	0.300	C(m)
	7.52	0.778	5.40	0.295	C
	7.50	0.775	5.33	0.291	C
	7.77	0.800	4.76	0.259	C
	2.94	0.283	2.99	0.152	C
	8.30	0.827	1.18	0.062	C
	8.57	0.846	0.00	--	C
50	--	--	11.97[b]	0.651	B
	4.59	0.484	9.95	0.555	B
	8.94	0.970	8.41	0.483	B
	11.97	1.332	7.67	0.451	B + C
	12.42	1.346	4.93	0.282	C
	13.05	1.404	3.73	0.212	C
	13.26	1.419	3.05	0.172	C
	13.27	1.414	2.64	0.149	C
	13.54	1.430	1.55	0.087	C
	13.49	1.400	0.00	--	C

[a] A = $Mg(IO_3)_2\cdot 10H_2O$; B = $Mg(IO_3)_2\cdot 4H_2O$;
C = $NaIO_3\cdot 5H_2O$; (m) = metastable.

[b] For binary systems the compiler computes the following:

Soly of $Mg(IO_3)_2$ = 0.122 mol kg^{-1} at 5°C,
= 0.250 mol kg^{-1} at 25°C,
= 0.3635 mol kg^{-1} at 50°C.

COMPONENTS:

(1) Magnesium iodate; $Mg(IO_3)_2$; [7790-32-1]

(2) Sodium iodate; $NaIO_3$; [7681-55-2]

(3) Water; H_2O; [7732-18-5]

ORIGINAL MEASUREMENTS:

Hill, A. E.; Ricci, J. E.

J. Am. Chem. Soc. 1931, *53*, 4305-15.

VARIABLES:

T/K = 278, 298 and 323
$NaIO_3$/mass % = 0 - 13.49

PREPARED BY:

Hiroshi Miyamoto

COMMENTS AND/OR ADDITIONAL DATA:

The phase diagram is given below (based on mass%).

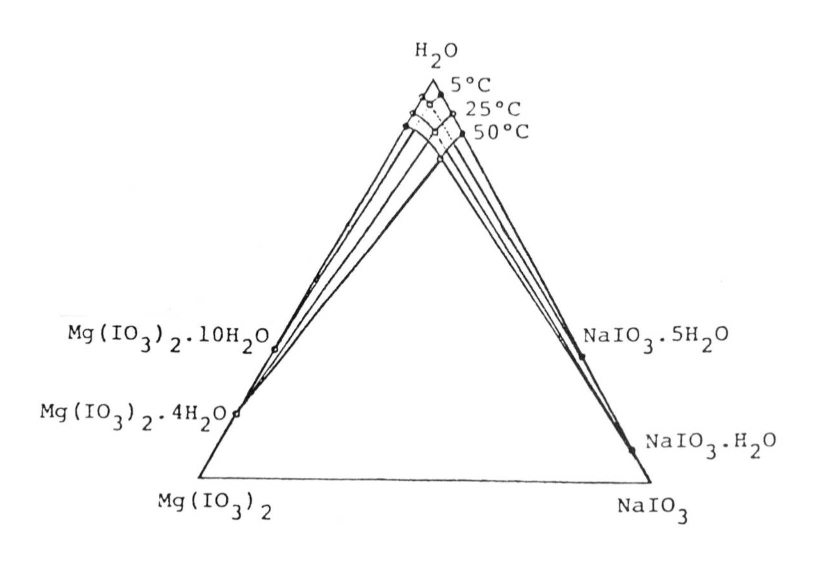

AUXILIARY INFORMATION

METHOD/APPARATUS/PROCEDURE:

The complex used for the ternary system were made up from weighed amounts of water, $NaIO_3$ and $Mg(IO_3)_2 \cdot 4H_2O$. For the 5°C isotherm, the solids were first dissolved by heating and the solutions were inoculated, after cooling with the expected solid phase. The materials were agitated in a thermostat at the desired temperature. The stirring times were about 14 days in most cases. Aliquots of saturated solution were withdrawn by pipet, weighed in a 100 ml volumetric flask. To this was added about 75 ml of 0.13 mol dm^{-3} NaOH and the solution brought up to the mark after shaking. After standing for 4 hours the precipitate of $Mg(OH)_2$ settled well. An aliquot sample of supernatant liquid was withdrawn, and titrated back with standard HCl using methyl orange as an indicator. Total iodate content was determined iodometrically. The $NaIO_3$ content was calculated from these values.

SOURCE AND PURITY OF MATERIALS:

Magnesium iodate was prepared by neutralizing an aqueous solution of HIO_3 and $MgCO_3$ and evaporating the slightly acidified solution at above 40-50°C for 4 days, which gave a copious crystalline of the tetrahydrate. The crystals were washed with, air-dried, ground and placed in a desiccator.
The source of $NaIO_3$ was not given.

ESTIMATED ERROR:

Soly: titrns upon 25 sets of duplicates showed that the average relative disagreement between duplicates was 0.5 %. Temp: not given.

ACKNOWLEDGEMENT:

The figure reprinted from the *J. Am. Chem. Soc.* by permission of the copyright owners, The American Chemical Society.

COMPONENTS:

(1) Magnesium iodate; $Mg(IO_3)_2$; [7790-32-1]

(2) Potassium iodate; KIO_3; [7758-05-6]

(3) Water; H_2O; [7732-18-5]

ORIGINAL MEASUREMENTS:

Vinogradov, E. E.; Karataeva, I. M.

Zh. Neorg. Khim. 1976, *21*, 1666-9; *Russ. J. Inorg. Chem. (Engl. Transl.)* 1976, *21*, 912-3.

VARIABLES:

T/K = 323

KIO_3/mass % = 0 - 11.75

PREPARED BY:

Hiroshi Miyamoto

EXPERIMENTAL VALUES:

t/°C	Composition of Saturated Solutions				Nature of the Solid Phase[a]
	Potassium Iodate		Magnesium Iodate		
	mass %	mol % (compiler)	mass %	mol % (compiler)	
50	--	--	11.85[b]	0.6432	A
	0.37	0.035	11.15	0.6030	A + C
	0.40	0.038	11.68	0.6354	A + C
	0.52	0.047	7.28	0.379	C
	1.52	0.139	6.74	0.352	C
	1.23	0.110	5.33	0.274	C
	1.08	0.0947	3.19	0.160	C
	1.76	0.154	2.43	0.122	C
	3.23	0.286	2.05	0.104	C
	3.66	0.324	1.59	0.0805	C
	4.11	0.364	1.31	0.0664	C
	12.95	1.240	0.20	0.011	B + C
	13.06	1.250	0.10	0.0055	B + C
	11.75[b]	1.108	--	--	B

[a] A = $Mg(IO_3)_2 \cdot 4H_2O$; B = KIO_3; C = $2KIO_3 \cdot Mg(IO_3)_2 \cdot 4H_2O$.

[b] For binary systems the compiler computes the following

Soly of $Mg(IO_3)_2$ = 0.3593 mol kg^{-1}

Soly of KIO_3 = 0.6222 mol kg^{-1}

AUXILIARY INFORMATION

METHOD/APPARATUS/PROCEDURE:

The mixtures of KIO_3, $Mg(IO_3)_2$ and H_2O were placed in glass vessels, which fitted with a magnetic stirrer and water jacket, through which water was circulated from a U-10 ultra-thermostat. Equilibrium was established in the system in 14-16 days.
The products were analyzed for all the ions present: for K^+ gravimetrically by precipitating with sodium tetraphenylborate, for Mg^{2+} by titrating with Trilon B, and for IO_3^- by titrating with $Na_2S_2O_3$.
The solid phases obtained were studied by thermogravimetric and X-ray diffraction methods.

SOURCE AND PURITY OF MATERIALS:

Magnesium iodate was prepared from the carbonate by the action of HIO_3 and chemical pure grade potassium iodate was used.

ESTIMATED ERROR:

Soly: nothing specified
Temp: ± 0.3°C (authors)

REFERENCES:

COMPONENTS:

(1) Magnesium iodate; $Mg(IO_3)_2$; [7790-32-1]

(2) Potassium iodate; KIO_3; [7758-05-6]

(3) Water; H_2O; [7732-18-5]

ORIGINAL MEASUREMENTS:

Vinogradov, E. E.; Karataeva, I. M.

Zh. Neorg. Khim. 1976, *21*, 1666-9; *Russ. J. Inorg. Chem. (Engl. Transl.)* 1976, *21*, 912-3.

COMMENTS AND/OR ADDITIONAL DATA:

The phase diagram is given below (based on mass%).

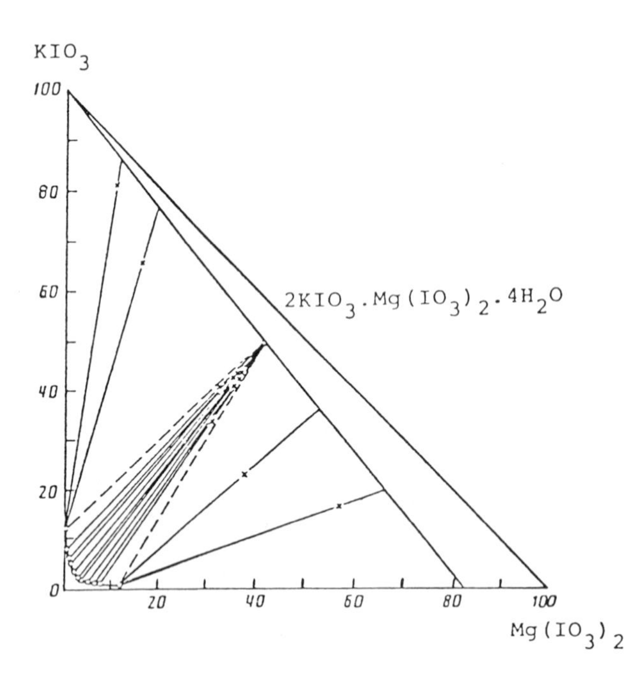

AUXILIARY INFORMATION

METHOD/APPARATUS/PROCEDURE:

SOURCE AND PURITY OF MATERIALS:

ESTIMATED ERROR:

ACKNOWLWDGEMENT:

The figure reprinted from *Zh. Neorg. Khim.* by permission of the copyright owners, VAAP, The Copyright Agency of the USSR.

COMPONENTS:

(1) Magnesium iodate; $Mg(IO_3)_2$; [7790-32-1]

(2) Rubidium iodate; $RbIO_3$; [13446-76-9]

(3) Water; H_2O; [7732-18-5]

ORIGINAL MEASUREMENTS:

Vinogradov, E. E.; Karataeva, I. M.

Zh. Neorg. Khim. 1976, *21*, 1666-9; *Russ. J. Inorg. Chem. (Engl. Transl.)* 1976, *21*, 912-3.

VARIABLES:

T/K = 323
composition

PREPARED BY:

Hiroshi Miyamoto

EXPERIMENTAL VALUES:

t/°C	Composition of Saturated Solutions				Nature of the Solid Phase[a]
	Rubidium Iodate		Magnesium Iodate		
	mass %	mol % (compiler)	mass %	mol % (compiler)	
50	--	--	11.85[b]	0.6432	A
	1.47	0.117	11.94	0.6589	A + B
	1.53	0.121	11.80	0.6506	A + B
	1.44	0.114	11.83	0.6518	A + B
	1.77	0.140	11.68	0.6448	A + B
	1.77	0.140	11.74	0.6485	A + B
	1.68	0.133	11.78	0.6504	A + B
	1.71	0.136	11.84	0.6543	A + B
	2.98	0.235	9.85	0.540	B
	2.82	0.214	6.73	0.356	B
	3.44	0.260	5.68	0.299	B
	3.50	0.264	5.42	0.285	B
	4.39[b]	0.317	--	--	B

[a] A = $Mg(IO_3)_2 \cdot 4H_2O$; B = $RbIO_3$.

[b] For binary systems the compiler computes the following

Soly of $Mg(IO_3)_2$ = 0.3593 mol kg^{-1}

Soly of $RbIO_3$ = 0.176 mol kg^{-1}

AUXILIARY INFORMATION

METHOD/APPARATUS/PROCEDURE:

The mixtures of $RbIO_3$, $Mg(IO_3)_2$ and H_2O were placed in glass vessels, which fitted with a magnetic stirrer and water jacket, through which water was circulated from a U-10 ultra-thermostat. Equilibrium was established in the system in 14-16 days.
The products were analyzed for all the ions present: for Rb^+ gravimetrically by precipitating with sodium tetraphenylborate, for Mg^{2+} by titrating with Trilon B, and for IO_3^- by titrating with $Na_2S_2O_3$.
The solid phases obtained were studied by thermogravimetric and X-ray diffraction methods.

SOURCE AND PURITY OF MATERIALS:

Magnesium iodate was prepared from the carbonate by the action of HIO_3 and chemical pure grade rubidium iodate was used.

ESTIMATED ERROR:

Soly: nothing specified
Temp: ± 0.3°C (authors)

REFERENCES:

COMPONENTS:

(1) Magnesium iodate; $Mg(IO_3)_2$; [7790-32-1]

(2) Rubidium iodate; $RbIO_3$; [13446-76-9]

(3) Water; H_2O; [7732-18-5]

ORIGINAL MEASUREMENTS:

Vinogradov, E. E.; Karataeva, I. M.

Zh. Neorg. Khim. 1976, *21*, 1666-9; *Russ. J. Inorg. Chem. (Engl. Transl.)* 1976, *21*, 912-3.

COMMENTS AND/OR ADDITIONAL DATA:

The phase diagram is given below. (based on mass%).

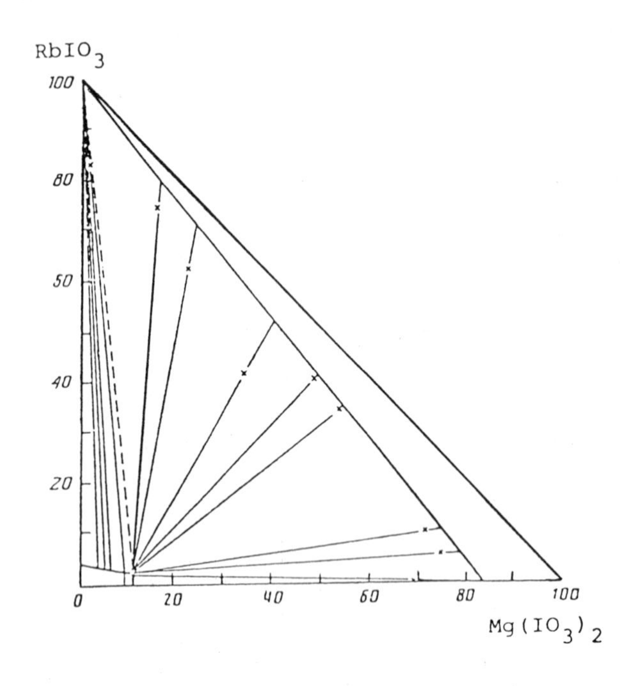

AUXILIARY INFORMATION

METHOD/APPARATUS/PROCEDURE:

SOURCE AND PURITY OF MATERIALS:

ESTIMATED ERROR:

ACKNOWLEDGEMENT:

The figure reprinted from *Zh. Neorg. Khim.* by permission of the copyright owners, VAAP, The Copyright Agency of the USSR.

COMPONENTS:

(1) Magnesium iodate; $Mg(IO_3)_2$; [7790-32-1]

(2) Magnesium nitrate; $Mg(NO_3)_2$; [10377-60-3]

(3) Water; H_2O; [7732-18-5]

ORIGINAL MEASUREMENTS:

Hill, A. E.; Moskowitz, S.

J. Am. Chem. Soc. 1931, *53*, 941-6.

EXPERIMENTAL VALUES:

t/°C	Composition of the Saturated Solutions				Nature of the Solid Phase[a]
	Magnesium Nitrate		Magnesium Iodate		
	mass %	mol % (compiler)	mass %	mol % (compiler)	
5	0.00	0.000	4.39[b]	0.221	A
	2.49	0.322	3.93	0.201	A
	5.45	0.724	3.92	0.206	A
	8.86	1.218	4.03	0.220	A
	13.10	1.885	4.26	0.243	A
	17.00	2.558	4.55	0.271	A + C
	18.52	2.830	4.44	0.269	C
	24.08	3.895	3.95	0.253	C
	34.21	6.178	2.82	0.202	C
	38.10	7.236	2.39	0.180	C + B
	39.02	7.260	0.46	0.034	B
	39.25	7.277	0.00	0.000	B
25	0.00	0.000	8.55[b]	0.448	C
	3.49	0.471	7.35	0.394	C
	7.31	1.018	6.66	0.368	C
	14.60	2.168	5.66	0.333	C
	18.48	2.846	5.14	0.314	C
	25.81	4.275	4.20	0.276	C
	33.50	6.030	3.25	0.232	C
	41.00	8.080	2.46	0.192	C + B
	41.60	8.072	0.90	0.069	B
	42.03	8.094	0.00	0.000	B
50	0.00	0.000	12.05[b]	0.655	C
	7.78	1.117	9.00	0.512	C
	15.17	2.312	7.31	0.442	C
	24.35	4.035	5.58	0.367	C
	32.15	5.776	4.36	0.311	C
	38.82	7.545	3.57	0.275	C
	44.41	9.300	3.13	0.260	C + B
	45.27	9.355	1.52	0.125	B
	46.09	9.408	0.00	0.000	B

[a] A = $Mg(IO_3)_2 \cdot 10H_2O$; B = $Mg(NO_3)_2$; C = $Mg(IO_3)_2 \cdot 4H_2O$.

[b] For binary systems the compiler computes the following

Soly of $Mg(IO_3)_2$ = 0.123 mol kg^{-1} at 5°C

= 0.250 mol kg^{-1} at 25°C

= 0.3662 mol kg^{-1} at 50°C

COMPONENTS:	ORIGINAL MEASUREMENTS:
(1) Magnesium iodate; $Mg(IO_3)_2$; [7790-32-1] (2) Magnesium nitrate; $Mg(NO_3)_2$; [10377-60-3] (3) Water; H_2O; [7732-18-5]	Hill, A. E.; Moskowitz, S. *J. Am. Chem. Soc.* 1931, *53*, 941-6.

COMMENTS AND/OR ADDITIONAL DATA:

The phase diagrams are given below (based on mass %).

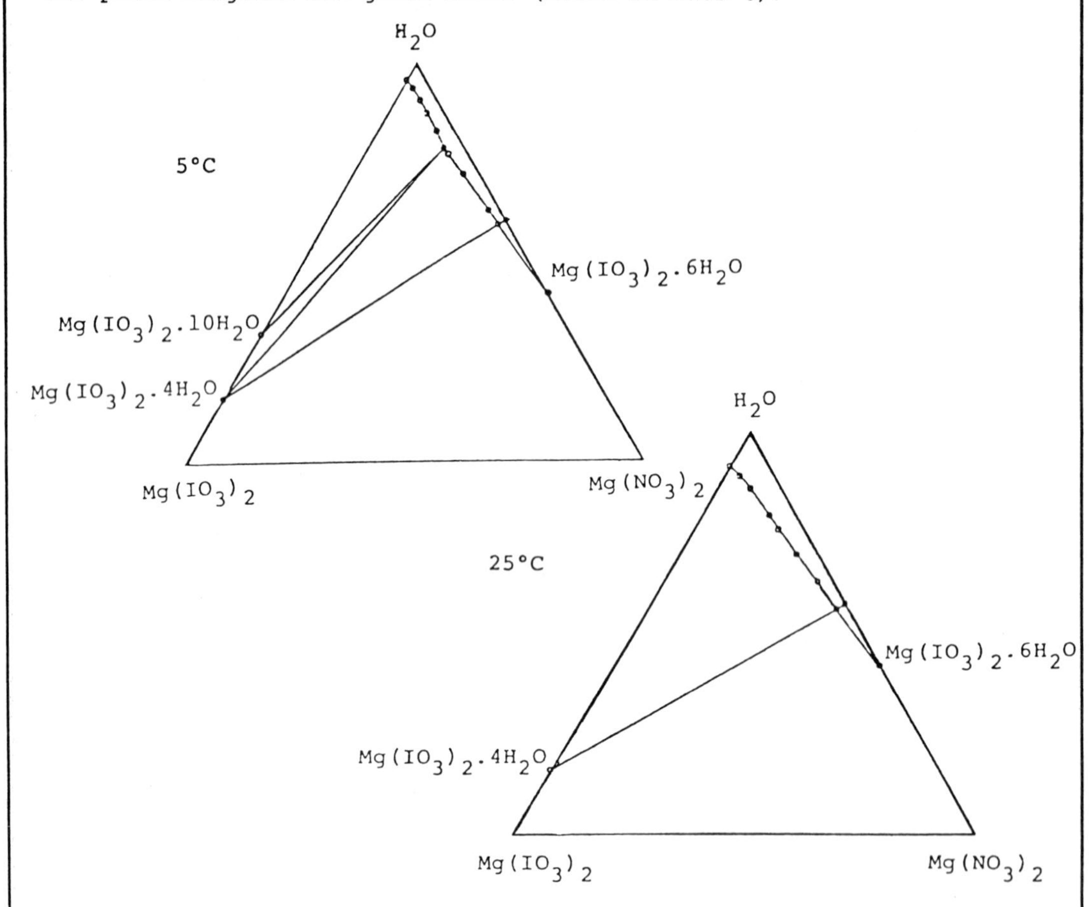

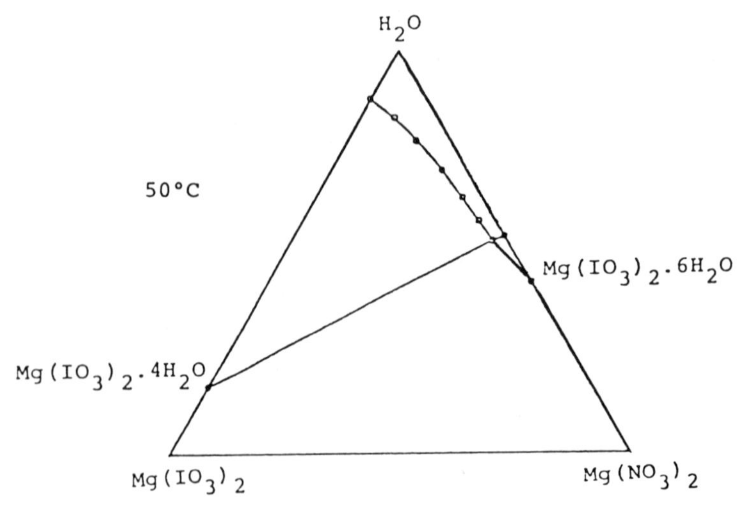

ACKNOWLEDGEMENT:

The figure reprinted from the *J. Am. Chem. Soc.* by permission of the copyright owners, The American Chemical Society.

COMPONENTS:

(1) Magnesium iodate; $Mg(IO_3)_2$; [7790-32-1]

(2) Magnesium nitrate; $Mg(NO_3)_2$; [10377-60-3]

(3) Water; H_2O; [7732-18-5]

ORIGINAL MEASUREMENTS:

Hill, A. E.; Moskowitz, S.

J. Am. Chem. Soc. 1931, *53*, 941-6.

VARIABLES:

T/K = 278, 298 and 323

$Mg(NO_3)_2$/mass % = 0 - 46

PREPARED BY:

Hiroshi Miyamoto

EXPERIMENTAL VALUES:

AUXILIARY INFORMATION

METHOD/APPARATUS/PROCEDURE:

The phase complexes used in the study of the ternary system were made from weighed amounts of magnesium iodate tetrahydrate, water and an analyzed solution of purified magnesium nitrate nearly saturated at room temperature. It was also necessary to use magnesium nitrate hexahydrate which had been partially dehydrated, in order to reduce the water content of the complex. The materials were agitated in a thermostat at the desired temperature for period of one to three days.
For the analysis of the iodate, the method of iodometry was used. The magnesium content was determined gravimetrically. After H_2SO_4 was added to samples of saturated solution, the solution was evaporated. The residue was weighed as $MgSO_4$.

SOURCE AND PURITY OF MATERIALS:

Magnesium iodate was prepared by neutralizing an aqueous solution of HIO_3 with $MgCO_3$ and evaporating the slightly acidified solution at above 40-50°C for 4 days, which gave a copious crystallization of the tetrahydrate. The crystals were washed with water, air-dried, ground and placed in a desiccator.
The source of magnesium nitrate was not given.

ESTIMATED ERROR:

Soly: the method gave results 0.25% low with a known solution.

Temp: not given

REFERENCES:

COMPONENTS:

(1) Magnesium iodate; $Mg(IO_3)_2$; [7790-32-1]

(2) Calcium iodate; $Ca(IO_3)_2$; [7789-80-2]

(3) Water; H_2O; [7732-18-5]

ORIGINAL MEASUREMENTS:

Shklovskaya, R. M.; Arkhipov, S. M.; Kidyarov, B. I.; Poleva, G. V.

Zh. Neorg. Khim. 1979, *24*, 1416-7; *Russ. J. Inorg. Chem.* (*Engl. Transl.*) 1979, *24*, 786-7.

VARIABLES:

T/K = 298
concentration

PREPARED BY:

Hiroshi Miyamoto

EXPERIMENTAL VALUES:

Composition of the Saturated Solutions[a]

t/°C	Magnesium Iodate			Calcium Iodate			Nature of the Solid Phase[b]
	mass %	mol %	m_1/mol kg^{-1}	mass %	mol %	$10^3 m_2$/mol kg^{-1}	
25	--	--	--	0.33	0.015	8.49	A
	0.94	0.046	0.025	0.065	0.0030	1.68	A
	1.79	0.0877	0.0487	0.028	0.0013	0.73	A
	3.15	0.156	0.0870	0.020	0.00095	0.53	A
	4.13	0.207	0.115	0.012	0.00058	0.32	A
	5.30	0.269	0.150	0.0097	0.00047	0.26	A
	6.64	0.341	0.190	0.0082	0.00040	0.23	A
	7.55	0.392	0.218	0.0075	0.00037	0.21	A
	8.27[c]	0.432	0.241	0.0073	0.00037	0.20	A + B
	8.49	0.445	0.248	--	--	--	B

[a] Molalities and mol % values calculated by the compiler.

[b] A = $Ca(IO_3)_2 \cdot 6H_2O$; B = $Mg(IO_3)_2 \cdot 4H_2O$

[c] Eutonic point

AUXILIARY INFORMATION

METHOD/APPARATUS/PROCEDURE:

Isothermal method used. Equilibrium was established in 15-20 days. The total concn of Mg in the coexisting phase was detd by complexometric titration at pH 10 using Acid Chrome Dark Blue indicator. The Ca content in liquid phase was detd by flame photometry. It was found that at the mass ratio Mg:Ca in solution up to 500 Mg does not hinder the determination of Ca. Solutions of that composition were analyzed by the method of boundary standard with standard solutions based on $CaCl_2$. For solutions which contained Mg and Ca in a ratio > 500 there was a decrease in the intensity of the Ca radiation. The author stated that such solutions were analysed by the "additions" method, but the details of the method were not described in the paper. The Mg content in the liquid phase was detd by difference. Mg compn in solid phase detd by atomic absorption, Ca by difference. The compositions of the solid phase were identified by the method of residues and checked by X-ray diffraction. The X-ray diffraction pattern of the specimens were recorded on URS-50 I diffractometer with filtered copper radiation.

SOURCE AND PURITY OF MATERIALS:

$Ca(IO_3)_2 \cdot 6H_2O$ was made from purified calcium nitrate and sodium iodate. $Mg(IO_3)_2 \cdot 4H_2O$ was made from "special purity grade" iodic acid and magnesium carbonate.

ESTIMATED ERROR:

Soly: the rel error in flame photometry measurement did not exceed 3-5%.
Temp: ± 0.1°C (authors)

COMPONENTS:

(1) Magnesium iodate; $Mg(IO_3)_2$; [7790-32-1]

(2) Barium iodate; $Ba(IO_3)_2$; [10567-69-8]

(3) Water; H_2O; [7732-18-5]

ORIGINAL MEASUREMENTS:

Ricci, J. E.; Freedman, A. J.

J. Am. Chem. Soc. 1952, *74*, 1769-73.

VARIABLES:

T/K = 298

PREPARED BY:

Hiroshi Miyamoto

EXPERIMENTAL VALUES:

The concentration of barium iodate in the saturated solution was found to be negligible.
The concentration of magnesium iodate in the solution saturated with both salt was the same as the pure solubility of $Mg(IO_3)_2 \cdot 4H_2O$.

The solubility of the pure $Mg(IO_3)_2 \cdot 4H_2O$ was 8.55 mass % or 0.250 mol kg^{-1} at 25°C.

AUXILIARY INFORMATION

METHOD/APPARATUS/PROCEDURE:

The details of the equilibrium procedure were not given.
Since H_2SO_4 gave no observable precipitate, the concentration of magnesium iodate in the saturated solution was determined iodometrically using H_2SO_4, KI and $Na_2S_2O_3$ solutions.

SOURCE AND PURITY OF MATERIALS:

$Ba(IO_3)_2 \cdot H_2O$ (commercial c.p. grade) used contained too little water, therefore it was leached with water until its solubility was constant, and it was then rinsed with some acetone and dried in air. The product contained 96.34% $Ba(IO_3)_2$ as compared with the theoretical 96.43%.
Magnesium iodate was prepared by neutralizing an aqueous solution of HIO_3 with $MgCO_3$ and evaporating the slightly acidified solution at above 40-50°C for 4 days, which gave a copious crystallization of the tetrahydrate. The crystals were washed with water, air-dried, ground and placed in a desiccator. The product contained 83.73% $Mg(IO_3)_2$ as compared with the theoretical 83.86%.

COMPONENTS:

(1) Magnesium iodate; $Mg(IO_3)_2$; [7790-32-1]

(2) Aluminium iodate; $Al(IO_3)_3$; [15123-75-8]

(3) Water; H_2O; [7732-18-5]

ORIGINAL MEASUREMENTS:

Tarasova, G. N.; Vinogradov, E. E.

Zh. Neorg. Khim. 1981, *26*, 2883-5; *Russ. J. Inorg. Chem. (Engl. Transl.)* 1981, *26*, 1544-5.

VARIABLES:

T/K = 298
composition

PREPARED BY:

Hiroshi Miyamoto

EXPERIMENTAL VALUES:

t/°C	Composition of Saturated Solutions				Nature of the Solid Phase[a]
	Aluminium Iodate		Magnesium Iodate		
	mass %	mol % (compiler)	mass %	mol % (compiler)	
25	5.70[b]	0.197	--	--	A
	4.51	0.156	1.60	0.0819	"
	4.68	0.167	4.31	0.227	"
	3.74	0.136	6.77	0.362	"
	3.42	0.124	7.24	0.388	A + B
	3.51	0.128	7.26	0.390	"
	3.66	0.133	7.24	0.389	"
	2.61	0.0950	8.15	0.437	B
	0.55	0.020	8.31	0.437	"
	--	--	8.55[b]	0.448	"

[a] A = $Al(IO_3)_3 \cdot 6H_2O$; B = $Mg(IO_3)_2 \cdot 4H_2O$

[b] For binary systems the compiler computes the following

Soly of $Mg(IO_3)_2$ = 0.250 mol kg^{-1}

Soly of $Al(IO_3)_3$ = 0.110 mol kg^{-1}

AUXILIARY INFORMATION

METHOD/APPARATUS/PROCEDURE:

The mixtures of $Al(IO_3)_3$, $Mg(IO_3)_2$ and water were stirred in a water thermostat. Equilibrium in the system was established in 10-14 days. The IO_3^- ion was determined by titration with sodium thiosulfate in the presence of sulfuric acid and KI. The Al^{3+} ion was determined complexometrically using Xylenol Orange indicator. The details of Mg analysis in the presence of Al^{3+} ions are given in ref 1.
The composition of the solid phases crystallizing in the system was determined by Schreinemakers' "residues" method.

SOURCE AND PURITY OF MATERIALS:

Magnesium iodate was synthesized from iodic acid and magnesium carbonate. Aluminium iodate was prepared at 80-90°C by neutralization of a saturated solution of HIO_3 with freshly precipitated aluminium hydroxide, taken in an equivalent amount, cooling the solution to room temperature, and drying the salt. The purity of the products was studied by thermal analysis, chemical analysis, and X-ray diffraction, the results were not given in the paper. The compiler assumes that the salts obtained are $Al(IO_3)_3 \cdot 6H_2O$ and $Mg(IO_3)_2 \cdot 4H_2O$.

ESTIMATED ERROR:

Soly: nothing specified
Temp: ± 0.1°C (authors)

REFERENCES:

1. Pribil, R. *Komplexony v Chemicke Analyse (Transl. into Russian)*, Inostr. Lit., Moscow, 1960, 491.

COMPONENTS:

(1) Magnesium iodate; $Mg(IO_3)_2$; [7790-32-1]

(2) Aluminium iodate; $Al(IO_3)_3$; [15123-75-8]

(3) Water; H_2O; [7732-18-5]

ORIGINAL MEASUREMENTS:

Tarasova, G. N.; Vinogradov, E. E.

Zh. Neorg. Khim 1981, *26*, 2883-5; *Russ. J. Inorg. Chem.(Engl. Transl.)* 1981, *26*, 1544-5.

COMMENTS AND/OR ADDITIONAL DATA:

The phase diagram is given below (based on mass %).

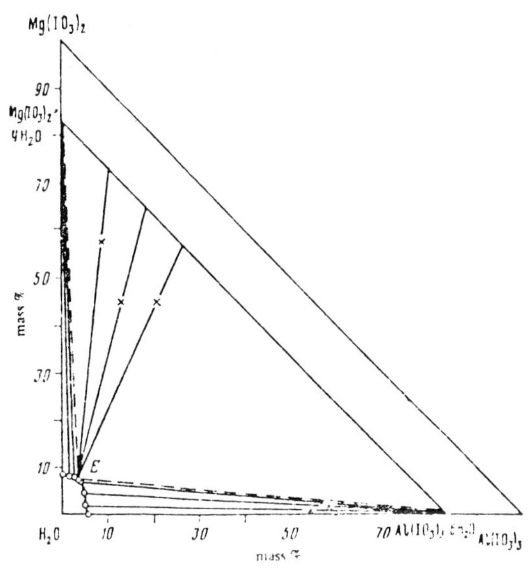

25 °C solubility isotherm for the $Al(IO_3)_3-Mg(IO_3)_2-H_2O$ system.

AUXILIARY INFORMATION

METHOD/APPARATUS/PROCEDURE:

SOURCE AND PURITY OF MATERIALS:

ESTIMATED ERROR:

ACKNOWLEDGEMENT:

The figure reprinted from *Zh. Neorg. Khim.* by permission of the copyright owners, VAAP, The Copyright Agency of the USSR.

COMPONENTS:

(1) Calcium chlorate; $Ca(ClO_3)_2$; [10137-74-3]

(2) Water; H_2O; [7732-18-5]

EVALUATOR:

Hiroshi Miyamoto
Department of Chemistry
Niigata University
Niigata, Japan

July, 1982

CRITICAL EVALUATION:

1. Solubility of calcium chlorate in water

Solubilities in the binary $Ca(ClO_3)_2$-H_2O system have been reported in 4 publications (1-4).

Mylius and Funk (1) measured the solubility of calcium chlorate dihydrate in water at 291.2 K.

Mazzetti (2) determined solubilities in ternary $Ca(ClO_3)_2$-$CaCl_2$-H_2O and $Ca(ClO_3)_2$-$KClO_3$-H_2O systems at 293.2K by the isothermal method.

Egorov (3) measured solubilities of calcium chlorate in pure water over a wide range of temperatures and reported the phase diagram of the binary $Ca(ClO_3)_2$-H_2O system: the synthetic method was employed.

Ehret (4) studied the solubility in ternary $Ca(ClO_3)_2$-$CaCl_2$-H_2O system at 298.2K by the isothermal method.

Depending upon temperature and composition, equilibrated solid phases of varying degrees of hydration have been reported. The following solid phases have been identified.

$Ca(ClO_3)_2$	[10137-74-3]
$Ca(ClO_3)_2 \cdot 2H_2O$	[10035-05-9]
$Ca(ClO_3)_2 \cdot 4H_2O$	[82808-59-1]
$Ca(ClO_3)_2 \cdot 6H_2O$	[82808-60-4]

The phase diagram reported by Egorov (3) is given in Figure 1.

The solubility data reported in (1) - (4) are given in Table 1, and the data are designated as tentative values.

The data in Table 1 were fitted to the following equations:

$$\ln(S_2/\text{mol kg}^{-1}) = -31.21447 + 44.22964/(T/100\text{K}) + 17.05511 \ln (T/100\text{K}) \quad : \quad \sigma = 0.032$$

$$\ln(S_0/\text{mol kg}^{-1}) = -21.36065 + 36.75924/(T/100\text{K}) + 10.90025 \ln (T/100\text{K}) \quad : \quad \sigma = 0.013$$

where S_2 and S_0 are the solubilities of calcium chlorate in equilibrium with the dihydrate and the anhydrous salt, respectively.

The results calculated from the smoothing equations are also given in Table 1.

COMPONENTS:

(1) Calcium chlorate; $Ca(ClO_3)_2$; [10137-74-3]

(2) Water; H_2O; [7732-18-5]

EVALUATOR:
Hiroshi Miyamoto
Department of Chemistry
Niigata University
Niigaga, Japan

July, 1982

CRITICAL EVALUATION:

Table 1 Tentative values for the solubility of calcium chlorate in water

T/K	m_1/mol kg^{-1}	ref	m_1'/mol kg^{-1}	Solid Phase
272.2	0.237	(3)		Ice
271.0	0.50	"		"
269.0	0.780	"		"
266.9	1.10	"		"
264.1	1.46	"		"
260.3	1.87	"		"
256.1	2.33	"		"
249.2	2.86	"		"
240.3	3.50	"		"
239.0	3.64	"		"
236.2	3.80	"		"
233.6	3.95	"		"
232.2	4.03	(3)		Ice + $Ca(ClO_3)_2\cdot 6H_2O$
235.7	4.12	(3)		$Ca(ClO_3)_2\cdot 6H_2O$
243.0	4.46	"		"
245.2	5.07	"		"
246.4	5.9	(3)		$Ca(ClO_3)_2\cdot 6H_2O$ + $Ca(ClO_3)_2\cdot 4H_2O$
246.9	5.90	(3)		$Ca(ClO_3)_2\cdot 4H_2O$
257.0	6.67	"		"
260.9	7.25	"		"
265.4	8.12	(3)	8.15	$Ca(ClO_3)_2\cdot 4H_2O$ + $Ca(ClO_3)_2\cdot 2H_2O$
268.2	8.23	(3)	8.19	$Ca(ClO_3)_2\cdot 2H_2O$
291.2	8.589	(1)	9.06	"
292.7	9.46	(3)	9.15	"
293.2	9.446	(2)	9.18	"
298.2	9.399	(4)	9.51	"
346.7	15.6	(3)	15.6	"
349.2	16	(3)	16	$Ca(ClO_3)_2\cdot 2H_2O$ + $Ca(ClO_3)_2$
366.2	17.1	(3)	16.9	$Ca(ClO_3)_2$
400.2	19.4	"	18.9	"
429.7	22.5	"	21.9	"
476.2	27.4	"	29.1	"
563.2	56	"	55	"

m_1: experimental value

m_1': calculated value

2. Ternary and quaternary systems

Solubility in ternary $Ca(ClO_3)_2$-$CaCl_2$-H_2O system. Mazzetti (2) and Ehret (4) reported solubilities in this ternary system at 293.2 and 298.2K, respectively. Double salts in this system were not formed.

Solubility in ternary $Ca(ClO_3)_2$-$KClO_3$-H_2O system. Mazzetti (2) reported the solubility in this system at 293.2K. No double salts were reported.

COMPONENTS:	EVALUATOR:
(1) Calcium chlorate; $Ca(ClO_3)_2$; [10137-74-3] (2) Water; H_2O; [7732-18-5]	Hiroshi Miyamoto Department of Chemistry Niigata University Niigata, JAPAN July, 1982

CRITICAL EVALUATION:

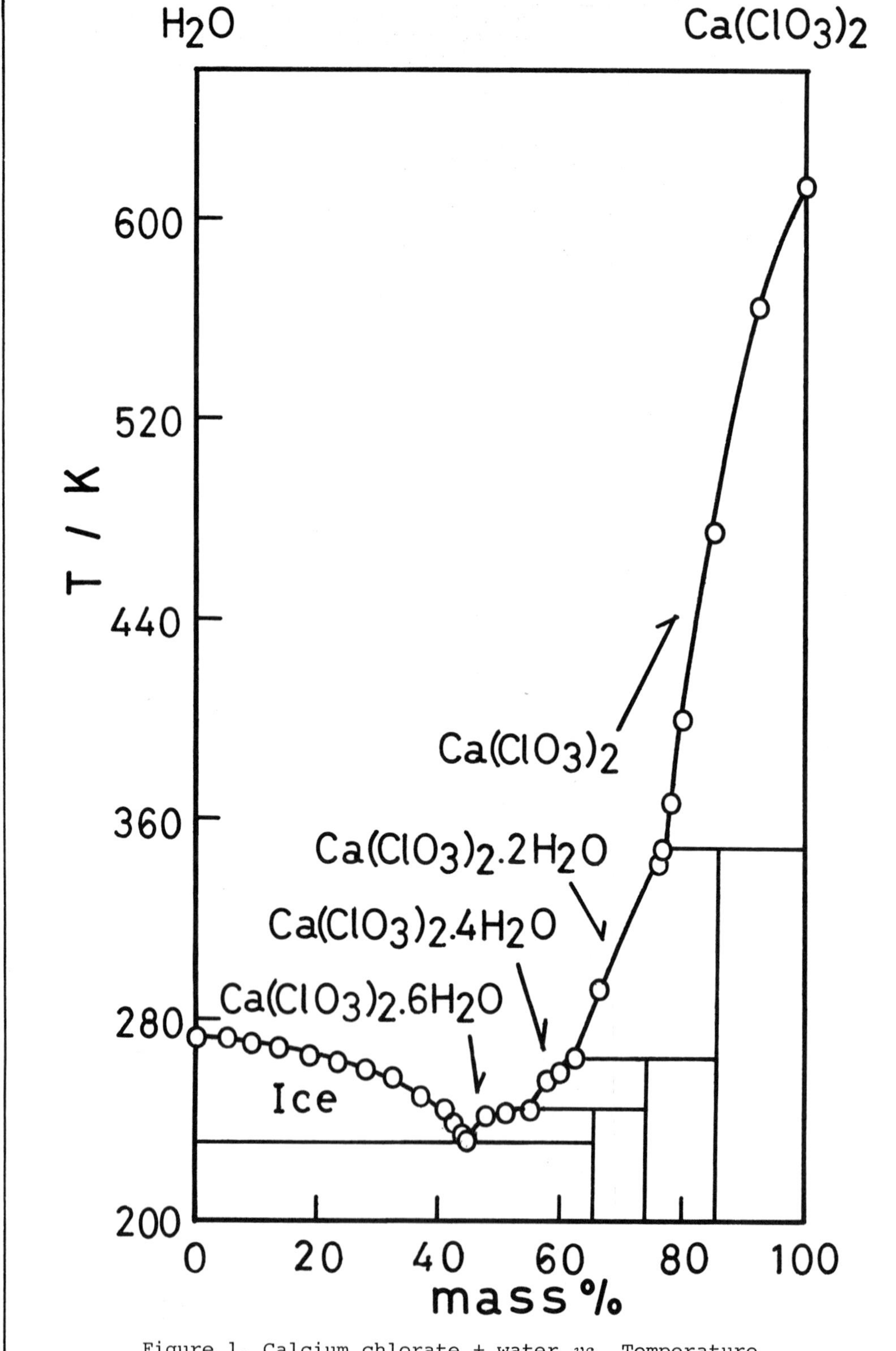

Figure 1. Calcium chlorate + water *vs*. Temperature

COMPONENTS:

(1) Calcium chlorate; $Ca(ClO_3)_2$; [10137-74-3]

(2) Water; H_2O; [7732-18-5]

EVALUATOR:

Hiroshi Miyamoto
Department of Chemistry
Niigata University
Niigata, Japan

July, 1982

CRITICAL EVALUATION:

Solubility in quaternary $Ca(ClO_3)_2$-$CaCl_2$-$KClO_3$-H_2O system. Mazzetti (2) has reported the solubility in this system at 293.2K. Compositions of saturated solutions with respect to two or three salts were reported. Double salts in this system were not formed.

REFERENCES:

1. Mylius, F.; Funk, R. *Ber. Dtsch. Chem. Ges.* 1897, *30*, 1716.

2. Mazzetti, C. *Ann. Chim. Appl.* 1929, *19*, 273.

3. Egorov, V. S. *J. Gen. Chem. (U.S.S.R.)* 1931, *1*, 1266.

4. Ehret, W. F. *J. Am. Chem. Soc.* 1932, *54*, 3126.

COMPONENTS:

(1) Calcium chlorate; $Ca(ClO_3)_2$; [10137-74-3]

(2) Water; H_2O; [7732-18-5]

ORIGINAL MEASUREMENTS:

Mylius, F.; Funk, R.

Ber. Dtsch. Chem. Ges. 1897, *30*, 1716-25.

VARIABLES:

T/K = 291

PREPARED BY:

Hiroshi Miyamoto

EXPERIMENTAL VALUES:

The solubility of $Ca(ClO_3)_2 \cdot 2H_2O$ in water at 18°C is given as below:

64 mass %	(authors)
177.8 g/100g[a] H_2O	(authors)
8.589 mol kg^{-1}	(compiler)

The density of the saturated solution at 18°C is also given:

1.729 g cm^{-3}

Based on this density, the compiler calculated the solubility in volume units as

5.346 mol dm^{-3}

[a] The compiler presumes that the first word in the fifth line from the end of page 1717 should read 100g.

AUXILIARY INFORMATION

METHOD/APPARATUS/PROCEDURE:

The salt and water were placed in a bottle and the bottle was shaken in a thermostat for a long time. After the saturated solution settled, an aliquot of solution was removed with a pipet. Calcium chlorate was determined by evaporation of the solution to dryness. The density of the saturated solution was also determined.

SOURCE AND PURITY OF MATERIALS:

The salt used was purchased as a "pure" chemical, and the traces of impurities were not present. The purity sufficed for the solubility determination.

ESTIMATED ERROR:

Soly: precision within 1 %
Temp: nothing specified

REFERENCES:

COMPONENTS:

(1) Calcium chlorate; $Ca(ClO_3)_2$; [10137-74-3]

(2) Water; H_2O; [7732-18-5]

ORIGINAL MEASUREMENTS:

Egorov, V. S.

J. Gen. Chem. (U.S.S.R.) 1931, *1*, 1266-70.

EXPERIMENTAL VALUES:

t/°C	Calcium Chlorate			Nature of the Solid Phase
	mass %	mol % (compiler)	m_1/mol kg^{-1} (compiler)	
- 1.0	4.67	0.425	0.237	Ice
- 2.2	9.3	0.88	0.50	"
- 4.2	13.9	1.39	0.780	"
- 6.3	18.6	1.95	1.10	"
- 9.1	23.2	2.56	1.46	"
-12.9	27.9	3.26	1.87	"
-17.1	32.5	4.02	2.33	"
-24.0	37.2	4.90	2.86	"
-32.9	42.0	5.93	3.50	"
-34.2	43.0	6.16	3.64	"
-37.0	44.0	6.40	3.80	"
-39.6	45.0	6.65	3.95	"
-41.	45.5	6.77	4.03	Ice + $Ca(ClO_3)_2\cdot 6H_2O$
-37.5	46.0	6.90	4.12	$Ca(ClO_3)_2\cdot 6H_2O$
-30.2	48.0	7.44	4.46	"
-28	51.2	8.37	5.07	"
-26.8	55	9.6	5.9	$Ca(ClO_3)_2\cdot 6H_2O$ + $Ca(ClO_3)_2\cdot 4H_2O$
-26.3	55.0	9.62	5.90	$Ca(ClO_3)_2\cdot 4H_2O$
-16.2	58.0	10.7	6.67	"
-12.3	60.0	11.5	7.25	"
- 7.8	62.7	12.8	8.12	$Ca(ClO_3)_2\cdot 4H_2O$ + $Ca(ClO_3)_2\cdot 2H_2O$
- 5.0	63.0	12.9	8.23	$Ca(ClO_3)_2\cdot 2H_2O$
19.5	66.2	14.6	9.46	"
73.5	76.3	21.9	15.6	"
76	77	23	16	$Ca(ClO_3)_2\cdot 2H_2O$ + $Ca(ClO_3)_2$
93	78.0	23.6	17.1	$Ca(ClO_3)_2$
127	80.1	25.9	19.4	"
156.5[a]	82.3	28.8	22.5	"
203	85.0	33.0	27.4	"
290	92	50	56	"

[a] The compiler assumes that 56.5 in the original paper should read 156.5.

COMPONENTS:

(1) Calcium chlorate; $Ca(ClO_3)_2$; [10137-74-3]

(2) Water; H_2O; [7732-18-5]

ORIGINAL MEASUREMENTS:

Egorov, V. S.

J. Gen. Chem. (U.S.S.R.) 1931, *1*, 1266-70.

VARIABLES:

T/K = 232 - 563

PREPARED BY:

Hiroshi Miyamoto

EXPERIMENTAL VALUES:

AUXILIARY INFORMATION

METHOD/APPARATUS/PROCEDURE:

Synthetic method used with visual observation of temperature of crystallization. A copper-constant thermocouple and a millivoltometer were used for the temperature measurements. At lower temperature, $Ca(ClO_3)_2 \cdot 2H_2O$ crystals and water were weighed and placed in a sealed vessel. But at higher temperature the anhydrous $Ca(ClO_3)_2$ was used. The sealed vessel was fixed to a large test tube and the test tube was placed in a Dewar vessel. The mixture of chloroform, tetrachlorocarbon and bromobenzene was used to cool the test tube and liquid air was also used.
Below 150°C the sealed vessel was placed in a cylinder covered by an asbestos, and the cylinder was heated using an oil bath. The sample solution in the sealed vessel was heated and the temperature recorded when the crystals disappeared. Next, the solution was cooled and the temperature measured when the crystals appeared. The processes were repeated.

SOURCE AND PURITY OF MATERIALS:

$Ca(ClO_3)_2 \cdot 2H_2O$ was prepared as follows: 20% sulfuric acid solution was added, with cooling, to aqueous $Ba(ClO_3)_2$ solution. The remaining barium ions were precipitated with sulfuric acid. The precipitated barium sulfate was filtered off and the filtrate added to c.p. grade $CaCO_3$. The barium carbonate was filtered off, and the $Ca(ClO_3)_2$ solution obtained was evaporated to dryness to obtain $Ca(ClO_3)_2 \cdot 2H_2O$ crystals. Analysis for chlorate ions showed 100% purity.

ESTIMATED ERROR:

Temp: precision of ± 0.2 K

COMPONENTS:

(1) Potassium chlorate; $KClO_3$; [3811-04-9]

(2) Calcium chlorate; $Ca(ClO_3)_2$; [10137-74-3]

(3) Water; H_2O; [7732-18-5]

ORIGINAL MEASUREMENTS:

Mazzetti, C.

Ann. Chim. Appl. 1929, *19*, 273-83.

VARIABLES:

T/K = 293
composition

PREPARED BY:

Bruno Scrosati
Hiroshi Miyamoto

EXPERIMENTAL VALUES:

t/°C	Composition of Saturated Solutions				Nature of the Solid Phase[a]
	Potassium Chlorate		Calcium Chlorate		
	mass %	mol % (compiler)	mass %	mol % (compiler)	
20	6.76	1.05	--	--	A
	2.97	0.500	11.15	1.112	"
	2.45	0.502	28.60	3.467	"
	2.39	0.547	36.89	4.995	"
	2.40	0.647	47.52	7.579	"
	1.10	0.401	64.31	13.87	A + B
	--	--	66.16[b]	14.54	B

[a] A = $KClO_3$; B = $Ca(ClO_3)_2 \cdot 2H_2O$

[b] For binary systems the compiler computes the following

Soly of $Ca(ClO_3)_2$ = 9.446 mol kg^{-1}

Soly of $KClO_3$ = 0.592 mol kg^{-1}

AUXILIARY INFORMATION

METHOD/APPARATUS/PROCEDURE:

The details of the method and the procedure for preparing the saturated solutions are not reported in the original paper.
The chlorate was determined volumetrically after reduction with FeSO4. Chlorine was determined volumetrically with the Volhard method. The calcium contents were determined volumetrically in the oxalate and potassium was tested by weight as potassium perchlorate. It was very difficult to establish the composition of the saturated solutions due to the high viscosity of the liquid.
A prolonged stirring was necessary to achieve equilibrium.

SOURCE AND PURITY OF MATERIALS:

Not reported

ESTIMATED ERROR:

Not possible to estimate due to insufficient details.

COMPONENTS:	ORIGINAL MEASUREMENTS:
(1) Potassium chlorate; $KClO_3$; [3811-04-9] (2) Calcium chlorate; $Ca(ClO_3)_2$; [10137-74-3] (3) Water; H_2O; [7732-18-5]	Mazzetti, C. *Ann. Chim. Appl.* 1929, *19*, 273-83.

COMMENTS AND/OR ADDITIONAL DATA:

The phase diagram is given as below (based on mass %).

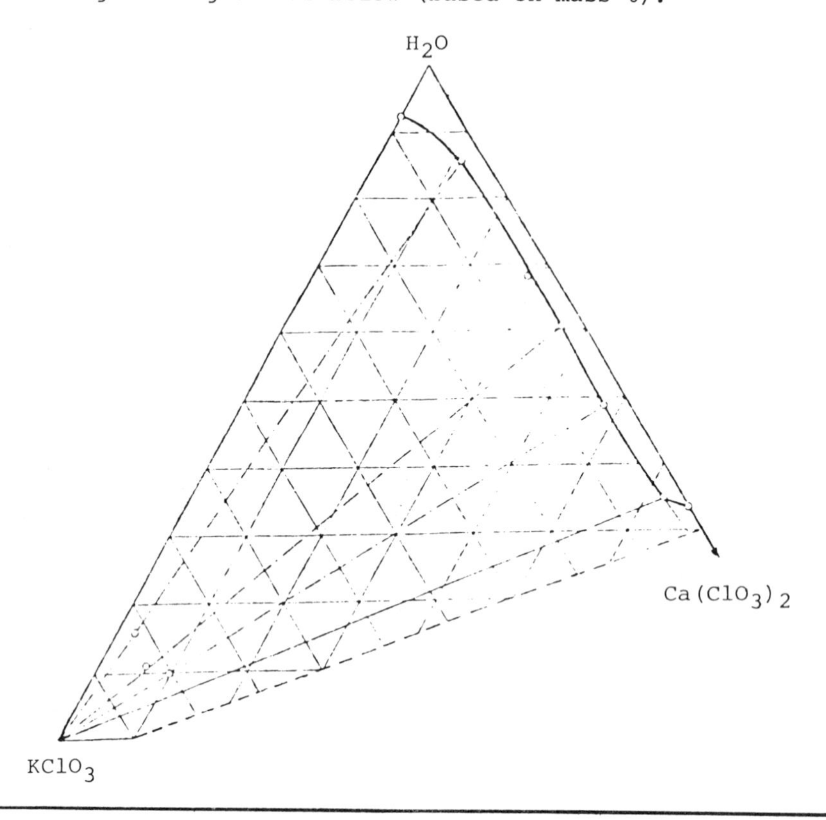

AUXILIARY INFORMATION

METHOD/APPARATUS/PROCEDURE:

SOURCE AND PURITY OF MATERIALS:

ESTIMATED ERROR:

REFERENCES:

COMPONENTS:

(1) Calcium chlorate; $Ca(ClO_3)_2$; [10137-74-3]

(2) Calcium chloride; $CaCl_2$; [10043-52-4]

(3) Water; H_2O; [7732-18-5]

ORIGINAL MEASUREMENTS:

Mazzetti, C.

Ann. Chim. Appl. 1929, *19*, 273-83.

VARIABLES:

T/K = 293
composition

PREPARED BY:

Bruno Scrosati
Hiroshi Miyamoto

EXPERIMENTAL VALUES:

t/°C	Composition of Saturated Solutions				Nature of the Solid Phase[a]
	Calcium Chloride		Calcium Chlorate		
	mass %	mol % (compiler)	mass %	mol % (compiler)	
20	42.7	10.8	--	--	A
	39.49	10.55	6.74	0.965	"
	37.35	10.26	10.56	1.556	"
	35.83	10.35	15.13	2.344	"
	33.84	10.82	22.88	3.923	A + B
	31.46	10.87	29.23	5.418	B
	28.29	10.04	33.47	6.368	B + C
	19.67	7.191	42.85	8.399	C
	12.83	4.737	49.61	9.822	"
	10.34	3.956	53.57	10.99	"
	7.16	2.88	58.74	12.66	"
	--	--	66.16[b]	14.54	"

[a] A = $CaCl_2 \cdot 6H_2O$; B = $CaCl_2 \cdot 4H_2O$; C = $Ca(ClO_3)_2 \cdot 2H_2O$

[b] For the binary system the compiler computes the following

Soly of $Ca(ClO_3)_2$ = 9.446 mol kg^{-1}

AUXILIARY INFORMATION

METHOD/APPARATUS/PROCEDURE:

The details of the method and the procedure for preparing the saturated solutions are not reported in the original paper.
Chloric acid was determined volumetrically after reduction with $FeSO_4$. Chlorine was determined volumetrically with the Volhard method. The calcium contents were determined volumetrically in the oxalate state.
It was very difficult to establish the composition of the saturated solutions due to the high viscosity of the liquids. Prolonged stirring was necessary to achieve equilibrium.

SOURCE AND PURITY OF MATERIALS:

Not reported

ESTIMATED ERROR:

Not possible to estimate due to insufficient details.

COMPONENTS:	ORIGINAL MEASUREMENTS:
(1) Calcium chlorate; $Ca(ClO_3)_2$; [10137-74-3] (2) Calcium chloride; $CaCl_2$; [10043-52-4] (3) Water; H_2O; [7732-18-5]	Mazzetti, C. *Ann. Chim. Appl.* 1929, *19*, 273-83.

COMMENTS AND/OR ADDITIONAL DATA:

The phase diagram is given as below (based on mass %).

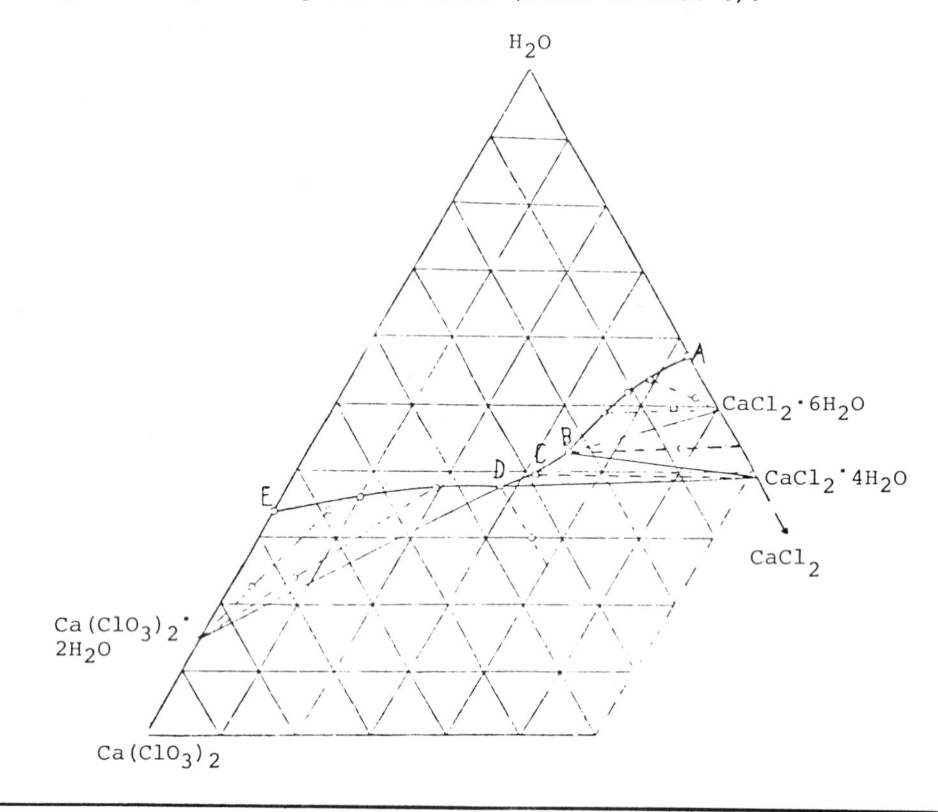

AUXILIARY INFORMATION

METHOD/APPARATUS/PROCEDURE:

See figure above:

Line AB equilibrium between $CaCl_2 \cdot 6H_2O(s)$ and aqueous $Ca(ClO_3)_2$.

Line BCD equilibrium between $CaCl_2 \cdot 4H_2O(s)$ and aqueous $Ca(ClO_3)_2$.

Line DE equilibrium between $Ca(ClO_3)_2 \cdot 2H_2O(s)$ and aqueous $CaCl_2$.

SOURCE AND PURITY OF MATERIALS:

ESTIMATED ERROR:

REFERENCES:

COMPONENTS:

(1) Calcium chloride; $CaCl_2$; [10043-52-4]

(2) Calcium chlorate; $Ca(ClO_3)_2$; [10137-74-3]

(3) Water; H_2O; [7732-18-5]

ORIGINAL MEASUREMENTS:

Ehret, W. F.

J. Am. Chem. Soc. 1932, *54*, 3126-34.

EXPERIMENTAL VALUES:

Composition of Saturated Solutions					
Calcium Chloride		Calcium Chlorate		Density	Nature of the Solid Phase[a]
mass %	mol % (Compiler)	mass %	mol % (Compiler)	ρ/g cm^{-3}	
0.00	0.00	66.05[b]	14.48	1.781	A
5.86	2.29	58.65	12.29	1.767	"
10.31	3.932	53.45	10.93	1.751	"
19.19	7.208	44.59	8.980	1.731	"
28.15	10.50	36.02	7.201	1.730	"
29.21	10.81	34.71	6.891	1.733	"
30.83	11.44	33.35	6.638	1.733	"
30.43	11.13	33.01	6.475	1.735	"
30.56	11.49	34.21	6.898	1.743	A + B
30.69	11.28	32.99	6.500	1.742	"
30.70	11.24	32.81	6.442	1.748	"
31.49	11.51	32.01	6.275	1.725	B
31.51	11.09	30.11	5.683	1.718	"
34.69	11.75	25.20	4.576	1.659	"
35.59	11.56	22.14	3.856	1.641	"
36.91	12.12	21.52	3.789	1.625	_[c]
36.89	12.00	21.02	3.666	1.629	B
37.75	12.15	19.65	3.391	1.618	"
38.16	11.97	17.85	3.003	1.606	"
38.17	11.94	17.69	2.968	1.607	_[c]
37.92	11.66	16.91	2.788	1.614	B
38.82	12.03	16.55	2.750	1.603	"
40.15	11.82	12.29	1.940	1.578	"
40.53	12.23	13.41	2.169	1.580	"
41.70	12.53	12.12	1.953	1.565	"
41.22	12.01	10.71	1.672	1.570	"
42.82	12.37	8.70	1.35	1.562	B + C
42.41	12.24	9.01	1.39	1.526	"
44.06	12.00	3.85	0.562	1.480	C
43.26	11.40	2.36	0.333	1.490	"
44.92	11.69	0.00	0.00	1.453	"

[a] A = $Ca(ClO_3)_2 \cdot 2H_2O$; B = $CaCl_2 \cdot 4H_2O$; C = $CaCl_2 \cdot 6H_2O$

[b] For binary system the compiler computes the following

Soly of $Ca(ClO_3)_2$ = 9.399 mol kg^{-1} at 25°C

[c] Nature of the solid phase is not given.

COMPONENTS:	ORIGINAL MEASUREMENTS:
(1) Calcium chloride; $CaCl_2$; [10043-52-4] (2) Calcium chlorate; $Ca(ClO_3)_2$; [10137-74-3] (3) Water; H_2O; [7732-18-5]	Ehret, W. F. *J. Am. Chem. Soc.* 1932, *54*, 3126-34.
VARIABLES:	**PREPARED BY:**
T/K = 298	Hiroshi Miyamoto

COMMENTS AND/OR ADDITIONAL DATA:
The phase diagram is given below (based on mass %)

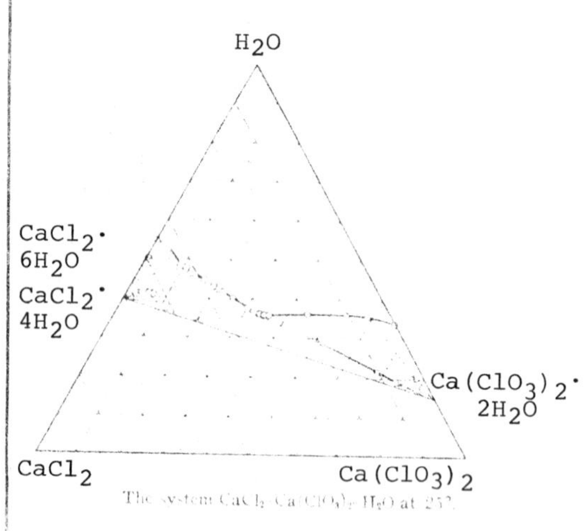

The system $CaCl_2$-$Ca(ClO_3)_2$-H_2O at 25°.

METHOD/APPARATUS/PROCEDURE:
Isothermal method was used. "Non-Sol" glass bottles containing the salt solutions were rotated in the thermostat. The mixtures of salts and water were made by starting with a solution containing only one salt and water, and in subsequent runs increments of the second salt were added to the solution. Two days of rotation in the bath was found by experience to be minimum length of time necessary for equilibrium. The same results were obtained when equilibrium was approached from above 25°C.
Samples of the liquid were withdrawn by means of a pipet provided with a folded filter paper at its lower end. The suction, approaching within 2-3mm of complete vacuum, was necessary. The saturated solution was immediately deposited in a specific gravity bottle at 25°C, and weighed. The weighed samples were diluted in a volumetric flask and aliquots taken for analysis. Calcium was first precipitated as the oxalate and then determined volumetrically with potassium permangante. Chloride was determined volumetrically and chlorate by difference. The composition of the solid phases in equilibrium with saturated solutions were determined by the direct (1) or "residue" method.

AUXILIARY INFORMATION

SOURCE AND PURITY OF MATERIALS:
Calcium chlorate was prepared as follows: about a liter of boiling hot solution of $(NH_4)_2SO_4$ was added, with stirring, to two liters of 1.4 mol dm^{-3} $Ba(ClO_3)_2$ solution. The remaining barium ions were precipitated with a dilute solution of $(NH_4)_2CO_3$. The precipitated sulfate and carbonate were allowed to settle and the ammonium chlorate solution removed by filteration. To the solution was added, with stirring, a calcium lime paste. The mixture was then boiled to remove ammonia. After cooling and filtering, the slight excess of lime was neutralized by means of chloric acid. The resulting solution containing calcium chlorate was evaporated to crystallization. The salt was recrystallized and contained only a trace of chloride. C.P. grade $CaCl_2 \cdot 6H_2O$ was melted in its own water of recrystallization and then recrystallized. $CaCl_2$ was prepared from the pure hexahydrate by heating in air at 110°C for several days. The salt was kept in a vacuum desiccator.

ESTIMATED ERROR:
Soly: nothing specified
Temp: ± 0.01°C (author)

REFERENCES:
1. Bancroft, W. D.
J. Phys. Chem. 1902, *6*, 179.

COMPONENTS:

(1) Potassium chlorate; $KClO_3$; [3811-04-9]

(2) Potassium chloride; KCl; [7447-40-7]

(3) Calcium chlorate; $Ca(ClO_3)_2$; [10137-74-3]

(4) Calcium chloride; $CaCl_2$; [10043-52-4]

(5) Water; H_2O; [7732-18-5]

ORIGINAL MEASUREMENTS:

Mazzetti, C.

Ann. Chim. Appl. 1929, *19*, 273-83.

VARIABLES:

T/K = 293

composition

PREPARED BY:

Bruno Scrosati

Hiroshi Miyamoto

EXPERIMENTAL VALUES:

Composition of Saturated Solutions at t/°C = 20

Calcium Chloride		Potassium Chlorate		Calcium Chlorate		Potassium Chloride		Nature of the Solid Phase[a]
mass %	mol %[b]	mass %	mol %[b]	mass %	mol %[b]	mass %	mol %[b]	
33.84	10.82	--	--	22.88	3.923	--	--	A + B
34.07	10.87	0.14	0.040	22.42	3.836	--	--	A + B + C
28.29	10.04	--	--	33.47	6.368	--	--	B + D
28.10	10.41	1.66	0.557	34.22	6.799	--	--	B + C + D
--	--	1.83	0.347	--	--	27.65	8.625	E + C
41.43	11.02	3.09	0.744	--	--	2.12	0.839	A + C + E
43.5	11.9	4.9	1.2	--	--	--	--	A + C

[a] A = $CaCl_2 \cdot 6H_2O$; B = $CaCl_2 \cdot 4H_2O$; C = $KClO_3$;

D = $Ca(ClO_3)_2 \cdot 2H_2O$; E = KCl

[b] mol % values calculated by the compiler.

AUXILIARY INFORMATION

METHOD/APPARATUS/PROCEDURE:

The details of the method and the procedure for preparing the saturated solutions are not reported in the original paper.
The ClO_3^- contents were determined volumetrically after reduction with $FeSO_4$. The Cl^- contents were determined volumetrically with the Volhard method.
The calcium contents were determined volumetrically in the oxalate and potassium was tested by weight as potassium chlorate and chloride.
It was very difficult to establish the composition of the saturated solutions due to the high viscosity of liquids.
Prolonged stirring was necessary to achieve equilibrium.

SOURCE AND PURITY OF MATERIALS:

Not reported

ESTIMATED ERROR:

Not possible to estimate due to insufficient details.

COMPONENTS:	ORIGINAL MEASUREMENTS:
(1) Potassium chlorate; $KClO_3$; [3811-04-9] (2) Potassium chloride; KCl; [7447-40-7] (3) Calcium chlorate; $Ca(ClO_3)_2$; [10137-74-3] (4) Calcium chloride; $CaCl_2$; [10043-52-4] (5) Water; H_2O; [7732-18-5]	Mazzetti, C. *Ann. Chim. Appl.* 1929, *19*, 273-83.

COMMENTS AND/OR ADDITIONAL DATA:

The phase diagram is given as below (based on mass %).

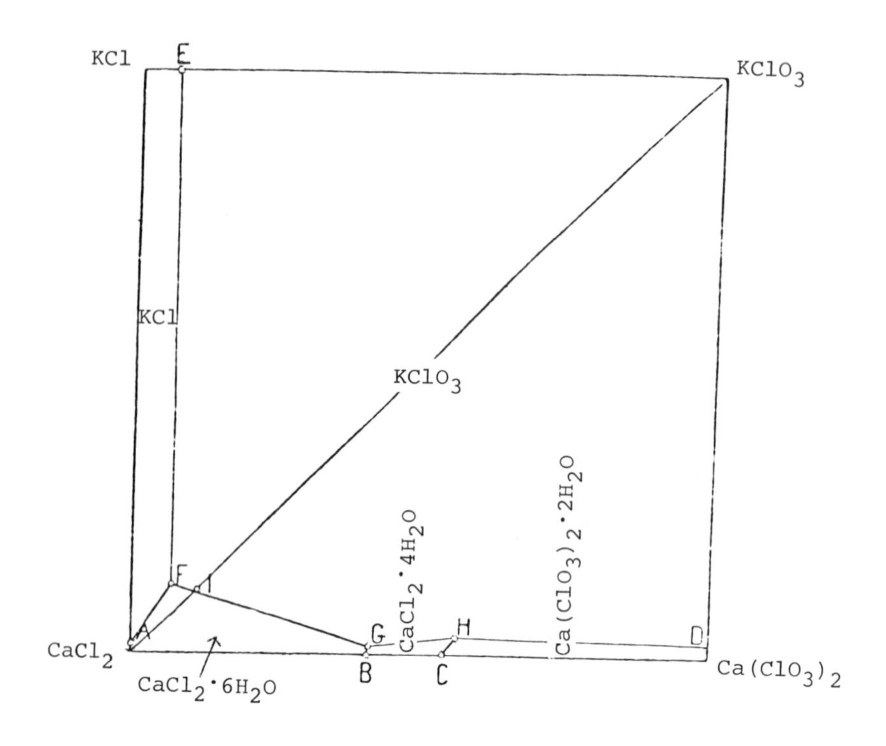

AUXILIARY INFORMATION

METHOD/APPARATUS/PROCEDURE:

SOURCE AND PURITY OF MATERIALS:

ESTIMATED ERROR:

REFERENCES:

COMPONENTS:	EVALUATOR:
(1) Calcium iodate; $Ca(IO_3)_2$; [7789-80-2] (2) Water; H_2O; [7732-18-5]	Hiroshi Miyamoto Department of Chemistry Niigata University Niigata, Japan April, 1982

CRITICAL EVALUATION:

1. The binary $Ca(IO_3)_2-H_2O$ system

Solubilities of calcium iodate in pure water have been reported in 25 publications (1-8, 10-19, 22, 24-29), and the solubility products of calcium iodate at zero ionic strength have been described in 12 publications (5, 7, 8, 15-21, 24, 25). The studies are summarized in Table I.

Table 1 Solubility studies of calcium iodate in water

Reference	T/K	Solid Phase	Soly and/or Soly Product	Method of Analysis
Mylius;	273-333	$Ca(IO_3)_2 \cdot 6H_2O$	Soly	gravimetric
Funk(1)	294-373	$Ca(IO_3)_2 \cdot H_2O$	Soly	"
Gross; Klinghoffer(2)	298	$Ca(IO_3)_2 \cdot 6H_2O$	Soly	volumetric (with $AgNO_3$)
Hill; Brown(3)	278-313	$Ca(IO_3)_2 \cdot 6H_2O$	Soly	iodometric(IO_3^-) gravimetric(Ca^{2+})
	308-343	$Ca(IO_3)_2 \cdot H_2O$	Soly	"
	330.7-363	$Ca(IO_3)_2$	Soly	"
Chloupek; Daneš; Danešova(4)	298	$Ca(IO_3)_2$	Soly	iodometric
Kilde(5)	291,298, 303	$Ca(IO_3)_2 \cdot 6H_2O$	Soly, K°_{s0}	iodometric
Kolthoff; Stenger(6)	298	$Ca(IO_3)_2 \cdot 6H_2O$	Soly	iodometric
Kilde(7)	291, 298, 303	$Ca(IO_3)_2 \cdot 6H_2O$	Soly, K°_{s0}	iodometric
Wise; Davies(8)	298	$Ca(IO_3)_2 \cdot 6H_2O$	Soly, K°_{s0}	iodometric
Davies(9)	Solubility of $Ca(IO_3)_2$ in pure water was not reported			
Keefer; Reiber; Bisson(10)	298	*	Soly	iodometric
Pedersen(11)	291	$Ca(IO_3)_2 \cdot 6H_2O$	Soly	iodometric
Derr; Vosburgh(12)	298	*	Soly	iodometric
Davies; Wyatt(13)	298	$Ca(IO_3)_2 \cdot 6H_2O$	Soly	iodometric
Davies; Waind(14)	298	$Ca(IO_3)_2 \cdot 6H_2O$	Soly	iodometric
Davies; Hoyle(15)	298	$Ca(IO_3)_2 \cdot 6H_2O$	Soly, K°_{s0}	iodometric
Monk(16)	298	$Ca(IO_3)_2 \cdot 6H_2O$	Soly, K°_{s0}	iodometric
Monk(17)	298	$Ca(IO_3)_2 \cdot 6H_2O$	Soly, K°_{s0}	iodometric

COMPONENTS:	EVALUATOR:
(1) Calcium iodate; $Ca(IO_3)_2$; [7789-80-2] (2) Water; H_2O; [7732-18-5]	Hiroshi Miyamoto Department of Chemistry Niigata University Niigata, Japan April, 1982

CRITICAL EVALUATION:

Table 1 (CONTINUED)

Reference	T/K	Solid Phase	Soly and/or Soly Product	Method of Analysis
Bell; George(18)	273,298, 313	$Ca(IO_3)_2 \cdot 6H_2O$ $Ca(IO_3)_2 \cdot H_2O$	Soly, K°_{s0} "	iodometric "
Rens(19)	298	$Ca(IO_3)_2 \cdot 6H_2O$	Soly, K°_{s0}	iodometric(IO_3^-) chelatometric(Ca^{2+})
Nezzal; Popiel; Vermande(20)	287-311 313-328	$Ca(IO_3)_2 \cdot 6H_2O$ $Ca(IO_3)_2 \cdot H_2O$	K°_{s0} "	iodometric "
Bousquet; Mathurin; Vermande(21)	287-311 313-328 333-359	$Ca(IO_3)_2 \cdot 6H_2O$ $Ca(IO_3)_2 \cdot H_2O$ $Ca(IO_3)_2$	K°_{s0} " "	iodometric " "
Miyamoto(22)	298	$Ca(IO_3)_2 \cdot 6H_2O$	Soly	iodometric
Fedorov; Robov; Shmyd'ko; Vorontsova; Mironov(23)	The solubility of $Ca(IO_3)_2$ in pure water was not given			
Das;Nair(24)	298,303, 308,313	*	Soly, K°_{s0}	iodometric
Das;Nair(25)	298,303, 308,313, 318	*	Soly, K°_{s0}	iodometric
Azarova; Vinogradov(26)	323	$Ca(IO_3)_2 \cdot 6H_2O$	Soly	iodometric(IO_3^-) complexometric(Ca^{2+})
Miyamoto; Suzuki; Yanai(27)	293,298, 303	$Ca(IO_3)_2 \cdot 6H_2O$	Soly	iodometric
Arkhipov; Kashina; Kidyarov(28)	298	*	Soly	iodometric(IO_3^-) complexometric(Ca^{2+})
Shklovskaya; Arkhipov; Kidyarov; Poleva(29)	298	$Ca(IO_3)_2$	Soly	flame photomotric (Ca^{2+})

* The number of water molecules was not given in the original paper.

COMPONENTS:

(1) Calcium iodate; $Ca(IO_3)_2$; [7789-80-2]

(2) Water; H_2O; [7732-18-5]

EVALUATOR:

Hiroshi Miyamoto
Department of Chemistry
Niigata University
Niigata, Japan

April, 1982

CRITICAL EVALUATION:

Davies (9) measured solubilities of calcium iodate in various aqueous solutions containing organic acids and sodium hydroxide, and Fedorov, Robov, Shmyd'ko, Vorontsova and Mironov (23) determined the solubility of calcium iodate in aqueous nitrate solutions. However, neither the solubility of calcium iodate in pure water nor the solubility product of calcium iodate at zero ionic strength was given in either publication (9,23).

Many of these studies deal with ternary systems, and two authors (1,3) have studied the solubility only in the binary $Ca(IO_3)_2$-H_2O system.

Depending upon temperature and composition, equilibrated solid phases of varying degrees of hydration have been reported. The following solid phases have been identified:

$Ca(IO_3)_2 \cdot 12H_2O$	[34992-36-4]
$Ca(IO_3)_2 \cdot 6H_2O$	[10031-33-1]
$Ca(IO_3)_2 \cdot H_2O$	[10031-32-0]
$Ca(IO_3)_2$	[7789-80-2]

The temperature dependence of the solubility of calcium iodate in pure water has been studied by Mylius and Funk (1) over the range from 273 to 373K and by Hill and Brown (3) 278 to 363 K. The temperature dependence of the solubility product of calcium iodate at zero ionic strength has also been reported by Bousquet, Mathurin and Vermande (21).

Fig. 1 reported by Hill and Brown (3) shows the relation between the solubility of calcium iodate in pure water and temperature. The graph shows clearly the existence of the three solid phases, with the transition temperatures of about 308 and 330.7K. The transition points could not be determined by direct measurements of the melting points due to the marked metastability of the various hydrated forms. Bousquet, Mathurin and Vermande (21) also found the three solid phases, and the transition temperatures were determined graphically. These authors (21) reported the hexahydrate→monohydrate transition temperature to be 309K, and the monohydrate→anhydrous salt transition temperature to be 325K. These results are in good agreement with those reported by Hill and Brown (3).

The evaluator found using the recommended and tentative values that the monohydrate→anhydrous salt transition temperature is 322.5K and the hexahydrate→monohydrate transition temperature is 307.9K.

The anhydrous form, having a nearly flat solubility curve, was not detected by Mylius and Funk (1); their data for temperatures above 333K were doubtless due to the presence of the metastable monohydrate.

A plot of the logarithm of the activity solubility product against the reciprocal of the absolute temperature was shown in (20) by Nazzal, Popiel and Vermande. They report that the point of intersection corresponding to 309K represents the temperature of transition between the hexahydrate and the monohydrate, but the temperature of transition between the monohydrate and the anhydrous salt was not reported.

The data to be considered in this critical evaluation are summarized in Table 2. The solubility is based on mol dm^{-3} units.

COMPONENTS:	EVALUATOR:
(1) Calcium iodate; $Ca(IO_3)_2$; [7789-80-2] (2) Water; H_2O; [7732-18-5]	Hiroshi Miyamoto Department of Chemistry Niigata University Niigata, JAPAN April, 1982

CRITICAL EVALUATION:

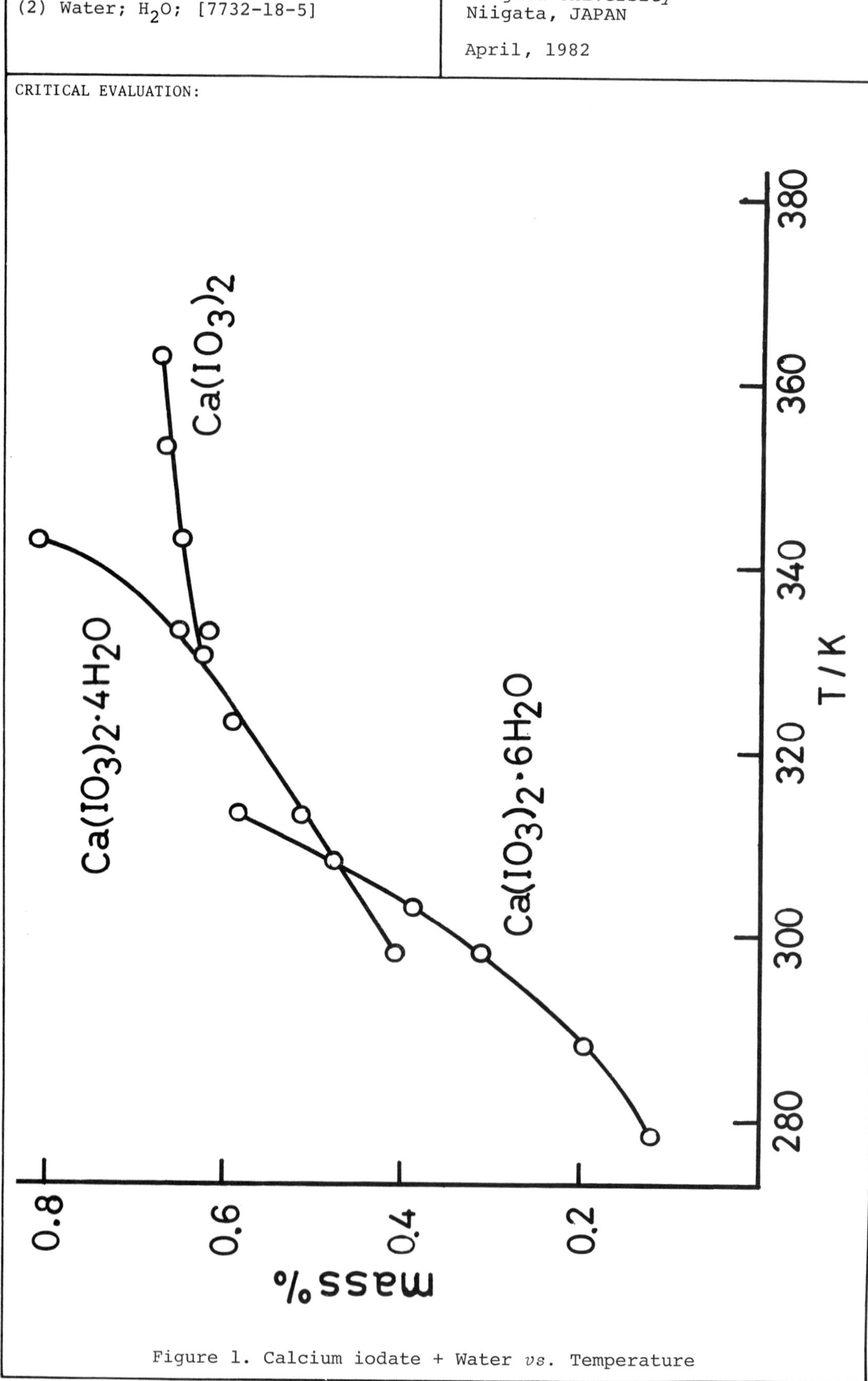

Figure 1. Calcium iodate + Water *vs*. Temperature

COMPONENTS:

(1) Calcium iodate; $Ca(IO_3)_2$; [7789-80-2]

(2) Water; H_2O; [7732-18-5]

EVALUATOR:

Hiroshi Miyamoto
Department of Chemistry
Niigata University
Niigata, Japan

April, 1982

CRITICAL EVALUATION:

Table 2 Summary of solubility data in binary $Ca(IO_3)_2$-H_2O system

Solid phase: $Ca(IO_3)_2 \cdot 6H_2O$

T/K	10^3c/mol dm^{-3}	ref
273.2	2.315	(18)
291.1	5.686	(11)
291.2	5.68	(7)
"	5.69	(5)
"	5.702	(11)
293.2	6.231	(19)
"	6.31	(27)
303.2	9.380	(25)
"	9.78	(7)
"	9.90	(27)
"	9.91	(5)
308.2	11.671	(25)
313.2	13.06	(18)
"	13.920	(25)

T/K	10^3c/mol dm^{-3}	ref
298.2	7.81	(7)
	7.820	(25)
	7.83	(12)
	7.838	(18)
	7.84	(5)
	7.840	(8)
	7.840	(13)
	7.84	(14)
	7.84	(15)
	7.84	(16)
	7.84	(17)
	7.84	(22)
	7.84	(27)
	7.85	(6)
	7.976	(2)

EVALUATION OF THE DATA (based on mol dm^{-3})

Solubility at 273.2K. Bell and George (18) made four independent analyses of the sample solution, and the average spread was about 0.3%. The value, 2.315 mmol dm^{-3}, is designated as a tentative value.

Solubility at 291.2K. Two values, 5.68 (5) and 5.69 (7) mmol dm^{-3}, have been reported by Kilde, and the temperature deviation is not given in either publication (5,7). The result (5.686 mmol dm^{-3}) reported in (11) obtained by Pedersen at 291.1K (19.9°C) is in good agreement with the values of Kilde. The arithmetic mean of four results is 5.69 mmol dm^{-3}, and the standard deviation, σ, is 0.009 mmol dm^{-3}. The mean value is designated as a recommended value.

Solubility at 293.2K. The solubility at this temperature has been reported by Rens (19) and Miyamoto, Suzuki and Yanai (27). The arithmetic mean of two results is 6.27 mmol dm^{-3}, and the standard deviation, σ, is 0.06 mmol dm^{-3}. The result is designated as a tentative value.

Solubility at 298.2K. This value has been reported in 15 publications. The hexahydrate is the stable solid phase in all solubility determinations at this temperature. The result reported in (2) obtained by Gross and Klinghoffer is certainly higher than that of other 14 publications, and is therefore rejected. The arithmetic mean of the 14 determinations is 7.836 mmol dm^{-3}, and the standard deviation is 0.010 mmol dm^{-3}. The value of 7.836 mmol dm^{-3} is designated as a recommended value.

COMPONENTS:

(1) Calcium iodate; $Ca(IO_3)_2$; [7789-80-2]

(2) Water; H_2O; [7732-18-5]

EVALUATOR:

Hiroshi Miyamoto
Department of Chemistry
Niigata University
Niigata, Japan

April, 1982

CRITICAL EVALUATION:

Solubility of 303.2K. The result reported by Miyamoto, Suzuki and Yanai (27) and that of Kilde (5) are almost identical. The value of Das and Nair is distinctly lower than that of others (5,7,27), and is therefore rejected. The arithmetic mean of the 3 determinations (5,7,27) is 9.86 mmol dm^{-3} and the standard deviation is 0.09 mmol dm^{-3}. The value of 9.86 mmol dm^{-3} is designated as a tentative value.

Solubility at 308.2K. Only Das and Nair have reported a solubility at this temperature. The value of 11.671 mmol dm^{-3} is designated as an approximate solubility value.

Solubility at 313.2K. The value of Das and Nair (25) is distinctly higher than that of Bell and George (18). Although the results of Das and Nair at other temperatures were rejected, the value at this temperature cannot be rejected because only one other study (18) is available for comparison. The arithmetic mean of two results (18,25) is 13.49 mmol dm^{-3}, and the standard deviation is 0.43 mmol dm^{-3}. The value of 13.49 mmol dm^{-3} is designated as a tentative value.

The recommended and tentative values for the solubility (mol dm^{-3}) of calcium iodate hexahydrate in water are given in Table 3.

Table 3 Recommended and tentative values for the solubility of calcium iodate in water

Solid phase: $Ca(IO_3)_2 \cdot 6H_2O$

T/K	$10^3 c_1$/mol dm^{-3}	$10^3 \sigma$/mol dm^{-3}	$10^3 c_1$(calcd)/ mol dm^{-3}
273.2	2.315	--	2.30
291.2	5.69[a]	0.009	5.78
293.2	6.27	0.06	6.33
298.2	7.836[a]	0.010	7.84
303.2	9.86	0.04	9.57
308.2	11.671	--	11.55
313.2	13.5	0.4	13.8

σ : standard deviation

a : recommended value

The recommended and tentative values in Table 3 were fitted to the following equation:

$$\ln(S/\text{mmol dm}^{-3}) = 64.26360 - 107.8724/(T/100\text{K}) - 23.82433 \ln(T/100\text{K}) \quad : \quad \sigma = 0.21$$

where S is the solubility of calcium iodate in water. The values calculated from the smoothing equation are also given in Table 3.

COMPONENTS:

(1) Calcium iodate; $Ca(IO_3)_2$; [7789-80-2]

(2) Water; H_2O; [7732-18-5]

EVALUATOR:

Hiroshi Miyamoto
Department of Chemistry
Niigata University
Niigata, Japan

April, 1982

CRITICAL EVALUATION:

The data to be considered in this critical evaluation are summarized in Table 4.

Table 4 Summary of solubility data for the binary $Ca(IO_3)_2$-H_2O system

T/K	10^3m/mol kg^{-1}	ref
Solid phase: $Ca(IO_3)_2 \cdot 6H_2O$		
273.2	3	(1)
278.2	3.06	(3)
283.2	4.4	(1)
288.2	5.01	(3)
291.2	6.4	(1)
298.2	7.774	(4)
"	7.82	(28)
"	7.86	(10)
"	7.87	(3)
"	7.87	(6)
"	8.49	(29)
303.2	9.89	(3)
"	11	(1)
308.2	12.27	(3)
313.2	15.07	(3)
"	16	(1)
323.2	23	(1)
327.2	27.0	(1)
333.2	35.4	(1)

T/K	10^3m/mol kg^{-1}	ref
Solid phase: $Ca(IO_3)_2 \cdot H_2O$		
294.2	9.5	(1)
298.2	10.43	(3)
308.2	12	(1)
313.2	13	(1)
"	13.33	(3)
318.2	14	(1)
323.2	14.0	(26)
"	15	(1)
"	15.22	(3)
330.7	16.03	(3)
333.2	16.83	(3)
"	17	(1)
343.2	20.97	(3)
353.2	20	(1)
373.2	24	(1)
Solid phase: $Ca(IO_3)_2$		
333.2	15.92	(3)
343.2	16.62	(3)
353.2	17.17	(3)
363.2	17.25	(3)

EVALUATION OF THE DATA (based on mol kg^{-1})

Solubility at 298.2K (solid phase: hexahydrate). The solubility at this temperature has been reported by Hill and Brown (3), Chloupek, Daneš and Danešova (4), Keefer, Reiber and Bisson (10), Arkhipov, Kashina and Kidyarov (28) and Shklovskaya, Arkhipov, Kidyarov and Poleva (29).

The result of Kolthoff and Stenger (6) was given in mol dm^{-3}, and the evaluator converted to mol kg^{-1} units using the density of the saturated solution given as 0.999 g cm^{-3} by authors.

COMPONENTS:	EVALUATOR:
(1) Calcium iodate; $Ca(IO_3)_2$; [7789-80-2] (2) Water; H_2O; [7732-18-5]	Hiroshi Miyamoto Department of Chemistry Niigata University Niigata, Japan April, 1982

CRITICAL EVALUATION:

The result reported in (29) obtained by Shklovskaya, Arkhipov, Kidyarov and Poleva is considerably higher than that of others (3,4,6,10,28), and therefore the result is rejected. The arithmetic mean of five results (3,4,6,10,28) is 7.84 mmol dm^{-3} and the standard deviation is 0.04 mmol dm^{-3}. The mean is a recommended value.

Solubility at other temperatures (solid phase: hexahydrate). Seidell and Linke (30) state that the result reported in (3) obtained by Hill and Brown agrees well with more recent investigators, but that the older work of Mylius and Funk is in error. The evaluator has also found the data of (1) to be in error as shown in the following analysis. The results of Mylius and Funk (1) and that of Hill and Brown with the value obtained from the result of critical evaluation at 298K are shown in Fig. 2.

As shown in Fig. 2, when the results of Mylius and Funk and that of Hill and Brown are plotted, two curves for the solubility are obtained. The curve drawn with the result of Mylius and Funk is clearly distinguished from that of Hill and Brown, and then the recommended value obtained from the critical evaluation at 298K can be plotted on the curve of Hill and Brown. Therefore, the result of Mylius and Funk is rejected.

The results reported in (3) obtained by Hill and Brown are designated as tentative values.

Solubility at 313.2K (solid phase: monohydrate). This result has been reported in 2 publications (1,3). The result reported in (1) obtained by Mylius and Funk is nearly equal to that of Hill and Brown (3). However, as in the above analysis, it is felt that the result of Hill and Brown is more accurate and reliable. The tentative value is based on the paper reported by Hill and Brown (3), and its value is 13.33 mmol kg^{-1}.

Solubility at 323.2K (solid phase: monohydrate). This result has been reported in 3 publications (1,3,26). The result reported in (26) obtained by Azarova and Vinogradov is considerably lower than that of others (1,3) and is therefore rejected. The result of Mylius and Funk is nearly equal to that of Hill and Brown. However, as in above analysis the result of Hill and Brown is designated as a tentative value, and its value is 15.22 mmol kg^{-1}.

Solubility at 333.2K (solid phase: monohydrate). This result has been reported in 2 publications (1,3). The result reported in (1) obtained by Mylius and Funk is nearly equal to that of Hill and Brown. However, as in the above analysis, it is felt that the result of Hill and Brown is more accurate and reliable. The tentative value is based on the paper reported by Hill and Brown (3), and is 16.83 mmol kg^{-1}.

Solubility at other temperatures (solid phase: monohydrate). Only one value has been reported at each temperature. The results at 298, 330.7 and 343K have been reported by Hill and Brown (3). These values are designated as tentative values. The results of Mylius and Funk (1) are rejected.

Solubility at 333.2, 343.2, 353.2 and 363.2K (solid phase: anhydrous form). Only one result at each temperature has been reported in (3). These values are designated as tentative values.

The recommended and tentative values for the solubility (mol kg^{-1} units) of calcium iodate in water over the temperature range of 273.2 to 363.2K are summarized in Table 5.

COMPONENTS:	EVALUATOR:
(1) Calcium iodate; $Ca(IO_3)_2$; [7789-80-2] (2) Water; H_2O; [7732-18-5]	Hiroshi Miyamoto Department of Chemistry Niigata University Niigata, JAPAN April, 1982

CRITICAL EVALUATION:

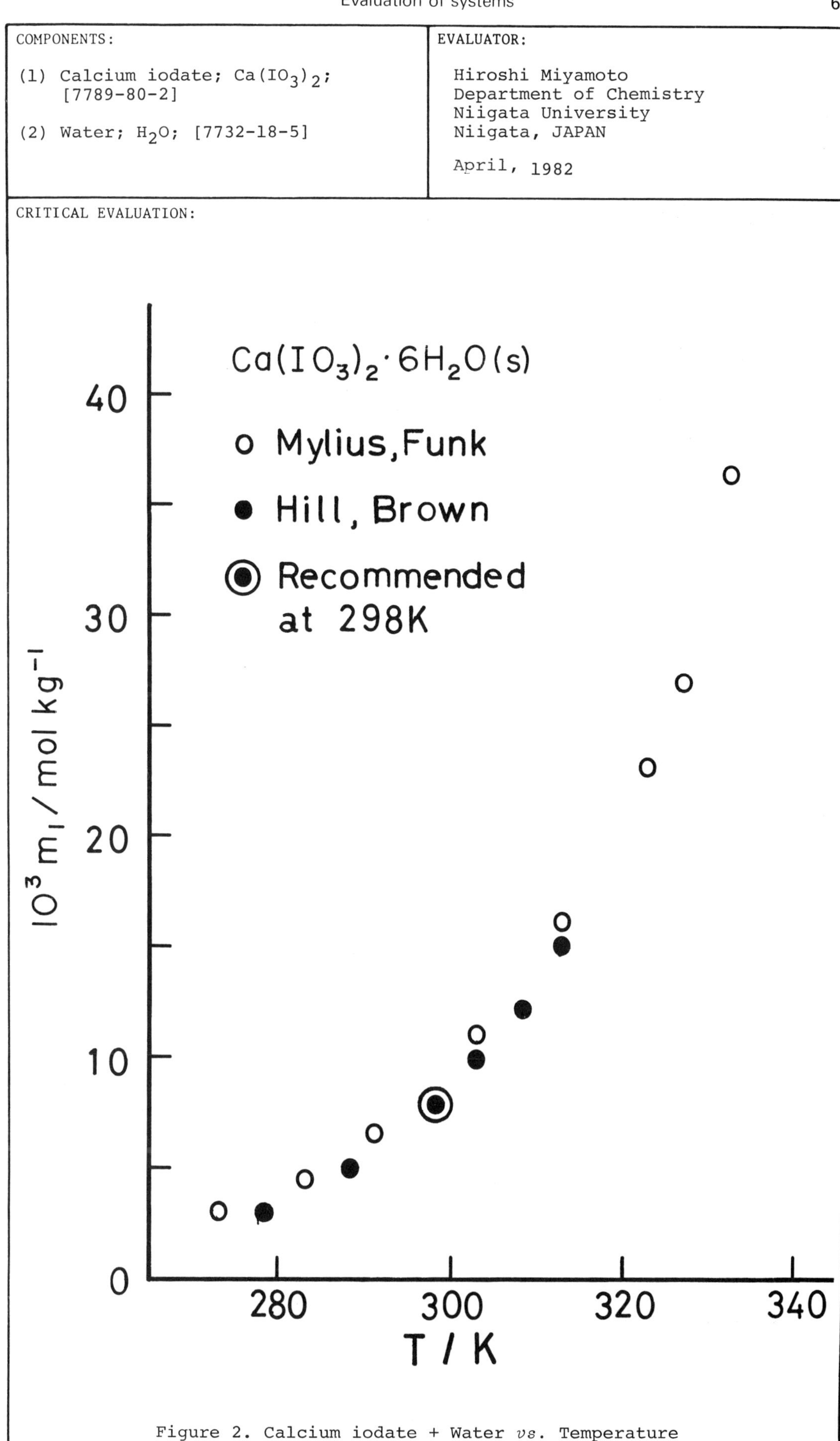

Figure 2. Calcium iodate + Water *vs.* Temperature

COMPONENTS:

(1) Calcium iodate; $Ca(IO_3)_2$; [7789-80-2]

(2) Water; H_2O; [7732-18-5]

EVALUATOR:

Hiroshi Miyamoto
Department of Chemistry
Niigata University
Niigata, Japan

April, 1982

CRITICAL EVALUATION:

Table 5 Recommended and tentative values for the solubility of calcium iodate in water

$Ca(IO_3)_2\cdot 6H_2O$		$Ca(IO_3)_2\cdot H_2O$		$Ca(IO_3)_2$	
T/K	Soly 10^3m/mol kg^{-1}	T/K	Soly 10^3m/mol kg^{-1}	T/K	Soly 10^3m/mol kg^{-1}
278.2	3.06	298.2	10.43	333.2	15.92
288.2	5.01	313.2	13.33	343.2	16.62
298.2	7.84[a]	323.2	15.22	353.2	17.17
303.2	9.89	330.2	16.03	363.2	17.25
308.2	12.27	333.2	16.83		
313.2	15.07	343.2	20.97		

a: recommended value

The data in Table 5 were fitted to the following equations:

$$\ln(S_6/\text{mmol kg}^{-1}) = 3.908264 - 23.44116/(T/100\text{K}) + 5.509761 \ln(T/100\text{K}) \quad : \quad \sigma = 0.050$$

$$\ln(S_1/\text{mmol kg}^{-1}) = -31.47739 + 42.45113/(T/100\text{K}) + 17.93882 \ln(T/100\text{K}) \quad : \quad \sigma = 0.69$$

$$\ln(S_0/\text{mmol kg}^{-1}) = 51.42688 - 77.07743/(T/100\text{K}) - 21.20998 \ln(T/100\text{K}) \quad : \quad \sigma = 0.0074$$

where S_6, S_1 and S_0 are the solubilities for the hexahydrate, the monohydrate and the anhydrous salt, respectively.

2. Solubility of calcium iodate in acid and alkali solutions

2-1. Solubility in HCl solution

Kilde (5) has reported the only result for the solubility in HCl solutions for HCl concentrations of 0.01 and 0.1 mol dm^{-3}. The solubility increases markedly with increasing the HCl concentration. The salt $Ca(IO_3)_2\cdot 6H_2O$ is very soluble in HCl solution. The dissociation constant of HIO_3 was calculated from the solubility data of $Ca(IO_3)_2$ in HCl solutions.

The results obtained by Kilde are tentative values.

2-2. Solubility in aqueous NaOH and KOH solutions

Only one result on the solubility in aqueous NaOH solutions at 291, 298, 303°K has been reported by Kilde (5). The solubility increases slightly with increasing the NaOH concentration. The dissociation constant of ion-pair $CaOH^+$ has been also reported.

Only one study on solubilities of calcium iodate in aqueous KOH solutions at 273, 298 and 303K has been reported by Bell and George (18). The solid phase is the hexahydrate at 273 and 298K, and the monohydrate at 303K. The KOH concentration was varied from 0 to 68.27 mmol dm^{-3} at 273K, to 54.29 mmol dm^{-3} at 298K and to 65.25 mmol dm^{-3} at 303K. The degree of the increase in solubility with increasing the KOH concentration is almost similar to that of the NaOH solution. The dissociation constant of $CaOH^+$ was calculated, and the thermodynamic solubility product was computed from the solubility data.

COMPONENTS:

(1) Calcium iodate; $Ca(IO_3)_2$; [7789-80-2]

(2) Water; H_2O; [7732-18-5]

EVALUATOR:

Hiroshi Miyamoto
Department of Chemistry
Niigata University
Niigata, Japan

April, 1982

CRITICAL EVALUATION:

The results reported in (5) obtained by Kilde and in (18) obtained by Bell and George are designated as tentative values.

2-3. Solubility in aqueous $Ca(OH)_2$ solution

The solubility of calcium iodate in aqueous $Ca(OH)_2$ solutions has been reported by Kilde (5) at 291, 298 and 303K, and by Davies and Holye (15) at 298K. Kilde reported only one datum in 0.0108 mol dm^{-3} $Ca(OH)_2$ solution at each temperature. In the study of Davies and Hoyle the $Ca(OH)_2$ concentration was varied from 0 to 20.93 mmol dm^{-3}. The solubility in the system decreases with increasing the $Ca(OH)_2$ concentration, and this observed behavior differs from that of the solubility in NaOH and KOH solutions. The dissociation constant, $K^{\circ}_{CaOH^+}$, obtained is in good agreement with that calculated from the solubility in NaOH and KOH solutions.

The results are designated as tentative values.

2-4. Solubility in aqueous NH_3 solution

The study on the solubility in aqueous NH_3 solution has been reported in 2 publications (6,12).

Derr and Vosburgh (12) state that the apparent solubility product of $Ca(IO_3)_2$ was found to be unaffected by the presence of 0.1 mol dm^{-3} of ammonia, but decreases at higher ammonia concentrations. However they did not report solubility data for NH_3 concentrations greater than 0.1 mol dm^{-3}.

Kolthoff and Stenger (6) have presented quantitative solubility determination in aqueous NH_3. The solubility of $Ca(IO_3)_2$ in this system decreases with increasing NH_3 concentration.

The results obtained by Kolthoff and Stenger are designated as tentative values.

3. Solubilities in solutions with alkali and alkaline earth iodates

The $Ca(IO_3)_2$-$LiIO_3$-H_2O system has been studied by Azarova and Vinogradov (26) at 323K, and by Arkhipov, Kashina and Kidyarov (28) at 298K. The $Ca(IO_3)_2$-$NaIO_3$-H_2O system has been studied by Hill and Brown (3) at 298K, and the $Ca(IO_3)_2$-$Mg(IO_3)_2$-H_2O system was studied by Shklovskaya, Arkhipov, Kidyarov and Poleva (29) at 298K.

The dominant feature in these studies reported in (3), (26), (28) and (29) is the existence of simplc eutonic type phase diagrams; the existence of double salts was not reported.

Kilde (5) determined the solubility of calcium iodate hexahydrate in dilute KIO_3 solution at 291.2, 298.2 and 303.2K, and Petersen measured the solubility in aqueous solution of magnesium iodate at 291.1K (17.9°C). The solubility of $Ca(IO_3)_2$ in these systems decreases with increasing the concentration of the iodate (KIO_3 or $Mg(IO_3)_2$).

4. Solubility of calcium iodate in aqueous salt solutions

Solubilities of calcium iodate in various aqueous salt solutions are summarized in Table 6.

COMPONENTS:	EVALUATOR:
(1) Calcium iodate; $Ca(IO_3)_2$; [7789-80-2] (2) Water; H_2O; [7732-18-5]	Hiroshi Miyamoto Department of Chemistry Niigata University Niigata, Japan April, 1982

CRITICAL EVALUATION:

Table 6 Summary of solubility of calcium iodate in aqueous salt solutions

Added electrolyte Salt	Concentration c_2/mol dm^{-3}	T/K	Calcium Iodate Concentration units	Reference
LiCl	0 - 1.510	298.2	mol dm^{-3}	(2)
"	0 - 0.15536	293.2	"	(19)
NaCl	0 - 1.988	298.2	mol dm^{-3}	(2)
"	0 - 1.000	291.2	"	(5)
"	0 - 1.000	298.2	"	"
"	0 - 1.000	303.2	"	"
"	0 - 0.1000	298.2	"	(8)
"	0 - 1.00021	293.2	"	(19)
KCl	0 - 1.946	298.2	mol dm^{-3}	(2)
"	0 - 0.1000	298.2	"	(8)
"	0 - 0.1008[a]	298.2	mol kg^{-1}	(10)
"	0 - 0.19002	293.2	mol dm^{-3}	(19)
$CaCl_2$	0 - 0.0500	291.2	mol dm^{-3}	(5)
"	0 - 0.0500	298.2	"	"
"	0 - 0.0500	303.2	"	"
"	0 - 0.0500	298.2	"	(8)
"	0 - 0.064134	293.2	"	(19)
$MgCl_2$	0 - 0.1[a]	298.2	mol kg^{-1}	(4)
"	0 - 0.505	291.2	mol dm^{-3}	(5)
"	0 - 0.505	298.2	"	"
"	0 - 0.505	303.2	"	"
"	0 - 0.062327	293.2	"	(19)
NH_4Cl	0 - 1.491	298.2	mol dm^{-3}	(2)
"	0 - 0.20006	293.2	"	(19)
KBr	0 - 0.18494	293.2	mol dm^{-3}	(19)
KI	0 - 0.19017	293.2	mol dm^{-3}	(19)
Na_2SO_4	0 - 0.025	298.2	mol dm^{-3}	(8)
K_2SO_4	0 - 0.1[a]	298.2	mol kg^{-1}	(4)
"	0 - 0.02371	273.2	mol dm^{-3}	(18)
"	0 - 0.02012	298.2	"	"
"	0 - 0.01409	313.2	"	"
$MgSO_4$	0 - 0.1[a]	298.2	mol kg^{-1}	(4)
"	0 - 0.025	"	mol dm^{-3}	(8)
$Na_2S_2O_3$	0 - 0.0015430	298.2	mol dm^{-3}	(13)
KNO_3	0 - 0.5[a]	298.2	mol kg^{-1}	(4)
$K_4[Fe(CN)_6]$	0 - 0.005	298.2	mol dm^{-3}	(8)

[a] mol kg^{-1} units

COMPONENTS:

(1) Calcium iodate; $Ca(IO_3)_2$; [7789-80-2]

(2) Water; H_2O; [7732-18-5]

EVALUATOR:

Hiroshi Miyamoto
Department of Chemistry
Niigata University
Niigata, Japan

April, 1982

CRITICAL EVALUATION:

4-1. Solubility in aqueous LiCl solutions

Gross and Klinghoffer (2) measured the solubility in this system at 298K. The molalities calculated by using density data from the Landolt-Börnstein-Roth tables have been reported in (2). No density data for $Ca(IO_3)_2$ solutions can be found in the tables. Therefore, the results are very approximate. The evaluator cannot check the results of Gross and Klinghoffer because no other investigations have been reported with the exception of Rens (19) who measured the solubility in this system at 293K.

COMPONENTS:

(1) Calcium iodate; $Ca(IO_3)_2$; [7789-80-2]

(2) Water; H_2O; [7732-18-5]

EVALUATOR:

Hiroshi Miyamoto
Department of Chemistry
Niigata University
Niigata, Japan

April, 1982

CRITICAL EVALUATION:

The solubility in this system increases with increasing the LiCl concentration.

The results reported in (19) obtained by Rens at 293K are designated as tentative values.

4-2. Solubility in aqueous NaCl solutions
This result has been reported in 4 publications. Gross and Klinghoffer (2) measured the solubility in this system and calculated the molalities using density data for NaCl solutions from Landolt-Börnstein-Roth tables, the results are therefore very approximate.

At 298K, the result in 0.100 mol dm^{-3} NaCl solution obtained by Kilde (5) is good agreement with that in 0.100 (0.0995) mol dm^{-3} reported by Gross and Klinghoffer, and is slightly higher than the corresponding result of Wise and Davies (8). The solubility in this system increases with increasing the NaCl concentration.

The results reported in (5) obtained by Kilde and in (8) obtained by Wise and Davies at 298K are designated as tentative values. The tentative value in 0.100 mol dm^{-3} NaCl solution is 10.4 mmol dm^{-3}, and the standard deviation is 0.2 mmol dm^{-3}. The results reported by Rens (19) at 293K are tentative values.

4-3. Solubility in aqueous KCl solutions
The solubility in this system at 298K has been reported in 3 publications (2,8,10). Rens (19) reports the only solubility result at a temperature of 293.2K.

Gross and Klinghoffer (2) measured the solubility in this system and calculated the molalities using density data for KCl solutions from Landolt Börnstein-Roth tables. No density data for KCl solutions can be found in the tables, and the results reported in (2) are therefore very approximate.

Wise and Davies (8) measured solubilities (mol dm^{-3} units) of calcium iodate hexahydrate in aqueous KCl solutions for KCl concentrations from 0 to 0.1000 mol dm^{-3} at 298K. The results of Gross and Klinghoffer are slightly higher than that of Wise and Davies. However as discussed above, it is felt that the results of Wise and Davies are more accurate and reliable.

Keefer, Reiber and Bisson (10) determined solubilities (mol kg^{-1} units) of calcium iodate iodometrically in aqueous KCl solutions for KCl concentrations from 0 to 0.1008 mol kg^{-1} at 298K, and they state that the densities of all solutions were determined. However we did not find them in the original paper. The degree of hydration of the salt used was, not reported, but the evaluator assumes that the hexahydrate was used because this is the stable phase at 298K.

The densities of the saturated solutions for this system has been reported by Wise and Davies, and the molalities reported in (8) can be converted into molarities. However, due to the absence of the reliable density data for the diluted solutions below 0.1 mol dm^{-3} KCl,the results of Wise and Davies are not directly comparable with those of Keefer, Reiber, and Bisson. The comparison of the solubility in pure water is possible, and both results are in good agreement with each other.
The results of Wise and Davies (8), Keefer, Reiber and Bisson (10), and Rens (19) under the stated experimental conditions are designated as tentative values.

COMPONENTS:
(1) Calcium iodate; $Ca(IO_3)_2$; [7789-80-2]
(2) Water; H_2O; [7732-18-5]

EVALUATOR:
Hiroshi Miyamoto
Department of Chemistry
Niigata University
Niigata, Japan

April, 1982

CRITICAL EVALUATION:

4-4. Solubility in aqueous $MgCl_2$ solutions
Solubilities of calcium iodate in aqueous $MgCl_2$ solutions have been reported in 3 publications (4,5,19).

Solubilities were measured by Chloupek, Daneš and Danešova (4) at 298K, by Kilde (5) at 291, 298 and 303K, and by Rens (19) at 293K. The results based on mol dm^{-3} units are given in (4), and those based on mol kg^{-1} are reported in (5) and (19).
The absence of reliable density data at 298K has not allowed the conversion of results reported in mol dm^{-3} into molal units. Therefore the results of Kilde cannot be directly compared with those of Chloupek, Daneš and Danešova.

4-5. Solubility in aqueous $CaCl_2$ solutions
This system has been reported in 3 publications (5,8,19).

Solubilities were measured by Kilde (5) at 291, 298 and 303K, by Wise and Davies (8) at 298K, and by Rens (19) at 293K.

Solubility at 298.2K. The results of Kilde are in good agreement with the corresponding values of Wise and Davies in aqueous solutions of same $CaCl_2$ concentration. The arithmetic mean of two results are designated as tentative values. The tentative values with the standard deviations are given in Table 7.

Table 7 Tentative values in aqueous $CaCl_2$ solutions

c_2/mol dm^{-3}	$10^3 c_1$/mol dm^{-3}	$10^3 \sigma$/mol dm^{-3}	ref
0	7.84	0.00	(5),(8)
0.0050	6.81	--	(5)
0.00625	6.69	--	(8)
0.0100	6.44	--	(5)
0.0250	5.51	0.09	(5),(8)
0.0500	4.91	0.005	(5),(8)

c_1: tentative value for $Ca(IO_3)_2$ solubility

c_2: concentration of $CaCl_2$

σ : standard deviation of c_1

Solubility at 293.2K. The results reported in (19) by Rens are designated as tentative values.

Solubility at 291.2 and 303.2K. The results reported in (5) obtained by Kilde are designated as tentative values.

4-6. Solubility in aqueous NH_4Cl solutions
This system has been reported in 2 publications (2,19).

Gross and Klinghoffer (2) measured solubilities in this system and calculated molalities using density data for NH_4Cl solutions from Landolt-Börnstein-Roth tables. No density data for solutions containing $Ca(IO_3)_2$ can be found in the tables, and the results reported are therefore very approximate. The evaluator cannot compare these results because no other studies have been reported at 298.2K. Rens (19) has investigated this system at 293.2K, and his results are designated as tentative.

COMPONENTS:	EVALUATOR:
(1) Calcium iodate; $Ca(IO_3)_2$; [7789-80-2] (2) Water; H_2O; [7732-18-5]	Hiroshi Miyamoto Department of Chemistry Niigata University Niigata, Japan April, 1982

CRITICAL EVALUATION:

4-7. Solubility in other salt solutions

Chloupek, Daneš and Danešov (4) determined solubilities of $Ca(IO_3)_2$ in $MgSO_4$ and KNO_3 solutions at 298K. The solubilities were determined iodometrically and reported in mol kg^{-1} units. In each of the salt solutions the solubility of $Ca(IO_3)_2$ increases with increasing salt concentration.

Wise and Davies (8) measured solubilities in Na_2SO_4 solutions at 298K and calculated the solubility product from the solubility data. For $K_2[Fe(CN)]$ solutions, the solubility of $Ca(IO_3)_2$ was also reported in (8). In each of the salt solutions, solubility increases with increasing salt concentration.

Davis and Wyatt (13) measured solubilities in $Na_2S_2O_3$ solutions at 298K, and calculated the dissociation constant of calcium thiosulfate from the solubility data.

Rens (19) studied solubilities in KBr and KI solutions at 293K, and calculated the activity product from the solubility data. The solubility increases with increasing salt concentration.

Only one article is published for solubilities of $Ca(IO_3)_2$ in each of the salt solutions. The results are designated as tentative values.

5. Solubility of calcium iodate in mixtures of various organic solvents and water

Pedersen (11) determined solubilities of calcium iodate hexahydrate in 1,4-dioxane-water mixtures at 293.1K and urea-water mixtures at 293.2K using iodometric titration for the analyses.

Monk (17) measured solubilities of calcium iodate hexahydrate in mixtures of various organic solvents and water at 298K. He used methanol, ethanol, 1-propanol, 1,2-ethandiol (ethylene glycol), 1,2,3-propanetriol (glycerol), 2-propanone (acetone), 1,4-dioxane and ethyl acetate.

Miyamoto (22) determined the solubility of calcium iodate hexahydrate in tetrahydrofuran-water mixtures at 298.2K and Miyamoto, Suzuki and Yanai (27) studied mixtures of N,N-dimethylformamide and water at 293, 298 and 303K. No stable solvates are formed in these systems.

They (17,22) studied solubilities in various mixed solvents with the purpose of testing the applicability of the simple electrostatic model, first given by Born (31), to solubility phenomena effects in different media. Solubilities in the various mixed solvents studied decrease with increasing concentration of organic solvents, and decrease with decreasing the dielectric constant of the mixed solvent. Monk (17) concludes that these results indicate that the chemical character of the solvent is of major importance in influencing decrease in solubility with decreasing the dielectric constant of solution.

The results reported in (11) by Pedersen, in (17) obtained by Monk, in (22) by Miyamoto, and in (27) by Miyamoto, Suzuki and Yanai are designated as tentative values.

COMPONENTS:	EVALUATOR:
(1) Calcium iodate; $Ca(IO_3)_2$; [7789-80-2] (2) Water; H_2O; [7732-18-5]	Hiroshi Miyamoto Department of Chemistry Niigata University Niigata, Japan April, 1982

CRITICAL EVALUATION:

6. Solubility of calcium iodate in amino acid solutions

Solubilities of calcium iodate both in aqueous glycine solutions and in aqueous alanine solutions have been reported by Keefer, Reiber, and Bisson (10), and Monk (16) at 298K. Monk also measured solubilities in aqueous glycyl glycine solutions.
Monk used the hexahydrate in solubility determination, but Keefer, Reiber, and Bisson did not report the degree of hydration for the salt used. The result of Monk is based on mol dm^{-3} units and that of Keefer, Reiber, and Bisson is based on molal units. The absence of applicable density data has not allowed the conversion of results reported in mol dm^{-3} units into molal units and vice versa. The result of Monk cannot directly compare with that of Keefer, Reiber, and Bisson.
Solubilities of calcium iodate in amino acid solutions increase with increasing concentration of amino acids. The dipolar ions formed by amino acids, being more polar than water, more or less replace water in the sphere of ion-hydration, and in addition the dielectric constant of the solution is increased in proportion to the acid concentration.

Monk calculated the thermodynamic solubility product from solubility data in amino acid solutions with or without the added electrolyte (KIO_3, $CaCl_2$, and KCl). In the calculations, the amounts of amino acid anions present are so small that the complex formation between calcium ion and amino acid anion was neglected. Keefer, Reiber, and Bisson did not calculate the thermodynamic solubility product from the solubility data.

The results reported in (10) obtained by Keefer, Reiber, and Bisson and in (16) obtained by Monk are designated as tentative values.

7. Solubility of calcium iodate in sodium salt of carboxylic acids.

The solubilities of calcium iodate in sodium salts of carboxylic acid solutions have been reported in 5 publications (7,8,9,24,25). Data summarizing the solubility studies on these systems are given in Table 8.

The solubilities of calcium iodate in solutions of the sodium salt of glycolic acid at 298K were reported by Davies (9), and by Das and Nair (24). Davies reports the solubilities in 0.020 and 0.040 mol dm^{-3} sodium salt solutions, and the interpolated results of the data obtained by Das and Nair agree very closely with the corresponding data of Davies.

The solubilities of calcium iodate in solutions of the sodium salt of mandelic acid were reported in (8) obtained by Wise and Davies, and in (24) obtained by Das and Nair. Wise and Davies report the solubilities in 0.020, 0.040, and 0.100 mol dm^{-3} sodium salt solutions. The interpolated results of the values obtained by Das and Nair are slightly lower than the corresponding data of Wise and Davies.

Results on other carboxylic acid systems have been reported in single publications (see Table 8).

The dissociation constant of the ion pair CaX^+ in all systmes has also been reported. The details are given on each compilation sheet.

COMPONENTS:

(1) Calcium iodate; $Ca(IO_3)_2$; [7789-80-2]

(2) Water; H_2O; [7732-18-5]

EVALUATOR:

Hiroshi Miyamoto
Department of Chemistry
Niigata University
Niigata, Japan

CRITICAL EVALUATION:

Table 8 Solubility studies of calcium iodate in sodium salt of carboxylate acids

T/K	Na Salt of Acids Acid	 Formula	Na Carboxylate concn c_2/mol dm^{-3}	ref
298.2	Glycolic	$HOCH_2COOH$	20.0 - 40.0	(9)
"	"	"	11.820 - 53.228	(24)
303.2	"	"	7.988 - 23.949	"
308.2	"	"	8.313 - 33.207	"
313.2	"	"	16.645 - 33.254	"
318.2	"	"	21.505 - 34.408	"
291.2	Lactic	$CH_3CHOHCOOH$	0 - 400.0	(7)
"	" (a)	"	0 - 99.2	"
298.2	"	"	0 - 400.0	"
"	" (a)	"	0 - 99.2	"
303.2	"	"	0 - 400.0	"
"	" (a)	"	0 - 99.2	"
298.2	β-Hydroxy butyric	$CH_3CH(OH)CH_2COOH$	21.48 - 43.04	(9)
298.2	Salicylic	o-HOC_6H_4COOH	20.0 - 40.0	(9)
298.2	Mandelic	$C_6H_5CH(OH)COOH$	0 - 100.0	(8)
"	"	"	10.970 - 56.148	(24)
303.2	"	"	27.069 - 72.340	"
308.2	"	"	7.042 - 56.336	"
313.2	"	"	30.340 - 60.680	"
318.2	"	"	37.921 - 60.680	"
298.2	Maleic	H-C-COOH ‖ H-C-COOH	0 - 36.04	(25)
303.2	"	"	0 - 51.03	"
308.2	"	"	0 - 59.08	"
313.2	"	"	0 - 63.87	"
298.2	Methoxyacetic	CH_3OCH_2COOH	20.0 - 40.0	(9)
298.2	Cyanoacetic	$NCCH_2COOH$	20.0 - 40.0	(9)
298.2	Pyruvic	$CH_3COCOOH$	20.0 - 40.0	(9)

(a): calcium salt

8. Solubility of calcium iodate in solutions of the sodium salt of amino acids and dipeptides

The calcium ion has a marked tendency to associate with the anions of α-amino acids. Davies and Waind measured the solubility of calcium iodate hexahydrate in aqueous solutions of some amino acids and dipeptides containing an equivalent volume of sodium hydroxide. The stabilities of a number of the association products of calcium with amino acids and dipeptides are reported in (14).

The amino acids and dipeptides used are glycine, aminopropionic acid, serine, D,L-glutamic acid, hipuric acid, 3,5-diodotyrosine, tyrosine, glycyl glycine, leucyl glycine and alanyl glycine.

COMPONENTS:

(1) Calcium iodate; $Ca(IO_3)_2$; [7789-80-2]

(2) Water; H_2O; [7732-18-5]

EVALUATOR:

Hiroshi Miyamoto
Department of Chemistry
Niigata University
Niigata, Japan

April, 1982

CRITICAL EVALUATION:

Davies (9) measured solubilities of calcium iodate hexahydrate in sodium glycinate solutions, and the results are in good agreement with those of Davies and Waind. In both publications the dissociation constant of the ion-pair ($CaH_2NCH_2COO^+$) was determined from the solubility data. The dissociation constants reported by Davies in (9) and in (14) are 0.037 and 0.042 mol dm^{-3} respectively.

Solubilities in solution of the sodium salt of amino acids or dipeptides increase with increasing concentration of amino acid or dipeptide.

The solubility data reported by Davies (9) and by Davies and Waind (14) are tentative values.

9. Solubility product of calcium iodate in aqueous solutions

Kilde (5) measured solubilities in different electrolyte solutions at 291, 298 and 303K, and the thermodynamic solubility product was calculated using the modified Debye-Hückel equation with the solubility data. The hexahydrate is in equilibrium with the saturated solution.

Wise and Davies (8) measured solubilities in water and numerous salt solutions at 298K, and reported the value for the solubility product extrapolated to zero ionic strength.

Monk (16) reported the thermodynamic solubility product in pure water and in amino acids solutions containing an inorganic salt from the solubility data at 298K. In the presence of amino-acids allowance for variations in dielectric constant must be made. For this purpose, he used the modified Davies expression (32). Monk (17) also reported the results for mixtures of various organic solvents and water.

Bell and George (18) measured the solubility of calcium iodate in salt solutions. The solid phase is the hexahydrate at 273 and 298K and the monohydrate at 313K. Combining the equilibrium constants of ion pairs with the experimental solubility data, they calculated the thermodynamic solubility product.

Rens (19) measured solubilities of calcium iodate hexahydrate in various salt solutions at 293K, and calculated the thermodynamic solubility product at zero ionic strength from the solubility data.

Nazzal, Popiel and Vermande (20) calculated the activity product of calcium iodate ranging from 287 to 328K from the solubility data in aqueous NaCl solutions, and found the transition temperature between the hexahydrate and the monohydrate.

Bousquet, Mathurin and Vermande (21) measured the solubility of the hydrate and anhydrate of calcium iodate in aqueous solutions over a wide temperature range from 287 to 359K, and calculated the activity solubility product from the experimental solubility data. The modified Debye-Hückel equation was used to calculate activity coefficients.

COMPONENTS:	EVALUATOR:
(1) Calcium iodate; $Ca(IO_3)_2$; [7789-80-2] (2) Water; H_2O; [7732-18-5]	Hiroshi Miyamoto Department of Chemistry Niigata University Niigata, Japan April, 1982

CRITICAL EVALUATION:

Das and Nair (24) reported solubilities of calcium iodate in buffer solutions of sodium glycolate and in mandelate buffers at 298, 303, 308, 313 and 318K, and calculated the activity products from the solubility data. They also measured solubilities in buffered solutions of sodium malate (25). In both publications, the degree of hydration of the salt was not specified, but the evaluator assumes that the hexahydrate was in equilibrium with saturated solutions at 298 and 303K, and the monohydrate at 313 and 318K. In calculations of activity coefficients, corrections were made for the incomplete dissociation of $CaIO_3^+$ and $NaIO_3$, and the activity coefficients were obtained from the modified Davies equation (33).

The data to be considered in this critical evaluation are summarized in Table 9.

Table 9 Summary of thermodynamic solubility product data for calcium iodate in aqueous solutions

T/K	10^6 K_{s0}°/mol^3 dm^{-9}	ref
	$Ca(IO_3)_2 \cdot 6H_2O$	
273.2	0.02859	(18)
287.2	0.211	(20)
"	0.211	(21)
291.2	0.324	(21)
"	0.327	(20)
"	0.329	(5)
293.2	0.4159	(19)
295.2	0.521	(20)
"	0.521	(21)
298.2	0.6953	(8)
"	0.698	(21)
"	0.708	(24)
"	0.7096	(25)
"	0.711	(16)
"	0.7119	(18)
"	0.736	(5)
"	0.794	(17)
299.2	0.822	(20)
301.2	0.971	(21)
303.2	1.119	(25)
"	1.12	(24)
"	1.271	(21)
"	1.291	(20)
"	1.35	(5)
305.2	1.542	(21)
307.2	1.928	(20)
308.2	1.91	(24)
"	1.919	(25)
"	2.005	(21)
311.2	2.832	(21)
	$Ca(IO_3)_2 \cdot H_2O$	
313.2	2.437	(18)
"	2.72	(21)
"	2.723	(20)
"	2.864	(25)
"	2.88	(24)
317.6	3.10	(21)
318.2	3.126	(20)
"	4.57	(24)
320.2	3.28	(21)
323.2	3.54	(20)
"	3.54	(21)
328.2	4.05	(21)
"	4.051	(20)
	$Ca(IO_3)_2$	
333.2	4.12	(21)
343.2	4.44	(21)
352.2	5.17	(21)
359.2	5.51	(21)

COMPONENTS:

(1) Calcium iodate; $Ca(IO_3)_2$; [7789-80-2]

(2) Water; H_2O; [7732-18-5]

EVALUATOR:

Hiroshi Miyamoto
Department of Chemistry
Niigata University
Niigata, Japan

April, 1982

CRITICAL EVALUATION:

EVALUATION OF THE SOLUBILITY PRODUCT DATA

Solubility product at 287.2K. This value has been reported in 2 publications (20,21). The result reported by Nazzal, Popiel and Vermande (20) is in excellent agreement with that reported by Bousquet, Mathurin and Vermande (21). The recommended value of the solubility product at 287.2K is therefore 0.211×10^{-6} mol^3 dm^{-9}.

Solubility product at 291.2K. This value has been reported in 3 publications (5,20,21). The arithmetic mean of 3 results is 0.327×10^{-6} mol^3 dm^{-9}, and the standard deviation is 0.003×10^{-6} mol^3 dm^{-9}.

Solubility product at 295.2K. Two data are reported. The result reported by Nazzal, Popiel and Vermande (20) is in excellent agreement with that reported by Bousquet, Mathurin and Vermande (21). The recommended value of solubility product at 295.2K is 0.521×10^{-6} mol^3 dm^{-9}.

Solubility product at 298.2K. This result has been reported in 8 publications (5,8,16,17,18,21,24,25). The result reported by Monk (17) is considerably higher than that of others (5,8,16,18,21,24,25), and is therefore rejected. The arithmetic mean of 7 results is 7.1×10^{-7} mol dm^{-9}, and the standard deviation is 0.1×10^{-7} mol dm^{-7}. The mean is designated as a recommended value.

Solubility product at 303.2K. This value has been reported in 5 publications (5,20,21,24,25). The data are widely distributed and range from 1.119×10^{-6} to 1.35×10^{-6} mol^3 dm^{-9}. The result by Kilde (5) is the highest value, and the data reported in (24) and (25) obtained by Das and Nair are lower than that of others (5,20,21). The arithmetic mean of 5 results is 1.23×10^{-6} mol^3 dm^{-9} and the standard deviation is 0.05×10^{-6} mol^3 dm^{-9}. The mean is designated as a tentative value.

Solubility product at 308.2K. This value has been reported in 3 publications (21,24,25). All results are in good agreement with each other. The arithmetic mean of the 3 results (21,24,25) is 1.94×10^{-6} mol^3 dm^{-9}, and the mean is designated as a recommended value.

Solubility product at 313.2K. This value has been reported in 5 publications (18,20, 21,24,25). The result reported by Bell and George (18) is lower than that of others (20,21,24,25), and therefore this value is rejected. The arithmetic mean of the remaining 4 results is 2.80×10^{-6} mol^3 dm^{-9}, and the standard deviation is 0.09×10^{-6} mol^3 dm^{-9}. The mean value is recommended.

Solubility product at 318.2K. The result reported by Das and Nair (24) is very high, and its value is higher than the results reported by Nezzal, Popiel and Vermande (20) at 323K, and by Bousquet, Mathurin and Vermande (21) at 320, 323 and 328K. Therefore, the result by Das and Nair is rejected. The tentative value is based on the result of Nazzal, Popiel and Vermande (20) and is 3.126×10^{-6} mol^3 dm^{-9}.

Solubility product at 323.2K Identical values have been reported in 2 publications (20,21). The tentative value is based on the results reported by Nazzal, Popiel and Vermande (20) and that of Bousquet, Mathurin and Vermande (21), and is 3.54×10^{-6} mol^3 dm^{-9}.

Solubility products at the other temperatures. Data based on single measurements are given in Table 9.

The recommended and tentative values of the solubility products along with the values calculated from the best fit equations are given in Table 10. In Table 10 K_{s0}°(exptl) is a recommended or tentative value.

COMPONENTS:

(1) Calcium iodate; $Ca(IO_3)_2$; [7789-80-2]

(2) Water; H_2O; [7732-18-5]

EVALUATOR:
Hiroshi Miyamoto
Department of Chemistry
Niigata University
Niigata, Japan

April, 1982

CRITICAL EVALUATION:

Table 10 Recommended and tentative values of the thermodynamic solubility product

T/K	$10^6 K^\circ_{s0}$(exptl)/mol^{3}dm^{-9}	$10^6 \sigma$/mol^{3}dm^{-9}	R or T^a	$10^6 K^\circ_{s0}$(calcd)/mol^{3}dm^{-9}
		Solid phase: $Ca(IO_3)_2 \cdot 6H_2O$		
273.2	0.02859	--	T	0.0290
287.2	0.211	--	R	0.200
291.2	0.327	0.003	R	0.327
293.2	0.4159	--	T	0.415
295.2	0.521	--	R	0.522
298.2	0.710	0.01	R	0.731
299.2	0.822	--	T	0.815
301.2	0.971	--	T	1.01
303.2	1.23	0.11	T	1.24
305.2	1.542	--	T	1.52
307.2	1.928	--	T	1.86
308.2	1.94	0.05	R	2.05
311.2	2.832	--	T	2.72
		Solid phase: $Ca(IO_3)_2 \cdot H_2O$		
313.2	2.80	0.09	R	2.80
317.6	3.10	--	T	3.09
318.2	3.126	--	T	3.13
320.2	3.28	--	T	3.28
323.2	3.54	--	T	3.54
328.2	4.05	--	T	4.05
		Solid phase: $Ca(IO_3)_2$		
333.2	4.12	--	T	4.10
343.2	4.44	--	T	4.51
352.2	5.17	--	T	5.08
359.2	5.51	--	T	5.55

a: R = recommended value, T = tentative value

The data in Table 10 were fitted to the following equations:

$$\ln K^\circ_{s0}(1) = 140.6256 - 271.5007/(T/100\text{K}) - 58.30910 \ln (T/100\text{K}) : \quad \sigma = 0.055 \times 10^{-6}$$

$$\ln K^\circ_{s0}(2) = -140.4491 + 175.7183/(T/100\text{K}) + 62.67982 \ln (T/100\text{K}) : \quad \sigma = 0.008 \times 10^{-6}$$

$$\ln K^\circ_{s0}(3) = -60.72781 + 67.05006/(T/100\text{K}) + 23.43024 \ln (T/100\text{K}) : \quad \sigma = 0.12 \times 10^{-6}$$

where $K^\circ_{s0}(1)$, $K^\circ_{s0}(2)$ and $K^\circ_{s0}(3)$ are the solubility products for the hexahydrate, the monohydrate and the anhydrous salt, respectively.

The values calculated from the smoothing equations are also given in Table 10.

COMPONENTS:

(1) Calcium iodate; $Ca(IO_3)_2$; [7789-80-2]

(2) Water; H_2O; [7732-18-5]

EVALUATOR:

Hiroshi Miyamoto
Department of Chemistry
Niigata University
Niigata, Japan

April, 1982

CRITICAL EVALUATION:

REFERENCES:

1. Mylius, F.; Funk, R. *Ber. Dtsch. Chem. Ges.* 1897, *30*, 1716.
2. Gross, F.; Klinghoffer, St. S. *Monatsh. Chem.* 1930, *55*, 338.
3. Hill, A. E.; Brown, S. F. *J. Am. Chem. Soc.* 1931, *53*, 4316.
4. Chloupek, J. B.; Daneš, VL. Z.; Danešova, B. A. *Collect. Czech. Chem. Commun.* 1933, *5*, 339.
5. Kilde, G. *Z. Anorg. Allg. Chem.* 1934, *218*, 113.
6. Kolthoff, I. M.; Stenger, V. A. *J. Phys. Chem.* 1934, *38*, 639.
7. Kilde, G. *Z. Anorg. Allg. Chem.* 1936, *229*, 321.
8. Wise, W. C. A.; Davies, C. W. *J. Chem. Soc.* 1938, 273.
9. Davies, C. W. *J. Chem. Soc.* 1938, 277.
10. Keefer, R. M.; Reiber, H. G.; Bisson, C. S. *J. Am. Chem. Soc.* 1940, *62*, 2951.
11. Pedersen, K. J. *K. Dan. Videns. Selsk. Math-Fys. Medd.* 1941, *18*, 1.
12. Derr, P. F.; Vosburgh, W. C. *J. Am. Chem. Soc.* 1943, *65*, 2408.
13. Davies, C. W.; Wyatt, P. A. H. *Trans. Faraday Soc.* 1949, *45*, 770.
14. Davies, C. W.; Waind, G. M. *J. Chem. Soc.* 1950, 301.
15. Davies, C. W.; Hoyle, B. E. *J. Chem. Soc.* 1951, 223.
16. Monk, C. B. *Trans. Faraday Soc.* 1951, *47*, 1233.
17. Monk, C. B. *J. Chem. Soc.* 1951, 2723.
18. Bell, R. P.; George, J. H. B. *Trans. Faraday Soc.* 1953, *49*, 619.
19. Rens, G. *Sucr. Belge* 1958, *77*, 193-208.
20. Nezzal, G.; Popiel, W. J.; Vermande, P. *Chem. Ind. (London)* 1967, *(30)*, 1294.
21. Bousquet, J.; Mathurin, D.; Vermande, P. *Bull. Soc. Chem. Fr.* 1969, 1111.
22. Miyamoto, H. *Nippon Kagaku Kaishi* 1972, 659.
23. Fedorov, V. A.; Robov, A. M.; Shmyd'ko, I. I.; Vorontsova, N. A.; Mironov, E. *Zh. Neorg. Khim.* 1974, *19*, 1746; *Russ. J. Inorg. Chem. (Engl. Transl.)* 1974, *19*, 950.
24. Das, A. R.; Nair, V. S. K. *J. Inorg. Nucl. Chem.* 1975, *37*, 991.
25. Das, A. R.; Nair, V. S. K. *J. Inorg. Nucl. Chem.* 1975, *37*, 2121.

COMPONENTS:

(1) Calcium iodate; $Ca(IO_3)_2$; [7789-80-2]

(2) Water; H_2O; [7732-18-5]

EVALUATOR:

Hiroshi Miyamoto
Department of Chemistry
Niigata University
Niigata, Japan

April, 1982

CRITICAL EVALUATION:

26. Azarova, L. A.; Vinogradov, E. E. *Zh. Neorg. Khim.* 1977, *22*, 273; *Russ. J. Inorg. Chem. (Engl. Transl.)* 1977, *22*, 153.

27. Miyamoto, H.; Suzuki, K.; Yanai, K. *Nippon Kagaku Kaishi* 1978, 1150.

28. Arkhipov, S. M.; Kashina, N. I.; Kidyarov, B. I. *Zh. Neorg. Khim.* 1978, *23*, 1422; *Russ. J. Inorg. Chem. (Engl. Transl.)* 1978, *23*, 784.

29. Shklovskaya, R. M.; Arkhipov, S. M.; Kidyarov, B. I.; Poleva, G. V. *Zh. Neorg. Khim.* 1979, *24*, 1416; *Russ. J. Inorg. Chem. (Engl. Transl.)* 1979, *24*, 786.

30. Linke, W. F. *Solubilities. Inorganic and Metal-Organic compounds.* American Chemical Society. Washington. 1965, 4th ed.

31. Born, M. *Z. Phys.* 1920, *1*, 45.

32. Davies, C. W. *J. Chem. Soc.* 1938, 2093.

33. Davies, C. W. *Ionic Association.* Butterworths. London. 1960. Ch. 3.

COMPONENTS:

(1) Calcium iodate; $Ca(IO_3)_2$; [7789-80-2]

(2) Water; H_2O; [7732-18-5]

ORIGINAL MEASUREMENTS:

Mylius, F.; Funk, R.

Ber. Dtsch. Chem. Ges. 1897, *30*, 1716-25.

VARIABLES:

T/K = 273 to 373

PREPARED BY:

Hiroshi Miyamoto

EXPERIMENTAL VALUES:

t/°C	Calcium Iodate mass %	m_1/mol kg^{-1} (compiler)	Nature of the Solid Phase
0	0.1	0.003	$Ca(IO_3)_2 \cdot 6H_2O$
10	0.17	0.0044	"
18[a]	0.25	0.0064	"
30	0.42	0.011	"
40	0.61	0.016	"
50	0.89	0.023	"
54	1.04	0.0270	"
60	1.36	0.0354	"
21	0.37	0.0095	$Ca(IO_3)_2 \cdot H_2O$
35	0.48	0.012	"
40	0.52	0.013	"
45	0.54	0.014	"
50	0.59	0.015	"
60	0.65	0.017	"
80	0.79	0.020	"
100	0.94	0.024	"

[a] The solubility, 0.25 g/100g H_2O, and the density of the saturated solution, 1, were also reported.

The compiler presumes that the first word in the fifth line from the end of page 1717 should read 100g.

AUXILIARY INFORMATION

METHOD/APPARATUS/PROCEDURE:

$Ca(IO_3)_2$ crystals and water were placed in bottles. The bottles were shaken in a constant temperature bath for a long time.

After the saturated solution settled, an aliquot of solution was removed with a pipet. Calcium iodate was determined by evaporation of the solution to dryness. The density of the saturated solution was also determined.

SOURCE AND PURITY OF MATERIALS:

The salt used was purchased as "pure" chemical, and traces of impurities were not present. The purity sufficed for the solubility determination.

ESTIMATED ERROR:

Soly: precision within 1% (compiler)
Temp: nothing specified

REFERENCES:

COMPONENTS:

(1) Calcium iodate; $Ca(IO_3)_2$; [7789-80-2]

(2) Water; H_2O; [7732-18-5]

ORIGINAL MEASUREMENTS:

Hill, A. E.; Brown, S. F.

J. Am. Chem. Soc. 1931, *53*, 4316-20.

VARIABLES:

T/K = 278 to 363

PREPARED BY:

Hiroshi Miyamoto

EXPERIMENTAL VALUES:

Solubility of Calcium Iodate in Water

t/°C	From under-saturation mass %	From super-saturation mass %	Average mass %	Average Molality[a] $10^2 m_1$/mol kg^{-1}	Nature of the Solid Phase[b]
5	0.118	0.120	0.119	0.306	A
15	0.194	0.196	0.195	0.501	A
25	0.306	0.307	0.306	0.787	A
30	0.384	0.384	0.384	0.989	A
35	0.475	0.477	0.476	1.227	A + B
40	0.584	---	0.584	1.507	A(m)
25	---	0.405	0.405	1.043	B(m)
40	0.514	0.519	0.517	1.333	B
50	0.589	0.590	0.590	1.522	B
57.5	0.621	---	0.621	1.603	B + C
60	0.652	---	0.652	1.683	B(m)
70	0.811	---	0.811	2.097	B(m)
60	---	0.617	0.617	1.592	C
70	0.643	0.645	0.644	1.662	C
80	0.665	0.665	0.665	1.717	C
90	0.668	0.668	0.668	1.725	C

[a] molalities calculated by compiler.

[b] A = $Ca(IO_3)_2 \cdot 6H_2O$; B = $Ca(IO_3)_2 \cdot H_2O$; C = $Ca(IO_3)_2$; (m) = metastable.

AUXILIARY INFORMATION

METHOD/APPARATUS/PROCEDURE:

In carrying out the solubility determinations, the specified hydrate in each temperature range was used. The time allowed for equilibrium varied from one day at the highest temperatures to two or three weeks at the lower temperatures. By using metastable phases, several points for metastable equilibrium were obtained, and in which the solubility of the metastable hydrate remained constant for as long a period as two weeks. The equilibrium between the liquid and solid phases was approached from the side of supersaturations and/or undersaturation. The concentration of calcium iodate in liquid phases was determined iodometrically.

SOURCE AND PURITY OF MATERIALS:

Calcium iodate was prepared by double decomposition of $Ca(NO_3)_2$ and KIO_3 in water, washed, and purified by recrystallization. The hexahydrate obtained by slow cooling within the temperature range below 30°C, the monohydrate below 100°C. Each sample was dried in a desiccator over the next lower hydrate as desicant, and was analyzed verifying the correct composition within a few tenths of a percent. The anhydrate was prepared by dehydration in an oven at 100°C.

ESTIMATED ERROR:

nothing specified

REFERENCES:

COMPONENTS:	ORIGINAL MEASUREMENTS:
(1) Calcium iodate; $Ca(IO_3)_2$; [7789-80-2]	Bousquet, J.; Mathurin, D.; Vermande, P.
(2) Water; H_2O; [7732-18-5]	*Bull. Soc. Chim. Fr.* 1969, 1111-5.

EXPERIMENTAL VALUES:

t/°C	Calcium Iodate Activity Product $10^7\ K^{\circ}_{s0}/mol^3\ dm^{-9}$	Nature of the Solid Phase
14	2.11	hexahydrate
18	3.24	"
22	5.21	"
25	6.98	"
28	9.71	"
30	12.71	"
32	15.42	"
35	20.05	"
38	28.32	"
40	27.2	monohydrate[a]
44.4	31.0	"
47	32.8	"
50	35.4	"
55	40.5	"
60	41.2	anhydrate[a]
70	44.4	"
79	51.7	"
86	55.1	"

[a] There are apparent misprintings of these values in the original article, but the correct values as printed here can be calculated from other data in Table in original article.

The solubility product, K°_{s0} of $Ca(IO_3)_2 \cdot xH_2O$ was defined as

$$K^{\circ}_{s0} = (c_{Ca^{2+}} \times c^2_{IO_3^-})(y_{Ca^{2+}} \times y^2_{IO_3^-})$$

$$= 4S^3 y_{\pm}^3 \qquad (1)$$

where S represents the solubility, $y_{\pm}$ the activity coefficient given by the modified Debye-Hückel equation

$$-\log y_{\pm} = Z_+Z_-\ A\sqrt{I} - BI \qquad (2)$$

From (1) and (2)

$$Y = -BI + 1/3 \log K^{\circ}_{s0} \qquad (3)$$

where $Y = 1/3 \log (4S^3) - Z_+Z_-A\sqrt{I}$, and A = 0.5115 at 25°C.

The solubility product (K°_{s0}) and constant B were evaluated from the intercept and the slope of Y vs I plots. The solubilities of $Ca(IO_3)_2$ in NaCl aqueous solutions were determined in order to obtain Y vs I plots, but these data were not given in the paper.

COMPONENTS:

(1) Calcium iodate; $Ca(IO_3)_2$; [7789-80-2]

(2) Water; H_2O; [7732-18-5]

ORIGINAL MEASUREMENTS:

Bousquet, J.; Mathurin, D.; Vermande, P.

Bull. Soc. Chim. Fr. 1969, 1111-5.

VARIABLES:

T/K = 287 - 359

PREPARED BY:

Hiroshi Miyamoto

EXPERIMENTAL VALUES:

AUXILIARY INFORMATION

METHOD/APPARATUS/PROCEDURE:

Aqueous NaCl solutions and the specified hydrate crystals were placed into glass-stoppered Erlenmeyer flasks. The flasks were stirred in a thermostat for 1-15 hours. The iodate content was determined iodometrically.

SOURCE AND PURITY OF MATERIALS:

$Ca(IO_3)_2 \cdot 6H_2O$ was prepared by mixing aqueous solutions of calcium nitrate and HIO_3. The product was washed. The monohydrate and anhydrate were prepared from the hexahydrates, which were furnished from BDH and prepared by authors, by hydration.

ESTIMATED ERROR:

Soly: nothing specified
Temp: ± 0.05°C (authors)

REFERENCES:

COMPONENTS:

(1) Calcium iodate; $Ca(IO_3)_2$; [7789-80-2]

(2) Ammonia; NH_3; [7664-41-7]

(3) Water; H_2O; [7732-18-5]

ORIGINAL MEASUREMENTS:

Kolthoff, I. M.; Stenger, V. A.

J. Phys. Chem. 1934, *38*, 639-43.

VARIABLES:

$T/K = 298$

$c_2/mol\ dm^{-3} = 0$ to 1.966

PREPARED BY:

Hiroshi Miyamoto and Mark Salomon

EXPERIMENTAL VALUES:

t/°C	Ammonia	Calcium Iodate			Density
	$c_2/mol\ dm^{-3}$	$c_1/mol\ dm^{-3}$	g/100 cm^3 sln.	g/100g sln.	$\rho/g\ cm^{-3}$
25	0	0.00785	0.306	0.306	0.999
	0.489	0.00779	0.304	0.305	0.995
	0.986	0.00756	0.295	0.298	0.991
	1.422	0.00733	0.285	0.289	0.987
	1.966	0.00715	0.279	0.284	0.983

The solid phase in equilibrium with the saturated solution is $Ca(IO_3)_2 \cdot 6H_2O$.

AUXILIARY INFORMATION

METHOD/APPARATUS/PROCEDURE:

Isothermal method. Solutions were equilibrated by shaking in paraffined containers for 14 to 20 h. The ppts were allowed to settle and samples of the supernatant liquid analysed. The soly of $Ca(IO_3)_2$ was found from iodometric titrn of IO_3. All analyses were made in duplicate or triplicate. The densities of the supernatant liquids were determined at 25°C.
In parallel studies on the NH_4ClO_4-NH_3-H_2O system, the authors state that duplicate analysis from a given bottle agreed to within 0.1 %, but that analysis from different bottles lead to differences as high as 1%, and that the solubility data for the NH_4ClO_4 systems are not as exact as those for $Ca(IO_3)_2$. The compilers assume this to mean that the reproducibility of the data for the $Ca(IO_3)_2$ analysis is about ± 0.1%.

SOURCE AND PURITY OF MATERIALS:

$Ca(IO_3)_2 \cdot 6H_2O$ prepd by treating a hot sln of c.p. grade $CaCl_2$ with a slight excess of KIO_3 (source not specified). The ppt was washed and twice recrystallized from conductivity water. Carbonate free ammonia obtained by distillation of 20% ammonia in the presence of excess $Ba(OH)_2$. The distillate was kept in a paraffined container protected from atmospheric CO_2.

ESTIMATED ERROR:

Soly: reproducibility probably around ± 0.1% (see discussion under METHOD).
Temp. of soly detns: precision ± 0.02 K.
Temp. of density detns: unknown.

REFERENCES:

COMPONENTS:

(1) Calcium iodate; $Ca(IO_3)_2$; [7789-80-2]

(2) Ammonia; NH_3; [7664-41-7]

(3) Water; H_2O; [7732-18-5]

ORIGINAL MEASUREMENTS:

Derr, P. F.; Vosburgh, W. C.

J. Am. Chem. Soc. 1943, *65*, 2408-11.

VARIABLES:

$T/K = 298$

$c_2/\text{mol dm}^{-3} = 0.1$

PREPARED BY:

Hiroshi Miyamoto

EXPERIMENTAL VALUES:

(1) The solubility of $Ca(IO_3)_2$ in pure water at 25°C is

7.83×10^{-3} mol dm^{-3}.

(2) The authors stated that the apparent solubility product of calcium iodate was found to be unaffected by the presence of 0.1 mol dm^{-3} of ammonia but decreased when the ammonia concentration was larger. However, the authors did not report the solubility of $Ca(IO_3)_2$ in solutions for which the NH_3 concentration was greater than 0.1 mol dm^{-3}.

AUXILIARY INFORMATION

METHOD/APPARATUS/PROCEDURE:

Method not stated, but the compiler assumes that the procedure is similar to that adopted for the silver iodate system given in same paper. A roughly measured volume of ammonia stock solutions was isothermally saturated with calcium iodate. Samples for analyses were withdrawn by forcing the solution through a filter and into a pipet by air pressure to avoid loss of NH_3. The total ammonia was determined by titration with standard acid. The total iodate was determined iodometrically.

SOURCE AND PURITY OF MATERIALS:

Calcium iodate was precipitated from solutions of reagent grade chemicals, digested at a high temperature, carefully washed, and preserved under water. The names of the chemicals used were not given. The number of hydrated waters in calcium iodate hydrate was not given.

ESTIMATED ERROR:

nothing specified

REFERENCES:

COMPONENTS:

(1) Calcium iodate; $Ca(IO_3)_2$; [7789-80-2]

(2) Ammonium chloride; NH_4Cl; [12125-02-9]

(3) Water; H_2O; [7732-18-5]

ORIGINAL MEASUREMENTS:

Gross, P.; Klinghoffer, St. S.

Monatsh. Chem. 1930, *55*, 338-41.

VARIABLES:

$T/K = 298$

$10^3 c_2/\text{mol dm}^{-3} = 0$ to 1491

PREPARED BY:

J. W. Lorimer and H. Miyamoto

EXPERIMENTAL VALUES:

t/°C	Ammonium Chloride[a] $10^3 c_2/\text{mol dm}^{-3}$	Calcium Iodate[b] $10^3 c_1/\text{mol dm}^{-3}$
25	0	7.976
	50.44	9.732
	99.07	10.68
	149.8	11.49
	298.7	13.44
	500.0	15.57
	747.4	17.75
	1048	19.66
	1491	22.92

[a] Concentrations of NH_4Cl appear to be initial values.

[b] Solid phase is $Ca(IO_3)_2 \cdot 6H_2O$.

COMMENTS:

See compilation of the authors' work in the system $Ca(IO_3)_2$ - LiCl - H_2O.

AUXILIARY INFORMATION

METHOD/APPARATUS/PROCEDURE:

Salts and water were shaken in sealed flasks in a thermostat for 24-48 h. One sample (two for determinations in pure water) was heated above 25°C and shaken for a short time to produce a solution which was supersaturated at 25°C. After settling, the solutions were filtered through cotton wool filters which were found to be inert to calcium iodate. Two samples were removed for analysis, presumably (compiler) by titration with $AgNO_3$.

SOURCE AND PURITY OF MATERIALS:

Calcium iodate hexahydrate was made from solutions of Merck p.a. $CaCl_2$ and KIO_3. NH_4Cl was Merck p.a.

ESTIMATED ERROR:

Temperature: control to ± 0.005 K, accuracy within ± 0.05 K.
Analyses for IO_3: precision within ± 0.3 %.

REFERENCES:

COMPONENTS:

(1) Calcium iodate; $Ca(IO_3)_2$; [7789-80-2]

(2) Ammonium chloride; NH_4Cl; [12125-02-9]

(3) Water; H_2O; [7732-18-5]

ORIGINAL MEASUREMENTS:

Rens, G.

Sucr. Belge 1958, *77*, 193-208.

VARIABLES:

T/K = 293

$10^3 c_2$/mol dm^{-3} = 0.000 to 200.06

PREPARED BY:

Hiroshi Miyamoto

EXPERIMENTAL VALUES:

t/°C	Ammonium Chloride $10^3 c_2$/mol dm^{-3}	1/2 Iodate Ion $10^3 (c/2)$/mol dm^{-3}	Calcium Ion $10^3 c$/mol dm^{-3}	Calcium Iodate $10^3 c_1$/mol dm^{-3} [a]
20	0.000	6.231	6.230	6.231
	5.010	6.395	6.423	6.409
	9.995	6.610	6.607	6.609
	20.01	6.903	6.932	6.918
	40.01	7.420	7.409	7.415
	79.99	8.197	8.166	8.182
	99.99	8.490	8.478	8.484
	150.02	9.167	9.144	9.156
	200.06	9.784	9.762	9.773

[a] Average value calculated by the compiler.

pK°_{s0} was calculated from the equation:

1) $I < 0.07$ mol dm^{-3}, $pK_{s0} = pK^\circ_{s0} - 6AI^{1/2} + 3CI$

where I is the ionic strength, and A = 0.5046 mol$^{-1/2}$ dm$^{3/2}$ (ref 1)

2) $0.07 < I < 0.25$ mol dm^{-3}, $pK_{s0} = pK^\circ_{s0} - 6AI^{1/2}/(1 + aBI^{1/2})$

where a is distance of closest approach.

For a = 3.4A°, the value of K°_{s0} was 4.159 x 10^{-7} mol^3 dm^{-9}

AUXILIARY INFORMATION

METHOD/APPARATUS/PROCEDURE:

Excess $Ca(IO_3)_2 \cdot 6H_2O$ and aqueous NH_4Cl solution were placed in sealed Erlenmeyer flasks. The flasks were rotated in a thermostat for 24 hours. Aliquots of saturated solution were filtered.
The concentration of iodate was determined iodometrically. The calcium content was determined by chelatometric titration using Eriochrome black T as an indicator.

SOURCE AND PURITY OF MATERIALS:

$Ca(IO_3)_2 \cdot 6H_2O$ was prepared by slowly adding the solution of KIO_3 (about 50g dm^{-3}) to an equivalent solution of $CaCl_2 \cdot 2H_2O$ at 20°C. The precipitate was washed by decantation, and was air-dried at room temperature. An analysis of the product gave the following values: IO_3 99.64% and Ca 99.95% of theoretical.

ESTIMATED ERROR:

Soly: nothing specified
Temp: ± 0.1°K (author)

REFERENCES:

1. Harned, H.; Owen, B. *The Physical Chemistry of Electrolytic Solutions.* Reinhold, New York, 1950, 447.

COMPONENTS:

(1) Calcium iodate; $Ca(IO_3)_2$; [7789-80-2]

(2) Lithium chloride; LiCl; [7447-41-8]

(3) Water; H_2O; [7732-18-5]

ORIGINAL MEASUREMENTS:

Gross, P.; Klinghoffer, St. S.

Monatsh. Chem. 1930, *55*, 338-41.

VARIABLES:

$T/K = 298$

$10^3c_2/mol\ dm^{-3} = 0$ to 1510

PREPARED BY:

J. W. Lorimer and H. Miyamoto

EXPERIMENTAL VALUES:

t/°C	Lithium Chloride[a] $10^3c_2/mol\ dm^{-3}$	Calcium Iodate[b] $10^3c_1/mol\ dm^{-3}$
25	0	7.976
	18.43	9.414
	96.32	10.28
	144.8	10.97
	288.6	12.35
	183.0	13.71
	784.3	15.50
	1126	17.20
	1510	19.10

[a] Concentrations of LiCl appear to be initial values.

[b] Solid phase is $Ca(IO_3)_2 \cdot 6H_2O$.

COMMENTS:

The authors claim to have calculated molalities using density data from Landolt-Börnstein-Roth tables. However, no density data on $Ca(IO_3)_2$ solutions can be found in these tables, and the authors' chloride concentrations are initial values before adding $Ca(IO_3)_2$, so the calculations are very approximate. The authors also plot, presumably for solubility data in NaCl, KCl, NH_4Cl as well as in LiCl,

$$\log m_1 - A\sqrt{I} + 2\log a(H_2O)$$

against ionic strength $I = 3m_1 + m_2$, where m is molality, $a(H_2O)$ the water activity in chloride solutions from (1), and $A = 1.01\ (kg\ mol^{-1})^{1/2}$ is the Debye-Hückel constant (corrected on original). The intercepts of such plots give -2.240 (sign inserted by compiler), which corresponds to $(1/3)\log(K_{s0}/4)$, or (compiler) $K_{s0} = a_{\pm}^3(Ca(IO_3)_2) = 8.513 \times 10^{-7}$. They use this value to calculate mean activity coefficients and solubilities (in mol kg^{-1}) at rounded ionic strengths. As noted above, these calculations are approximate.

AUXILIARY INFORMATION

METHOD/APPARATUS/PROCEDURE:

Salts and water were shaken in sealed flasks in a thermostat for 24-48h. One sample (two for determinations in pure water) was heated above 25°C and shaken for a short time to produce a solution which was supersaturated at 25°C. After settling, the solutions were filtered through cotton wool filters which were found to be inert to calcium iodate. Two samples were removed for analysis, presumably (compiler) by titration with $AgNO_3$.

SOURCE:

Calcium iodate hexahydrate was made from solutions of Merck p.a. $CaCl_2$ and KIO_3. LiCl was Merck p.a.

ESTIMATED ERROR:

Temperature: control to ± 0.005 K, accuracy within ± 0.05 K.
Analyses for IO_3: precision within ± 0.3 %.

REFERENCES:

1. Lewis, G. N.; Randall, M. *Thermodynamics*, McGraw-Hill, New York, 1923, Chap. XXIII, XXVIII.

COMPONENTS:

(1) Calcium iodate; $Ca(IO_3)_2$; [7789-80-2]

(2) Lithium chloride; $LiCl$; [7447-41-8]

(3) Water; H_2O; [7732-18-5]

ORIGINAL MEASUREMENTS:

Rens, G.

Sucr. Belge 1958, *77*, 193-208.

VARIABLES:

T/K = 293

$10^3 c_2$/mol dm^{-3} = 0.000 to 155.36

PREPARED BY:

Hiroshi Miyamoto

EXPERIMENTAL VALUES:

t/°C	Lithium Chloride $10^3 c_2$/mol dm^{-3}	1/2 Iodate Ion $10^3 (c/2)$/mol dm^{-3}	Calcium Ion $10^3 c$/mol dm^{-3}	Calcium Iodate $10^3 c_1$/mol dm^{-3} [a]
20	0.000	6.231	6.230	6.231
	3.717	6.375	6.385	6.380
	51.80	7.571	7.555	7.563
	113.88	8.447	8.450	8.449
	155.36	8.888	8.877	8.883

[a] Average value calculated by the compiler.

pK°_{s0} was calculated from the equation:

1) $I < 0.07$ mol dm^{-3}, $pK_{s0} = pK^\circ_{s0} - 6AI^{1/2} + 3CI$

where I is the ionic strength, and A = 0.5046 mol$^{-1/2}$ dm$^{3/2}$ (ref 1)

2) $0.07 < I < 0.25$ mol dm^{-3}, $pK_{s0} = pK^\circ_{s0} - 6AI^{1/2}/(1 + aBI^{1/2})$

where a is distance of closest approach.

For a = 3.9A°, the value of K°_{s0} was 4.159 x 10^{-7} mol^3 dm^{-9} .

AUXILIARY INFORMATION

METHOD/APPARATUS/PROCEDURE:

Excess $Ca(IO_3)_2 \cdot 6H_2O$ and aqueous LiCl solution were placed in sealed Erlenmeyer flasks. The flasks were rotated in a thermostat for 24 hours. Aliquots of saturated solution were filtered.
The concentration of iodate was determined iodometrically. The calcium content was determined by chelatometric titration using Eriochrome black T as an indicator.

SOURCE AND PURITY OF MATERIALS:

$Ca(IO_3)_2 \cdot 6H_2O$ was prepared by slowly adding the solution of KIO_3 (about 50g dm^{-3}) to an equivalent solution of $CaCl_2 \cdot 2H_2O$ at 20°C. The precipitate was washed by decantation, and was air-dried at room temperature. An analysis of the product gave the following values: IO_3 99.64% and Ca 99.95% of theoretical.

ESTIMATED ERROR:

Soly: nothing specified
Temp: ± 0.1°K (author)

REFERENCES:

1. Harned, H.; Owen, B. *The Physical Chemistry of Electrolytic Solutions.* Reinhold, New York, 1950, 447.

COMPONENTS:

(1) Calcium iodate; $Ca(IO_3)_2$; [7789-80-2]

(2) Lithium nitrate; $LiNO_3$; [7790-69-4]

(3) Lithium perchlorate; $LiClO_4$; [7791-03-9]

(4) Water; H_2O; [7732-18-5]

ORIGINAL MEASUREMENTS:

Fedorov, V. A.; Robov, A. M.; Shmyd'ko, I. I.; Vorontsova, N. A.; Mironov, V. E.

Zh. Neorg. Khim. 1974, *19*, 1746-50; *Russ. J. Inorg. Chem.* (*Engl. Transl.*) 1974, *19*, 950-3.

VARIABLES:

T/K = 298

Concentration of $LiNO_3$ and $LiClO_4$

PREPARED BY:

Hiroshi Miyamoto

EXPERIMENTAL VALUES:

t/°C	Lithium Nitrate c_2/mol dm^{-3}	Calcium Iodate, 10^2c_1/mol dm^{-3}				
		Ionic Strength[a]				
		0.5	1.0	2.0	3.0	4.0
25	0	1.29	1.48	1.71	1.53	1.29
	0.1	1.34	--	--	--	--
	0.2	1.38	1.57	1.82	--	--
	0.3	1.44	--	--	--	--
	0.4	1.49	1.65	--	1.75	1.50
	0.5	1.54	--	1.98	--	--
	0.6		1.74	--	--	--
	0.8		1.83	2.18	1.99	1.73
	1.0		1.92	2.28	--	--
	1.2			--	2.20	1.95
	1.3			2.45	--	--
	1.5			2.57	--	--
	1.6			--	2.45	2.20
	1.8			2.73	--	--
	2.0			2.83	2.70	2.43
	2.4				2.98	2.72
	2.8				3.18	2.95
	3.0				3.28	--
	3.2					3.21
	3.6					3.50
	4.0					3.79

[a] The ionic strength adjusted by addition of lithium perchlorate to the lithium nitrate given above.

AUXILIARY INFORMATION

METHOD/APPARATUS/PROCEDURE:

Equilibrium between $Ca(IO_3)_2$ crystals and $LiNO_3$ solution containing $LiClO_4$ was reached by vigorous agitation with a magnetic stirrer in stoppered vessels in a thermostat. Equilibrium was established after stirring 4-6 hours, and checked by removing specimens after equal intervals of time. The concentrations of $Ca(IO_3)_2$ in the saturated solutions were determined iodometrically with visual or amperometric indication of the equivalence points.

SOURCE AND PURITY OF MATERIALS:

$Ca(IO_3)_2 \cdot 6H_2O$ was prepared by mixing solutions of $CaCl_2$ and HIO_3 at room temperature. The product was washed with water.

Chemically pure grade $LiNO_3$ and $LiClO_4$ were recrystallized from twice-distilled water. Before recrystallization, the solutions were boiled with active carbon.

ESTIMATED ERROR:

Soly: the reproducibility of the results averages ± 1.5-2 %.

Temp: not given.

REFERENCES:

COMPONENTS:

(1) Calcium iodate; $Ca(IO_3)_2$; [7789-80-2]

(2) Sodium chloride; NaCl; [7647-14-5]

(3) Water; H_2O; [7732-18-5]

ORIGINAL MEASUREMENTS:

Gross, P.; Klinghoffer, St. S.

Monatsh. Chem. 1930, *55*, 338-41.

VARIABLES:

$T/K = 298$

$10^3 c_2/\text{mol dm}^{-3}$ = 0 to 1988

PREPARED BY:

J. W. Lorimer and H. Miyamoto

EXPERIMENTAL VALUES:

t/°C	Sodium Chloride[a] $10^3 c_2/\text{mol dm}^{-3}$	Calcium Iodate[b] $10^3 c_1/\text{mol dm}^{-3}$
25	0	7.976
	50.02	9.677
	99.47	10.52
	119.5	11.23
	323.4	13.03
	498.8	14.74
	747.4	16.48
	1048	18.67
	1491	21.15
	1988	23.65

[a] Concentrations of NaCl appear to be initial values.

[b] Solid phase is $Ca(IO_3)_2 \cdot 6H_2O$.

COMMENTS:

See compilation of the authors' work on the system $Ca(IO_3)_2$-LiCl-H_2O.

AUXILIARY INFORMATION

METHOD/APPARATUS/PROCEDURE:

Salts and water were shaken in sealed flasks in a thermostat for 24-48 h. One sample (two for determinations in pure water) was heated above 25°C and shaken for a short time to produce a solution which was supersaturated at 25°C. After settling, the solutions were filtered through cotton wool filters which were found to be inert to calcium iodate. Two samples were removed for analysis, presumably (compiler) by titration with $AgNO_3$.

SOURCE AND PURITY OF MATERIALS:

Calcium iodate hexahydrate was made from solutions of Merck p.a. $CaCl_2$ and KIO_3. NaCl was Merck p.a.

ESTIMATED ERROR:

Temperature: control to ± 0.005 K, accuracy within ± 0.05 K.
Analyses for IO_3: precision within ± 0.3 %.

REFERENCES:

COMPONENTS:	ORIGINAL MEASUREMENTS:
(1) Calcium iodate; $Ca(IO_3)_2$; [7789-80-2] (2) Sodium chloride; NaCl; [7647-14-5] (3) Water; H_2O; [7732-18-5]	Kilde, G. *Z. Anorg. Allg. Chem.* 1934, *218*, 113-28.

VARIABLES:	PREPARED BY:
T/K = 291, 298 and 303 $c_2/mol\ dm^{-3}$ = 0 to 1.000	Hiroshi Miyamoto

EXPERIMENTAL VALUES:

t/°C	Sodium Chloride $c_2/mol\ dm^{-3}$	Calcium Iodate $10^3 c_1/mol\ dm^{-3}$	$10^6 K_{s0}/mol^3\ dm^{-9}$
18	0	5.69	0.737
	0.0998	7.77	1.88
	0.250	9.24	3.16
	0.500	11.05	5.40
	1.000	13.8	10.5
25	0	7.84	1.93
	0.0998	10.5	4.63
	0.250	12.3	7.44
	0.500	14.6	12.5
	1.000	17.9	22.9
30	0	9.91	3.89
	0.0998	12.9	8.58
	0.250	15.0	13.5
	0.500	17.7	22.2
	1.000	21.6	40.3

AUXILIARY INFORMATION

METHOD/APPARATUS/PROCEDURE:

An excess of $Ca(IO_3)_2 \cdot 6H_2O$ was shaken with NaCl aqueous solutions for at least 24 hours in a thermostat at the desired temperature. Aliquots of saturated solutions were filtered through cotton wool, and the iodate content was determined iodometrically.

SOURCE AND PURITY OF MATERIALS:

Calcium iodate hexahydrate was prepared by mixing calcium chloride solution and KIO_3 solution. The precipitate was washed and dried at room temperature. Reagent grade NaCl was used.

ESTIMATED ERROR:

Soly: precision within 1%
Temp: nothing specified

Concentration solubility product

$K_{s0} = [Ca^{2+}][IO_3^-]^2$

K°_{s0} was calculated from the equation

$$1/3 \log K^\circ_{s0} = 1/3 \log K_{s0} - Z_1 Z_2 A I^{1/2} + BI$$

where $Z_1 Z_2 A$ = 0.998 at 18°C, 1.008 at 25°C and 1.018 at 30°C, and I is the ionic strength.

The values obtained were the following: $K^\circ_{s0} = 0.329 \times 10^{-6}$ at 18°C, 0.736×10^{-6} at 25°C and 1.35×10^{-6} at 30°C.

COMPONENTS:

(1) Calcium iodate; $Ca(IO_3)_2$; [7789-80-2]

(2) Sodium chloride; NaCl; [7647-14-5]

(3) Water; H_2O; [7732-18-5]

ORIGINAL MEASUREMENTS:

Wise, W. C. A.; Davies, C. W.

J. Chem. Soc. 1938, 273-7.

VARIABLES:

$T/K = 298$

$10^3c_2/\text{mol dm}^{-3} = 0 \text{ to } 100.0$

PREPARED BY:

Hiroshi Miyamoto
Mark Salomon

EXPERIMENTAL VALUES:

t/°C	Sodium Chloride $10^3c_2/\text{mol dm}^{-3}$	Calcium Iodate $10^3c_1/\text{mol dm}^{-3}$	Density $\rho/\text{g cm}^{-3}$
25	0	7.840	0.9998
	12.5	8.285	1.0010
	25.0	8.676	1.0015
	50.0	9.287	1.0025
	100.0	10.23	1.0050

COMMENTS AND/OR ADDITIONAL DATA:

The conductivity data at 18°C were used to evaluate the thermodynamic ion pair dissociation constant, K_D°; it was found that $K_D^\circ = 0.13$. Using this value for the ion pair dissociation constant, the concentration of the ion pair $CaIO_3^+$ was calculated from

$$\log [CaIO_3^+] = \log [Ca^{2+}][IO_3^-] - \log K_D^\circ - 2.02I^{1/2} + 2.0I$$

where I is the ionic strength. Utilizing this relation to compute the ionic concentrations of Ca^{2+} and IO_3^-, the authors plotted (1/3) log $[Ca^{2+}]$x$[IO_3^-]$ against the ionic strength and extrapolated to zero ionic strength to obtain the thermodynamic solubility product constant. The result of this extrapolation is $K_{s0}^\circ = 6.953 \times 10^{-7}$.

AUXILIARY INFORMATION

METHOD/APPARATUS/PROCEDURE:

Saturating column method as in (1) and modified as in (2). A bulb containing the solvent solution is attached to a column containing the slightly soluble salt, and the solvent is allowed to flow through the column at a rate sufficient to insure saturation (1). The modification (2) consisted of connecting the column by capillary tubing to a second parallel arm in which the saturated solution collected. The entire apparatus was placed in a thermostat. Weighed samples of the satd slns were taken for analysis by method described in (3): i.e. the satd slns added to acidified KI sln and the liberated I_2 titrd by weight against an approx 0.15N thiosulfate sln, 0.01N iodine sln being used for the back titrn. The densities of the satd slns were measured at 25°C, and the molar conductivities at 18°C for the binary $Ca(IO_3)_2$-H_2O system are also reported.

SOURCE AND PURITY OF MATERIALS:

$Ca(IO_3)_2 \cdot 6H_2O$ was prepared by dropwise addition of solutions of KIO_3 and $CaCl_2$ in equivalent amounts to a large volume of conductivity water. The precipitate was washed first by decantation and then in the solubility columns until a constant solubility was obtained.

ESTIMATED ERROR:

Soly: not specified, but reproducibility probably around ± 0.3% as in ref. (3).
Temp: ± 0.01°K (authors)

REFERENCES:

1. Brönsted, J. N.; La Mer, V. K. *J. Am. Chem. Soc.* 1924, *46*, 555.
2. Money, R. W.; Davies, C. W. *J. Chem. Soc.* 1934, 400.
3. Macdougall, G.; Davies, C. W. *J. Chem. Soc.* 1935, 1416.

COMPONENTS:

(1) Calcium iodate; $Ca(IO_3)_2$; [7789-80-2]

(2) Sodium chloride; NaCl; [7647-14-5]

(3) Water; H_2O; [7732-18-5]

ORIGINAL MEASUREMENTS:

Rens, G.

Sucr. Belge 1958, *77*, 193-208.

VARIABLES:

$T/K = 293$

$10^3 c_2/\text{mol dm}^{-3} = 0.00$ to 1000.21

PREPARED BY:

Hiroshi Miyamoto

EXPERIMENTAL VALUES:

t/°C	Sodium Chloride $10^3 c_2/\text{mol dm}^{-3}$	1/2 Iodate Ion $10^3 (c/2)/\text{mol dm}^{-3}$	Calcium Ion $10^3 c/\text{mol dm}^{-3}$	Calcium Iodate $10^3 c_1/\text{mol dm}^{-3}$ [a]
20	0.00	6.231	6.230	6.231
	10.00	6.577	6.607	6.592
	20.00	6.917	6.938	6.928
	40.00	7.389	7.409	7.399
	60.04	7.771	7.789	7.780
	100.03	8.436	8.476	8.456
	130.01	8.832	8.841	8.737
	180.06	9.393	9.438	9.416
	200.06	9.636	9.692	9.664
	500.00	12.041	12.085	12.063
	1000.21	14.902	14.924	14.913

[a] Average value calculated by the compiler

pK°_{s0} was calculated from the equation:

1) $I < 0.07 \text{ mol dm}^{-3}$, $pK_{s0} = pK^\circ_{s0} - 6AI^{1/2} + 3CI$

where I is the ionic strength, and $A = 0.5046 \text{ mol}^{-1/2} \text{ dm}^{3/2}$ (ref 1)

2) $0.07 < I < 0.25 \text{ mol dm}^{-3}$, $pK_{s0} = pK^\circ_{s0} - 6AI^{1/2}/(1 + aBI^{1/2})$

where a is distance of closest approach.

For $a = 3.5$A°, the value of K°_{s0} was $4.159 \times 10^{-7} \text{ mol}^3\text{dm}^{-9}$

AUXILIARY INFORMATION

METHOD/APPARATUS/PROCEDURE:

Excess $Ca(IO_3)_2 \cdot 6H_2O$ and aqueous NaCl solution were placed in sealed Erlenmeyer flasks. The flasks were rotated in a thermostat for 24 hours. Aliquots of saturated solution were filtered.
The concentration of iodate was determined iodometrically. The calcium content was determined by chelatometric titration using Eriochrome black T as an indicator.

SOURCE AND PURITY OF MATERIALS:

$Ca(IO_3)_2 \cdot 6H_2O$ was prepared by slowly adding the solution of KIO_3 (about 50g dm^{-3}) to an equivalent solution of $CaCl_2 \cdot 2H_2O$ at 20°C. The precipitate was washed by decantation, and was air-dried at room temperature. An analysis of the product gave the following values: IO_3 99.64% and Ca 99.95% of theoretical.

ESTIMATED ERROR:

Soly: nothing specified
Temp: ± 0.1°K (author)

REFERENCES:

1. Harned, H.; Owen, B. *The Physical Chemistry of Electrolytic Solutions.* Reinhold, New York, 1950, 447.

COMPONENTS:

(1) Calcium iodate; $Ca(IO_3)_2$; [7789-80-2]

(2) Sodium chloride; NaCl; [7647-14-5]

(3) Hydrochloric acid; HCl; [7647-01-0]

(4) Water; H_2O; [7732-18-5]

ORIGINAL MEASUREMENTS:

Kilde, G.

Z. Anorg. Allg. Chem. 1934, *218*, 113-28.

VARIABLES:

T/K = 291, 298, and 303

c_3/mol dm^{-3} = 0.0499 - 1.000

PREPARED BY:

Hiroshi Miyamoto

EXPERIMENTAL VALUES:

t/°C	Hydrochloric Acid c_3/mol dm^{-3}	Sodium Chloride c_2/mol dm^{-3}	Calcium Iodate $10^3 c_1$/mol dm^{-3}	K_D [a]/mol dm^{-3}
18	0.0499	0.0499	8.6	0.293
	0.0998	--	9.4	0.296
	1.000	--	31.7	0.378
25	0.0499	0.0499	11.5	0.284
	0.0998	--	12.6	0.278
	1.000	--	42.4	0.355
30	0.0499	0.0499	14.1	0.312
	0.0998	--	15.4	0.298
	1.000	--	52.3	0.338

[a] $K_D = [H^+][IO_3^-]/[HIO_3]$,

$K_D^\circ(HIO_3)$ at 25°C was calculated from the equation

$$-\log K_D^\circ(HIO_3) = -\log K_D - \log y_{H^+} - 0.504I^{1/2} + 0.38I$$

where y_{H^+} is the activity coefficient of hydrogen ion and the value is given in ref 1.

$K_D^\circ(HIO_3) = 0.180 \sim 0.184$ at 25°C.

AUXILIARY INFORMATION

METHOD/APPARATUS/PROCEDURE:

The compiler assumes that the method of the solubility determination in acidic solution was similar to that adopted in the case of neutral salt solution.

An excess of $Ca(IO_3)_2 \cdot 6H_2O$ was shaken with hydrochloric acid containing NaCl for at least 24 hours in a thermostat at the desired temperature. Aliquots of saturated solutions were filtered through cotton wool, and the iodate content was determined iodometrically.

SOURCE AND PURITY OF MATERIALS:

Calcium iodate hexahydrate was prepared by mixing calcium chloride solution and sodium iodate solution. The precipitate was washed and dried at room temperature. Reagent grade NaCl was used. The source of hydrochloric acid was not given.

ESTIMATED ERROR:

Soly: precision within 1 %.
Temp: nothing specified.

REFERENCES:

1. Bjerrum, N.; Unmack, A. *K. Dan. Vidensk. Selsk. Mat-Fys. Medd.* 1929, *9*, 1.

COMPONENTS:

(1) Calcium iodate; $Ca(IO_3)_2$; [7789-80-2]

(2) Sodium chloride; NaCl; [7647-14-5]

(3) Sodium hydroxide; NaOH; [1310-73-2]

(4) Water; H_2O; [7732-18-5]

ORIGINAL MEASUREMENTS:

Kilde, G.

Z. Anorg. Allg. Chem. 1934, *218*, 113-28.

VARIABLES:

T/K = 291, 298, and 303

$c_2/mol\ dm^{-3}$ = 0 to 1.000

$c_3/mol\ dm^{-3}$ = 0.0100 to 0.0500

PREPARED BY:

Hiroshi Miyamoto

EXPERIMENTAL VALUES:

Sodium Hydroxide $c_3/mol\ dm^{-3}$	Sodium Chloride $c_2/mol\ dm^{-3}$	Calcium Iodate $10^2 c_1/mol\ dm^{-3}$		
	t/°C	18	25	30
0.0100	0.0000	0.62	0.84	1.10
0.0250	0.0000	0.71	0.94	1.17
0.0250	0.100	0.85	1.14	1.38
0.0250	0.250	1.01	1.31	1.60
0.0250	0.500	1.17	1.53	1.84
0.0250	1.000	1.42	1.84	2.22
0.0500	0.0000	0.80	1.05	1.29
0.0500	0.0500	0.87	1.14	1.41
0.0500	0.100	0.92	1.23	1.50
0.0500	0.250	1.05	1.39	1.70
0.0500	0.500	1.21	1.60	1.93
0.0500	1.000	1.47	1.91	2.29

K°_D for $Ca(OH)^+$ was calculated from the equation

$$pK^\circ_D = pK_D + 2.0I^{1/2} - 1.1I$$

The value obtained was 0.040.

AUXILIARY INFORMATION

METHOD/APPARATUS/PROCEDURE:

An excess of $Ca(IO_3)_2 \cdot 6H_2O$ was shaken with NaOH aqueous solution containing NaCl for at least 24 hours in a thermostat at the desired temperature. Aliquots of saturated solutions were filtered through cotton wool, and the iodate content was determined iodometrically.

SOURCE AND PURITY OF MATERIALS:

Calcium iodate hexahydrate was prepared by mixing calcium chloride solution and KIO_3 solution. The precipitate was washed and dried at room temperature. Reagent grade NaCl was used. The source of NaOH was not given.

ESTIMATED ERROR:

Soly: precision within 1 %

Temp: nothing specified

REFERENCES:

COMPONENTS:

(1) Calcium iodate; $Ca(IO_3)_2$; [7789-80-2]
(2) Sodium chloride; NaCl; [7647-14-5]
(3) Potassium chloride; KCl; [7447-40-7]
(4) Water; H_2O; [7732-18-5]

ORIGINAL MEASUREMENTS:

Nezzal, G.; Popiel, W. J.; Vermande, P.

Chem. Ind. (London) 1967, (*30*) 1294-5.

EXPERIMENTAL VALUES:

t/°C	Activity Product $10^7\ K^\circ_{s0}/mol^3\ dm^{-9}$
14	2.11
18	3.27
22	5.21
26	8.22
30	12.91
34	19.28
40	27.23
45	31.26
50	35.40
55	40.51

The paper is concerned with the solubility study of calcium iodate in water and in aqueous KCl and NaCl solutions (0.125-0.10 mol dm^{-3}). The solubility data of calcium iodate in water and in the solutions were not reported, but the calculated activity solubility products were given in the paper.
The activity solubility product is given by,

$$K^\circ_{s0} = (C_{Ca^{2+}}\ C^2_{IO_3^-})(y_{Ca^{2+}}\ y^2_{IO_3^-}) = 4S^3 y^3_\pm$$

where S represents the solubility of the iodate, C being the concentration and y the activity coefficient was eliminated by introducing one of the following modified Debye-Hückel equations:

(i) Brönsted equation: $-\log y_\pm = Z_+Z_-A\sqrt{I} - BI$

where Z is the ionic valency, I the ionic strength, A values taken from Robinson-Stokes (1), and B an unknown constant;

(ii) Davies equation: $-\log y_\pm = Z_+Z_-A[(I/1 + \sqrt{I}) - 0.2I]$.

In the first case K°_{s0} was obtained by plotting ($\log S - 2A\sqrt{I}$) against I, and extrapolating to I = 0. In the second case $\log S$ was plotted against the expression in the square brackets. The authors say that the two equations gave substantially similar results. However, the first method gives a shorter extrapolation, and therefore more reliance was placed on the values of K°_{s0} obtained in this way. The values of B for calcium iodate were 0.84±0.04 in NaCl, and 0.80±0.02 in KCl solution. The authors do not discuss ion pairing.

COMPONENTS:

(1) Calcium iodate; $Ca(IO_3)_2$; [7789-80-2]

(2) Sodium chloride; NaCl; [7647-14-5]

(3) Potassium chloride; KCl; [7447-40-7]

(4) Water; H_2O; [7732-18-5]

ORIGINAL MEASUREMENTS:

Nezzal, G.; Popiel, W. J.; Vermande, P.

Chem. Ind. (*London*) 1967, (*30*) 1294-5.

VARIABLES:

T/K = 287 to 328

Concentration of NaCl and KCl

PREPARED BY:

Hiroshi Miyamoto

EXPERIMENTAL VALUES:

AUXILIARY INFORMATION

METHOD/APPARATUS/PROCEDURE:

The measurements of solubility were determined over two temperature range of 14 to 38°C for the hexahydrate, and 40 to 55°C for the monohydrate. Saturated solutions of the iodate were prepared in distilled water and in aqueous KCl and NaCl solutions, saturation was attained in the following two ways.

In the static or isothermal method, the chloride solutions were left in contact with an excess of the solid iodate in a thermostat with occasional shaking. Equilibrium was established in 24 hours. The solutions for analysis were removed with a filter stick. In the dynamic or saturating column method, the chloride solutions were allowed to percolate slowly (about 50 ml per hour) through columns of solid iodate (diameter 14mm, length 170mm) built into a thermostatic bath or at room temperature. The iodate contents in all saturated solutions were determined iodometrically.

SOURCE AND PURITY OF MATERIALS:

$Ca(IO_3)_2 \cdot 6H_2O$ and $Ca(IO_3)_2 \cdot H_2O$ crystals were used in the solubility determination.

The sources of the iodates, KCl and NaCl were not given.

ESTIMATED ERROR:

nothing specified

REFERENCES:

1. Robinson, R. A.; Stokes, R. H. *Electrolyte Solutions*. Butterworths, London, 1965.

COMPONENTS:

(1) Calcium iodate; $Ca(IO_3)_2$; [7789-80-2]

(2) Sodium chloride; NaCl; [7647-14-5]

(3) Calcium hydroxide; $Ca(OH)_2$; [1305-62-0]

(4) Water; H_2O; [7732-18-5]

ORIGINAL MEASUREMENTS:

Kilde, G.

Z. Anorg. Allg. Chem. 1934, *218*, 113-28.

VARIABLES:

T/K = 291, 298, and 303

c_2/mol dm^{-3} = 0 to 0.500

PREPARED BY:

Hiroshi Miyamoto

EXPERIMENTAL VALUES:

Calcium Hydroxide c_3/mol dm^{-3}	Sodium Chloride c_2/mol dm^{-3}	Calcium Iodate $10^2 c_1$/mol dm^{-3}		
		t/°C 18	25	30
0.0108	0.000	0.45	0.64	0.86
0.0108	0.100	0.59	0.85	1.08
0.0108	0.250	0.72	1.01	1.23
0.0108	0.500	0.88	1.22	1.54

K_D° for $Ca(OH)^+$ was calculated from the equation

$$pK_D^\circ = pK_D + 2.0I^{1/2} - 1.1I$$

The value obtained was 0.040.

AUXILIARY INFORMATION

METHOD/APPARATUS/PROCEDURE:

An excess of $Ca(IO_3)_2 \cdot 6H_2O$ was shaken with $Ca(OH)_2$ aqueous solutions containing NaCl for at least 24 hours in a thermostat at the desired temperature. Aliquots of saturated solutions were filtered through cotton wool, and the iodate content was determined iodometrically.

SOURCE AND PURITY OF MATERIALS:

Calcium iodate hexahydrate was prepared by mixing calcium chloride solution and KIO_3 solution. The precipitate was washed and dried at room temperature. Reagent grade NaCl was used, but the source of $Ca(OH)_2$ was not given.

ESTIMATED ERROR:

Soly: precision within 1 %

Temp: nothing specified

REFERENCES:

COMPONENTS:

(1) Calcium iodate; $Ca(IO_3)_2$; [7789-80-2]

(2) Sodium chloride; NaCl; [7647-14-5]

(3) Sucrose, $C_{12}H_{22}O_{11}$; [57-50-1]

(4) Water; H_2O; [7732-18-5]

ORIGINAL MEASUREMENTS:

Rens, G.

Sucr. Belge 1958, *77*, 193-208.

VARIABLES:

T/K = 293

10^3c_2/mol dm^{-3} = 0.000 to 180.35

10^3c_3/mol dm^{-3} = 0.00 to 100.01

PREPARED BY:

Hiroshi Miyamoto

EXPERIMENTAL VALUES:

Sodium Chloride 10^3c_2/mol dm^{-3}	Sucrose 10^3c_3/mol dm^{-3}	1/2 Iodate Ion $10^3(c/2)$/mol dm^{-3}	Calcium Ion 10^3c/mol dm^{-3}	Calcium Iodate c_1/mol dm^{-3} [a]
0.000	0.00	6.231	6.230	6.231
5.002	100.03	6.419	6.444	6.432
35.02	100.03	7.272	7.290	7.281
35.01	200.04	7.274	7.289	7.282
35.00	400.07	7.274	7.289	7.282
110.00	100.00	8.570	8.589	8.580
110.03	200.04	8.592	8.612	8.602
110.04	400.12	8.608	8.632	8.620
180.35	100.01	9.426	9.456	9.441

[a] Average value calculated by the compiler

pK°_{s0} was calculated from the equation:

1) I < 0.07 mol dm^{-3}, $pK_{s0} = pK^\circ_{s0} - 6AI^{1/2} + 3CI$

where I is the ionic strength, and A = 0.5046 mol$^{-1/2}$ dm$^{3/2}$ (ref 1)

2) 0.07 < I < 0.25 mol dm^{-3}, $pK_{s0} = pK^\circ_{s0} - 6AI^{1/2}/(1 + aBI^{1/2})$

where a is distance of closest approach.

For a = 3.5A°, the value of K°_{s0} was 4.159 x 10^{-7} mol^3 dm^{-9}

AUXILIARY INFORMATION

METHOD/APPARATUS/PROCEDURE:

Excess $Ca(IO_3)_2 \cdot 6H_2O$ and aqueous NaCl and sucrose solutions were placed in sealed Erlenmeyer flasks. The flasks were rotated in a thermostat for 24 hours. Aliquots of saturated solution were filtered.
The concentration of iodate was determined iodometrically. The calcium content was determined by chelatometic titration using Eriochrome black T as an indicator.

SOURCE AND PURITY OF MATERIALS:

$Ca(IO_3)_2 \cdot 6H_2O$ was prepared by slowly adding the solution of KIO_3 (about 50g dm^{-3}) to an equivalent the solution of $CaCl \cdot 2H_2O$ at 20°C. The precipitate was washed by decantation, and was air-dried at room temperature. The analysis of the product gave the following values: IO_3 99.64% and Ca 99.95% of theoretical.

ESTIMATED ERROR:

Soly: nothing specified

Temp: ± 0.1°K (author)

REFERENCES:

1. Harned, H.; Owen, B. *The Physical Chemistry of Electrolytic Solutions.* Reinhold, New York, 1950, 447.

COMPONENTS:

(1) Calcium iodate; $Ca(IO_3)_2$; [7789-80-2]

(2) Sodium sulfate; Na_2SO_4; [7757-82-6]

(3) Water; H_2O; [7732-18-5]

ORIGINAL MEASUREMENTS:

Wise, W. C. A.; Davies, C. W.

J. Chem. Soc. 1938, 273-7.

VARIABLES:

T/K = 298

10^3c_2/mol dm^{-3} = 0 to 25.0

PREPARED BY:

Hiroshi Miyamoto

Mark Salomon

EXPERIMENTAL VALUES:

t/°C	Sodium Sulfate 10^3c_2/mol dm^{-3}	Calcium Iodate 10^3c_1/mol dm^{-3}	Density ρ/g cm^{-3}
25	0	7.840	0.9998
	6.25	8.898	1.001
	12.5	9.745	1.002
	18.75	10.45	1.003
	25.0	11.05	1.004

COMMENTS AND/OR ADDITIONAL DATA:

The conductivity data at 18°C were used to evaluate the thermodynamic ion pair dissociation constant, K_D°; it was found that K_D° = 0.13. Using this value for the ion pair dissociation constant, the concentration of the ion pair $CaIO_3^+$ was calculated from

$$\log [CaIO_3^+] = \log [Ca^{2+}][IO_3^-] - \log K_D^\circ - 2.02I^{1/2} + 2.0I$$

where I is the ionic strength. Utilizing this relation to compute the ionic concentrations of Ca^{2+} and IO_3^-, the authors plotted (1/3) log $[Ca^{2+}]$x$[IO_3^-]$ against the ionic strength and extrapolated to zero ionic strength to obtain the thermodynamic solubility product constant. The result of this extrapolation is $K_{s0}^\circ = 6.953 \times 10^{-7}$.

AUXILIARY INFORMATION

METHOD/APPARATUS/PROCEDURE:

Saturating column method as in (1) and modified as in (2). A bulb containing the solvent solution is attached to a column containing the slightly soluble salt, and the solvent is allowed to flow through the column at a rate sufficient to insure saturation (1). The modification (2) consisted of connecting the column by capillary tubing to a second parallel arm in which the saturated solution collected. The entire apparatus was placed in a thermostat. Weighed samples of the satd slns were taken for analysis by method described in (3): i.e. the satd slns added to acidified KI sln and the liberated I_2 titrd by weight against an approx 0.15N thiosulfate sln, 0.01N iodine sln being used for the back titrn. The densities of the satd slns were measured at 25°C, and the molar conductivities at 18°C for the binary $Ca(IO_3)_2$-H_2O system are also reported.

SOURCE AND PURITY OF MATERIALS:

$Ca(IO_3)_2\cdot 6H_2O$ was prepared by dropwise addition of solutions of KIO_3 and $CaCl_2$ in equivalent amounts to a large volume of conductivity water. The precipitate was washed first by decantation and then in the solubility columns until a constant solubility was obtained.

ESTIMATED ERROR:

Soly: not specified, but reproducibility probably around ± 0.3% as in ref. (3).

Temp: ± 0.01°K (authors)

REFERENCES:

1. Brönsted, J. N.; La Mer, V. K. *J. Am. Chem. Soc.* 1924, *46*, 555.
2. Money, R. W.; Davies, C. W. *J. Chem. Soc.* 1934, 400.
3. Macdougall, G.; Davies, C. W. *J. Chem. Soc.* 1935, 1416.

COMPONENTS:

(1) Calcium iodate; $Ca(IO_3)_2$; [7789-80-2]

(2) Sodium thiosulfate; $Na_2S_2O_3$; [7772-98-7]

(3) Water; H_2O; [7732-18-5]

ORIGINAL MEASUREMENTS:

Davies, C. W.; Wyatt, P. A. H.

Trans. Faraday Soc. 1949, *45*, 770-3.

VARIABLES:

$T/K = 298$

$10^3 c_2/mol\ dm^{-3} = 0$ to 15.430

PREPARED BY:

Hiroshi Miyamoto

EXPERIMENTAL VALUES:

t/°C	Sodium Thiosulfate $10^3 c_2/mol\ dm^{-3}$	Calcium Iodate $10^3 c_1/mol\ dm^{-3}$
25	0	7.840
	7.235	8.907
	10.853	9.303
	11.566	9.367
	14.148	9.613
	15.430	9.749

COMMENTS AND/OR ADDITIONAL DATA:

The concentration of the individual ionic species and ion pairs were calculated by successive approximation from the relation

$$\log[Ca^{2+}][S_2O_3^{2-}]/[CaS_2O_3] = \log K_D^\circ + 2I^{1/2}(1 + I^{1/2}) - 0.40I$$

where I = ionic strength and K_D° is the dissociation constant of the ion pairs. For ion pairs $Ca(S_2O_3)$, the value taken for K_D° was 0.0089.

AUXILIARY INFORMATION

METHOD/APPARATUS/PROCEDURE:

Saturating column method used as in (1): see compilation of ref 2 for details.
The saturator was immersed in a thermostat regulated 25 ± 0.01°C.
Samples of saturated solution were withdrawn in warmed pipets and analyzed by iodometric titration for the iodate.
Each solubility value is the mean of two determinations.

SOURCE AND PURITY OF MATERIALS:

$Ca(IO_3)_2 \cdot 6H_2O$ was prepared by slow dropwise addition of solutions of A.R. grade $CaCl_2$ and KIO_3 to a large volume of water.
$Na_2S_2O_3$ was an A.R. grade sample that had been recrystallized and dried over a saturated $CaCl_2$ solution.

ESTIMATED ERROR:

Soly: duplicate determinations of the solubilities agreed to within 0.1%.
Temp: ± 0.01°C (authors)

REFERENCES:

1. Money, R. W.; Davies, C. W. *J. Chem. Soc.* 1934, 400.
2. Macdougall, G.; Davies, C. W. *J. Chem. Soc.* 1935, 1416.

COMPONENTS:

(1) Calcium iodate; $Ca(IO_3)_2$; [7789-80-2]

(2) Sodium hydroxyacetate (sodium glycolate)[a]; $C_2H_3O_3Na$; [2836-32-0]

(3) Water; H_2O; [7732-18-5]

ORIGINAL MEASUREMENTS:

Davies, C. W.

J. Chem. Soc. 1938, 277-81.

VARIABLES:

$T/K = 298$

$10^3 c_2/mol\ dm^{-3} = 20.0$ and 40.0

PREPARED BY:

Hiroshi Miyamoto

EXPERIMENTAL VALUES:

t/°C	Sodium Glycolate $10^3 c_2/mol\ dm^{-3}$	Calcium Iodate $10^3 c_1/mol\ dm^{-3}$	Dissociation constant $K_D^\circ/mol\ dm^{-3}$
25	20.0	9.315	0.0257
	40.0	10.56	0.0257

K_D° was calculated from the equation

$$\log K_D^\circ = \log[Ca^{2+}][X^-]/[CaX^+] - 2.02I^{1/2} + 2.8I$$

where $X^- = HOCH_2COO^-$ and I is the ionic strength.

AUXILIARY INFORMATION

METHOD/APPARATUS/PROCEDURE:

Saturated column method as described in the preceding paper (1). The saturated solutions were passed several times through the saturating column.

The author stated that the concentration of $Ca(IO_3)_2$ in the saturated solution was determined by the ordinary volumetric method. The method is probably similar to that reported in the preceding paper (1), which has been compiled elsewhere in this volume.

[a] Acetic acid,hydroxy-,monosodium salt.

SOURCE AND PURITY OF MATERIALS:

The source of calcium iodate was not given, but probably same as in the preceding paper (1): see compilation of the paper for details. Glycolic acid was a commercial sample, the acid equivalent determined to be within 0.1% of theoretical. The acid and AnalaR NaOH used to prepare sodium glycolate.

ESTIMATED ERROR:

Soly: the mean of two titrations agreed within at least 0.2%.

Temp: ± 0.01°C (author)

REFERENCES:

1. Wise, W. C. A.; Davies, C. A. *J. Chem. Soc.* 1938, 273.

COMPONENTS:	ORIGINAL MEASUREMENTS:
(1) Calcium iodate; $Ca(IO_3)_2$; [7789-80-2]	Das, A. R.; Nair, V. S. K.
(2) Sodium glycolate; $C_2H_3O_3Na$; [2836-32-0]	*J. Inorg. Nucl. Chem.* 1975, *37*, 991-3.
(3) Water; H_2O; [7732-18-5]	

EXPERIMENTAL VALUES:

t/°C	Sodium Glycolate 10^3c_2/mol dm^{-3}	Calcium Iodate 10^3c_1/mol dm^{-3}	log K_A [a]
25	11.820	8.739	1.65
	17.740	9.094	1.60
	23.985	9.551	1.64
	25.650	9.795	1.68
	32.860	10.176	1.66
	40.070	10.512	1.64
	47.314	11.058	1.68
	53.228	11.358	1.68
			mean 1.65 ± 0.05
30	7.988	10.251	1.77
	11.975	10.595	1.76
	15.996	10.945	1.72
	19.957	11.217	1.72
	23.949	11.636	1.76
			mean 1.75 ± 0.02
35	8.313	12.733	1.84
	12.470	13.276	1.85
	16.626	13.751	1.85
	19.957	14.021	1.87
	24.900	14.530	1.84
	29.056	14.903	1.83
	33.207	15.260	1.82
			mean 1.84 ± 0.03
40	16.645	16.021	1.90
	19.960	16.680	1.97
	24.939	17.304	1.97
	29.095	17.789	1.96
	33.254	18.245	1.94
			mean 1.95 ± 0.06
45	21.505	20.21	1.98
	25.806	20.82	1.98
	30.107	21.892	2.07
	34.408	22.473	2.04
			mean 2.02 ± 0.03

[a] K_A is the association constant for $CaCH_2(OH)CO_2^+$, K_A/dm^3 mol^{-1}.

COMMENTS AND/OR ADDITIONAL DATA:

The authors state that a plot of log K_A against $(T/K)^{-1}$ is linear and have calculated free energies, enthalpies, and entropies for the association reaction. Individual ionic entropies have also been calculated, and the original paper should be consulted for these details.

The authors do not discuss the nature of the solid phases in equilibrium with the saturated solutions. At 40° and 45°C, the solid phase is probably the monohydrate, $Ca(IO_3)_2 \cdot H_2O$, and below these temperatures, the solid phase is probably the hexahydrate $Ca(IO_3)_2 \cdot 6H_2O$ (compilers).

The results of the calculations for the thermodynamic solubility product constant for $Ca(IO_3)_2$ as a function of temperature are:

t/°C	25	30	35	40	45
$-\log K^0_{s0}$	6.15	5.95	5.72	5.54	5.34

COMPONENTS:

(1) Calcium iodate; $Ca(IO_3)_2$; [7789-80-2]

(2) Sodium glycolate; $C_2H_3O_3Na$; [2836-32-0] [a]

(3) Water; H_2O; [7732-18-5]

ORIGINAL MEASUREMENTS:

Das, A. R.; Nair, V. S. K.

J. Inorg. Nucl. Chem. 1975, *37*, 991-3.

VARIABLES:

T/K = 298 - 318

$10^3 c_2$/mol dm^{-3} = 8 - 53

PREPARED BY:

Hiroshi Miyamoto

Mark Salomon

METHOD/APPARATUS/PROCEDURE:

The saturating column method was used as described by Davies (1) (see the compilation for this reference for details). All solutions were made by dilution of the stock solution, and the pH was measured before and after percolation through the column. The pH of all solutions were in the range of 3.6 - 3.8 (pK_A of glycolic acid = 3.80 at 25°C). Other details same as in ref. (2): i.e. the slns were passed through the column at least four times (twice was sufficient for saturation), and satd slns were analysed for iodate by iodometric titrn in a nitrogen atmosphere.

The thermodynamic soly product constant was calcd from

$$K^\circ_{s0} = [Ca^{2+}][IO_3^-]^2 y_\pm(Ca^{2+}) y_\pm(IO_3^-)^2$$

In these calculations the ion pairing constant for $NaIO_3$ was taken as 3.0 at 25°C (3) and the association constant for $CaIO_3^+$ was estimated for 30-45°C by the electrostatic theory of Bjerrum (4): values for these association constants are not given in the paper. The association constant for $CaGL^+$ (K_A in the tables, GL^- = $CH_2(OH)COO^-$) was calcd from

$$K_A(CaGL^+) = [CaGL^+]/[Ca^{2+}][GL^-]y_\pm(Ca^{2+})$$

All activity coefficients were calculated from the Davies equation (5).

AUXILIARY INFORMATION

SOURCE AND PURITY OF MATERIALS:

$Ca(IO_3)_2$ prepd by stoichiometric addn of $CaCl_2$ and KIO_3 slns over a 2 h period to a large volume of water at 65°C (authors do not state which hydrate is formed). The ppt was washed 6 times by decantation and digested in double dist water for 4h at 80°C. Glycolic acid was "repeatedly" recrystallized from water and dried in vacuo before use. Stock slns of Na glycolate were prepd from the acid and CO_2-free NaOH sln so as to obtain a 1:1 buffer ratio. Conductivity water prepared by mixed-bed ionization (6) was used in all preparations and solubility experiments.

[a] Acetic acid,hydroxy-,monosodium salt

ESTIMATED ERROR:

Soly: nothing specified

Temp: Precision probably ± 0.1% as in (2).

REFERENCES:

1. Davies, C. W. *J. Chem. Soc.* 1930, 2471.
2. Ghosh, R.; Nair, V. S. K. *J. Inorg. Nucl. Chem.* 1970, *32*, 3025.
3. Wise, W. C. A.; Davies, C. W. *J. Chem. Soc.* 1938, 273.
4. Bjerrum, N. *K. Danske Vidensk. Selsk. Math-Fys. Medd.* 1926, *7*, No. 9.
5. Davies, C. W. *Ion Association.* Butterworths. London. 1960.
6. Davies, C. W.; Nancollas, G. H. *Chem. Ind.* 1950, *7*, 129.

COMPONENTS:

(1) Calcium iodate; $Ca(IO_3)_2$; [7789-80-2]

(2) Glycine, monosodium salt (sodium aminoacetate); $C_2H_4NO_2Na$; [6000-44-8]

(3) Water; H_2O; [7732-18-5]

ORIGINAL MEASUREMENTS:

Davies, C. W.

J. Chem. Soc. 1938, 277-81.

VARIABLES:

$T/K = 298$

$10^3 c_2/\text{mol dm}^{-3} = 20.0$ and 40.0

PREPARED BY:

Hiroshi Miyamoto

EXPERIMENTAL VALUES:

t/°C	Sodium Aminoacetate $10^3 c_2/\text{mol dm}^{-3}$	Calcium Iodate $10^3 c_1/\text{mol dm}^{-3}$	Dissociation constant $K_D^\circ/\text{mol dm}^{-3}$
25	20.0	9.113	0.038
	40.0	10.01	0.045

K_D° was calculated from the equation

$$\log K_D^\circ = \log[Ca^{2+}][X^-]/[CaX^+] - 2.02I^{1/2} + 2.8I$$

where $X^- = H_2NCH_2COO^-$ and I is the ionic strength.

AUXILIARY INFORMATION

METHOD/APPARATUS/PROCEDURE:

Saturated column method as described in the preceding paper (1). The saturated solutions were passed several times through the saturating column.

The author stated that the concentration of $Ca(IO_3)_2$ in the saturated solution was determined by the ordinary volumetric method. The method is probably similar to that reported in the preceding paper (1), which has been compiled elsewhere in this volume.

SOURCE AND PURITY OF MATERIALS:

The source of calcium iodate was not given, but probably same as in the preceding paper (1): see compilation of the paper for details.
AnalaR glycine was used without further purification.
The acid and AnalaR NaOH used to prepared sodium salt of glycine.

ESTIMATED ERROR:

Soly: the mean of two titrations agreed within at least 0.2%
Temp: ± 0.01°C (author)

REFERENCES:

1. Wise, W. C. A.; Davies, C. A. *J. Chem. Soc.* 1938, 273.

COMPONENTS:

(1) Calcium iodate; $Ca(IO_3)_2$; [7789-80-2]

(2) Monosodium glycinate; $C_2H_5NO_2Na$; [6000-44-8]

(3) Water; H_2O; [7732-18-5]

ORIGINAL MEASUREMENTS:

Davies, C. W.; Waind, G. M.

J. Chem. Soc. 1950, 301-3.

VARIABLES:

$T/K = 298$

$10^3c_2/\text{mol dm}^{-3} = 0$ to 149.4

PREPARED BY:

Hiroshi Miyamoto

EXPERIMENTAL VALUES:

t/°C	Monosodium Glycinate $10^3c_2/\text{mol dm}^{-3}$	Calcium Iodate $10^3c_1/\text{mol dm}^{-3}$
25	0	7.84
	28.35	9.55
	56.70	10.81
	74.70	11.40
	149.4	13.77

NaA was assumed to be completely dissociated.

$K_D^\circ(CaA^+) = 0.037$

AUXILIARY INFORMATION

METHOD/APPARATUS/PROCEDURE:

The saturating column method was employed (see ref (1) for details). The saturation was ensured by passing a portion of the solution through the saturating column a second time. The analyses were effected by withdrawing 10 cm^3 of the saturating solution in a calibrated pipette, and running it into an acidic KI solution. The liberated iodine was titrated by weight against $Na_2S_2O_3$ solution, iodine solution being used for the back titration.

SOURCE AND PURITY OF MATERIALS:

$Ca(IO_3)_2\cdot6H_2O$ was prepared by dropwise addition of solutions of KIO_3 and $CaCl_2$ in equivalent amounts to a large volume of conductivity water. The hexahydrate separated, and was washed (see also compilation of ref (1)). The purity of the acid was checked by potentiometric titration of the solution against standard NaOH. The sodium salt was then made up in accordance with the observed end-point.

ESTIMATED ERROR:

Soly: nothing specified

Temp: ± 0.01°C (authors)

REFERENCES:

1. Wise, W. C. A.; Davies, C. W. *J. Chem. Soc.* 1938, 273.

COMPONENTS:

(1) Calcium iodate; $Ca(IO_3)_2$; [7789-80-2]

(2) 2-Oxopropanoic acid, sodium salt (sodium pyruvate); $C_3H_3O_3Na$; [113-24-6]

(3) Water; H_2O; [7732-18-5]

ORIGINAL MEASUREMENTS:

Davies, C. W.

J. Chem. Soc. 1938, 277-81.

VARIABLES:

$T/K = 298$

$10^3 c_2/\text{mol dm}^{-3} = 20.0$ and 40.0

PREPARED BY:

Hiroshi Miyamoto

EXPERIMENTAL VALUES:

t/°C	Sodium Pyruvate $10^3 c_2/\text{mol dm}^{-3}$	Calcium Iodate $10^3 c_1/\text{mol dm}^{-3}$	Dissociation constant $K_D^\circ/\text{mol dm}^{-3}$
25	20.0	8.837	0.079
	40.0	9.609	0.086

K_D° was calculated from the equation

$$\log K_D^\circ = \log[Ca^{2+}][X^-]/[CaX^+] - 2.02I^{1/2} + 2.8I$$

where $X^- = H_3CCOCOO^-$ and I is the ionic strength.

AUXILIARY INFORMATION

METHOD/APPARATUS/PROCEDURE:

Saturated column method as described in the preceding paper (1). The saturated solutions were passed several times through the saturating column.

The author stated that the concentration of $Ca(IO_3)_2$ in the saturated solution was determined by the ordinary volumetric method. The method is probably similar to that reported in the preceding paper (1), which has been compiled elsewhere in this volume.

SOURCE AND PURITY OF MATERIALS:

The source of calcium iodate was not given, but probably same as in the preceding paper (1): see compilation of the paper for details. Pyruvic acid (b.p. 73°C/11mmHg) was redistilled just before use; its acid equivalent was 87.5 (calcd: 88.0). The acid and NaOH used to prepare sodium salt of pyruvic acid.

ESTIMATED ERROR:

Soly: the mean of two titrations agreed within at least 0.2%

Temp: ± 0.01°C (author)

REFERENCES:

1. Wise, W. C. A.; Davies, C. A. *J. Chem. Soc.* 1938, 273.

COMPONENTS:

(1) Calcium iodate; $Ca(IO_3)_2$; [7789-80-2]

(2) Propanoic acid, 2-hydroxy-, monosodium salt (sodium lactate); $C_3H_5O_3Na$; [72-17-3]

(3) Water; H_2O; [7732-18-5]

ORIGINAL MEASUREMENTS:

Kilde, G.

Z. Anorg. Allg. Chem. 1936, *229*, 321-36.

VARIABLES:

T/K = 291, 298 and 303

c_2/mol dm^{-3} = 0 to 0.4000

PREPARED BY:

Hiroshi Miyamoto

EXPERIMENTAL VALUES:

t/°C	Sodium Lactate	Calcium Iodate		
	c_2/mol dm^{-3}	c_1/mol dm^{-3}	$10^6 K_{s0}$/mol^{3}dm^{-9} [a]	K_D/mol dm^{-3} [b]
18	0	5.68	--	--
	0.0100	6.31	0.875	0.063
	0.0200	6.93	1.01	0.057
	0.0400	7.81	1.26	0.072
	0.1000	10.03	1.94	0.087
	0.2000	13.29	2.79	0.081
	0.4000	17.81	4.32	0.091
25	0	7.81	--	--
	0.0100	8.53	2.25	0.089
	0.0200	9.18	2.58	0.094
	0.0400	10.32	3.16	0.095
	0.1000	13.15	4.54	0.093
	0.2000	16.89	6.21	0.090
	0.4000	22.87	9.73	0.098
30	0	9.78	--	--
	0.0100	10.58	4.17	0.062
	0.0200	11.31	4.59	0.069
	0.0400	12.74	5.50	0.071
	0.1000	16.09	8.28	0.091
	0.2000	20.42	11.8	0.099
	0.4000	27.56	17.8	0.102
				K_D°=0.040

[a] $K_{s0} = [Ca^{2+}][IO_3^-]^2$ [b] $K_D = [Ca^{2+}][C_3H_5O_3^-]/[CaC_3H_5O_3^+]$

AUXILIARY INFORMATION

METHOD/APPARATUS/PROCEDURE:

An excess of calcium iodate hexahydrate and sodium lactate solution were placed into bottles. The bottles were shaken in a thermostat for at least 24 hours. Samples of saturated solutions were withdrawn through a filter fitted with cotton wool, and the iodate estimation was made by addition of KI to about 25 cm^3 of the saturated solution, followed by sulfuric acid, and titrated with thiosulfate solution.

SOURCE AND PURITY OF MATERIALS:

$Ca(IO_3)_2 \cdot 6H_2O$ was prepared by dropwise addition of $CaCl_2$ solution to KIO_3 solution.
Sodium lactate solution was prepared by mixing of an equivalent of sodium hydroxide and lactic acid. The solution was boiled to remove lactylacetic acid.

ESTIMATED ERROR:

Soly: precision within 1%
Temp: nothing specified

REFERENCES:

COMPONENTS:

(1) Calcium iodate; $Ca(IO_3)_2$; [7789-80-2]

(2) Sodium methoxyacetate; $C_3H_5O_3Na$; [50402-70-5]

(3) Water; H_2O; [7732-18-5]

ORIGINAL MEASUREMENTS:

Davies, C. W.

J. Chem. Soc. 1938, 277-81.

VARIABLES:

T/K = 298

$10^3 c_2$/mol dm^{-3} = 20.0 and 40.0

PREPARED BY:

Hiroshi Miyamoto

EXPERIMENTAL VALUES:

t/°C	Sodium Methoxyacetate $10^3 c_2$/mol dm^{-3}	Calcium Iodate $10^3 c_1$/mol dm^{-3}	Dissociation constant K_D°/mol dm^{-3}
25	20.0	8.850	0.075
	40.0	9.664	0.077

K_D° was calculated from the equation

$$\log K_D^\circ = \log[Ca^{2+}][X^-]/[CaX^+] - 2.02I^{1/2} + 2.8I$$

where $X^- = H_3COCH_2COO^-$ and I is the ionic strength.

AUXILIARY INFORMATION

METHOD/APPARATUS/PROCEDURE:

Saturated column method as described in the preceding paper (1). The saturated solutions were passed several times through the saturating column.

The author stated that the concentration of $Ca(IO_3)_2$ in the saturated solution was determined by the ordinary volumetric method. The method is probably similar to that reported in the preceding paper (1), which has been compiled elsewhere in this volume.

SOURCE AND PURITY OF MATERIALS:

The source of calcium iodate was not given, but probably same as in the preceding paper (1): see compilation of the paper for details. The author stated that the methoxyacetic acid was prepared. But the method and the place of preparation were not described.
The acid and AnalaR NaOH used to prepare sodium methoxyacetate.

ESTIMATED ERROR:

Soly: the mean of two titrations agreed within at least 0.2%
Temp: ± 0.01°C (author)

REFERENCES:

1. Wise, W. C. A.; Davies, C. A. *J. Chem. Soc.* 1938, 273.

COMPONENTS:

(1) Calcium iodate; $Ca(IO_3)_2$; [7789-80-2]

(2) Sodium cyanoacetate; $C_3H_2NO_2Na$; [1071-36-9]

(3) Water; H_2O; [7732-18-5]

ORIGINAL MEASUREMENTS:

Davies, C. W.

J. Chem. Soc. 1938, 277-81.

VARIABLES:

T/K = 298

10^3c_2/mol dm^{-3} = 20 and 40

PREPARED BY:

Hiroshi Miyamoto

EXPERIMENTAL VALUES:

t/°C	Sodium Cyanoacetate 10^3c_2/mol dm^{-3}	Calcium Iodate 10^3c_1/mol dm^{-3}	Dissociation constant K_D°/mol dm^{-3}
25	20.0	8.652	0.24
	40.0	9.241	0.31

K_D° was calculated from the equation

$$\log K_D^\circ = \log[Ca^{2+}][X^-]/[CaX^+] - 2.02I^{1/2} + 2.8I$$

where $X^- = NCH_2CCOO^-$ and I is the ionic strength.

AUXILIARY INFORMATION

METHOD/APPARATUS/PROCEDURE:

Saturated column method as described in the preceding paper (1). The saturated solutions were passed several times through the saturating column.

The author stated that the concentration of $Ca(IO_3)_2$ in the saturated solution was determined by the ordinary volumetric method. The method is probably similar to that reported in the preceding paper (1), which has been compiled elsewhere in this volume.

SOURCE AND PURITY OF MATERIALS:

The source of calcium iodate was not given, but probably same as in the preceding paper (1): see compilation of the paper for details. Cyanoacetic acid was a commercial sample, the acid equivalent determined to be within 0.1% of theoretical. AnalaR NaOH was used. The acid and NaOH used to prepare sodium salt of cyanoacetic acid.

ESTIMATED ERROR:

Soly: the mean of two titrations agreed within at least 0.2%

Temp: ± 0.01°C (author)

REFERENCES:

1. Wise, W. C. A.; Davies, C. A. *J. Chem. Soc.* 1938, 273.

COMPONENTS:

(1) Calcium iodate; $Ca(IO_3)_2$; [7789-80-2]

(2) Sodium aminopropionate; $C_3H_6NO_2Na$; [23338-69-4]

(3) Water; H_2O; [7732-18-5]

ORIGINAL MEASUREMENTS:

Davies, C. W.; Waind, G. M.

J. Chem. Soc. 1950, 301-3.

VARIABLES:

$T/K = 298$

$10^3 c_2/mol\ dm^{-3} = 0$ to 109.6

PREPARED BY:

Hiroshi Miyamoto

EXPERIMENTAL VALUES:

t/°C	Sodium Aminopropionate $10^3 c_2/mol\ dm^{-3}$	Calcium Iodate $10^3 c_1/mol\ dm^{-3}$
25	0	7.84
	19.62	8.96
	21.08	8.99
	31.63	9.45
	42.17	9.85
	52.72	10.24
	63.27	10.61
	73.80	10.92
	84.35	11.26
	98.64	11.58
	109.6	11.70

NaA was assumed to be completely dissociated.

$K_D^\circ(CaA^+) = 0.058$

AUXILIARY INFORMATION

METHOD/APPARATUS/PROCEDURE:

The saturating column method was employed (see ref (1) for details). The saturation was ensured by passing a portion of the solution through the saturating column a second time. The analyses were effected by withdrawing 10 cm^3 of the saturating solution in a calibrated pipette, and running it into an acidic KI solution. The liberated iodine was titrated by weight against $Na_2S_2O_3$ solution, iodine solution being used for the back titration.

SOURCE AND PURITY OF MATERIALS:

$Ca(IO_3)_2 \cdot 6H_2O$ was prepared by drop-wise addition of solutions of KIO_3 and $CaCl_2$ in equivalent amounts to a large volume of conductivity water. The hexahydrate separated, and was washed (see also compilation of ref (1)). The purity of the acid was checked by potentiometric titration of the solution against standard NaOH. The sodium salt was then made up in accordance with the observed end-point.

ESTIMATED ERROR:

Soly: nothing specified

Temp: ± 0.01°C (authors)

REFERENCES:

1. Wise, W. C. A.; Davies, C. W. *J. Chem. Soc.* 1938, 273.

COMPONENTS:

(1) Calcium iodate; $Ca(IO_3)_2$; [7789-80-2]

(2) Serine, monosodium salt; $C_3H_6NO_3Na$; [41521-39-5]

(3) Water; H_2O; [7732-18-5]

ORIGINAL MEASUREMENTS:

Davies, C. W.; Waind, G. M.

J. Chem. Soc. 1950, 301-3.

VARIABLES:

$T/K = 298$

$10^3 c_2/\text{mol dm}^{-3} = 0$ to 56.10

PREPARED BY:

Hiroshi Miyamoto

EXPERIMENTAL VALUES:

t/°C	Serine, monosodium salt $10^3 c_2/\text{mol dm}^{-3}$	Calcium Iodate $10^3 c_1/\text{mol dm}^{-3}$
25	0	7.84
	11.22	8.62
	21.10	9.20
	56.10	10.68

NaA was assumed to be completely dissociated.

$K^{\circ}_{D}(CaA^{+}) = 0.037$

AUXILIARY INFORMATION

METHOD/APPARATUS/PROCEDURE:

The saturating column method was employed (see ref (1) for details). The saturation was ensured by passing a portion of the solution through the saturating column a second time. The analyses were effected by withdrawing 10 cm^3 of the saturating solution in a calibrated pipette, and running it into an acidic KI solution. The liberated iodine was titrated by weight against $Na_2S_2O_3$ solution, iodine solution being used for the back titration.

SOURCE AND PURITY OF MATERIALS:

$Ca(IO_3)_2 \cdot 6H_2O$ was prepared by dropwise addition of solutions of KIO_3 and $CaCl_2$ in equivalent amounts to a large volume of conductivity water. The hexahydrate separated, and was washed. (see also compilation of ref (1)). The purity of the acid was checked by potentiometric titration of the solution against standard NaOH. The sodium salt was then made up in accordance with the observed end-point.

ESTIMATED ERROR:

Soly: nothing specified
Temp: ± 0.01°C (authors)

REFERENCES:

1. Wise, W. C. A.; Davies, C. W. *J. Chem. Soc.* 1938, 273.

COMPONENTS:

(1) Calcium iodate; $Ca(IO_3)_2$; [7789-80-2]

(2) Sodium malate; $C_4H_3O_4Na$; [58214-38-3]

(3) Water; H_2O; [7732-18-5]

ORIGINAL MEASUREMENTS:

Das, A. R.; Nair, V. S. K.

J. Inorg. Nucl. Chem. 1975, *37*, 2121-3.

EXPERIMENTAL VALUES:

t/°C	Sodium Malate $10^3 c_2$/mol dm^{-3}	Calcium Iodate $10^3 c_1$/mol dm^{-3}	log K_A [a]
25	0	7.820	
	15.98	9.176	2.98
	21.31	9.582	3.00
	26.63	9.943	3.02
	31.07	10.159	3.00
	36.04	10.627	3.04
			mean 3.01 ± 0.04
30	0	9.380	
	10.74	10.611	3.14
	18.80	11.499	3.17
	26.86	12.159	3.16
	34.91	12.818	3.17
	51.03	13.903	3.17
			mean 3.16 ± 0.04
35	0	11.671	
	42.97	15.985	3.19
	51.03	16.769	3.20
	59.08	17.743	3.26
			mean 3.21 ± 0.04
40	0	13.920	
	18.40	16.684	3.38
	25.89	17.642	3.36
	33.56	18.503	3.36
	56.29	20.246	3.32
	63.87	21.083	3.32
			mean 3.34 ± 0.06

[a] K_A is the association constant for $CaHO_2CCH{=}CHCO_2^+$

$K_A/dm^3 mol^{-1}$

COMMENTS AND/OR ADDITIONAL DATA:

The authors state that a plot of log K_A against $(T/K)^{-1}$ is linear and have calculated free energies, enthalpies, and entropies for the association reaction. Individual ionic entropies have also been calculated, and the original paper should be consulted for these details.

The authors do not discuss the nature of the solid phase in equilibrium with the saturated solutions. At 40° and 45°C, the solid phase is probably the monohydrate, $Ca(IO_3)_2 \cdot H_2O$, and below these temperatures, the solid phase is probably the hexahydrate $Ca(IO_3)_2 \cdot 6H_2O$ (compilers).

The results of the calculations for the thermodynamic solubility product constant for $Ca(IO_3)_2$ as a function of temperature are:

t/°C	25	30	35	40
$-\log K^0_{s0}$	6.149	5.951	5.717	5.543

COMPONENTS:

(1) Calcium iodate; $Ca(IO_3)_2$; [7789-80-2]

(2) Sodium malate; $C_4H_3O_4Na$; [58214-38-3]

(3) Water; H_2O; [7732-18-5]

ORIGINAL MEASUREMENTS:

Das, A. R.; Nair, V. S. K.

J. Inorg. Nucl. Chem. 1975, *37*, 2121-3.

VARIABLES:

T/K = 298 to 313

$10^3 c_2$/mol dm^{-3}

PREPARED BY:

Hiroshi Miyamoto
Mark Salomon

METHOD/APPARATUS/PROCEDURE:

The saturating column method was used as described by Davies (1) (see the compilation for this reference for details). All solutions were made by dilution of the stock solution, and the pH was measured before and after percolation through the column. The pH of all solutions were in the range of 3.6 - 3.8. Other details same as in ref. (2): i.e. the slns were passed through the column at least four times (twice was sufficient for saturation), and satd slns were analysed for iodate by iodometric titrn in a nitrogen atmosphere.

The thermodynamic soly product constant was calcd from

$$K^\circ_{s0} = [Ca^{2+}][IO_3^-]^2 y_\pm(Ca^{2+}) y_\pm(IO_3^-)^2$$

In these calculations the ion pairing constant for $NaIO_3$ was taken as 3.0 at 25°C (3) and the association constant for $CaIO_3^+$ was estimated for 30-45°C by the electrostatic theory of Bjerrum (4): values for these association constants are not given in the paper. The association constant for $K_A(CaMal^+)$ (K_A in the tables, Mal^- = $HO_2CCH{=}CHCO_2^-$) was calcd. from

$$K_A(CaMal^+) = [CaMal^+]/[Ca^{2+}][Mal^-] y_\pm(Ca^{2+})$$

All activity coefficients were calculated from the Davies equation (5).

AUXILIARY INFORMATION

SOURCE AND PURITY OF MATERIALS:

$Ca(IO_3)_2$ prepd by stoichiometric addn of $CaCl_2$ and KIO_3 slns over a 2 h period to a large volume of water at 65°C (authors do not state which hydrate is formed). The ppt was washed 6 times by decantation and digested in double dist water for 4h at 80°C. Malic acid was "repeatedly" recrystallized from water and dried in vacuo before use. Stock slns of Na malate were prepd from the acid and CO_2-free NaOH sln so as to obtain a 1:1 buffer ratio. Conductivity water prepared by mixed-bed ionization (6) was used in all preparations and solubility experiments.

ESTIMATED ERROR:

Soly: nothing specified
Temp: Precision probably ± 0.1% as in (2).

REFERENCES:

1. Davies, C. W. *J. Chem. Soc.* 1930, 2471.
2. Ghosh, R.; Nair, V. S. K. *J. Inorg. Nucl. Chem.* 1970, *32*, 3025.
3. Wise, W. C. A.; Davies, C. W. *J. Chem. Soc.* 1938, 273.
4. Bjerrum, N. *K. Danske Vidensk. Selsk. Math-Fys. Medd.* 1926, *7*, No. 9.
5. Davies, C. W. *Ion Association.* Butterworths. London. 1960.
6. Davies, C. W.; Nancollas, G. H. *Chem. Ind.* 1950, *7*, 129.

COMPONENTS:

(1) Calcium iodate; $Ca(IO_3)_2$; [7789-80-2]

(2) 3-Hydroxybutanoic acid, mono-sodium salt (sodium β-hydroxy-butyrate); $C_4H_7O_3Na$; [150-83-4]

(3) Water; H_2O; [7732-18-5]

ORIGINAL MEASUREMENTS:

Davies, C. W.

J. Chem. Soc. 1938, 277-81.

VARIABLES:

$T/K = 298$

$10^3 c_2/\text{mol dm}^{-3} = 21.48$ and 43.04

PREPARED BY:

Hiroshi Miyamoto

EXPERIMENTAL VALUES:

t/°C	Sodium β-Hydroxybutyrate $10^3 c_2/\text{mol dm}^{-3}$	Calcium Iodate $10^3 c_1/\text{mol dm}^{-3}$	Dissociation constant $K_D^\circ/\text{mol dm}^{-3}$
25	21.48	8.791	0.13
	43.04	9.470	0.16

K_D° was calculated from the equation

$$\log K_D^\circ = \log[Ca^{2+}][X^-]/[CaX^+] - 2.02I^{1/2} + 2.8I$$

where $X^- = CH_3CH(OH)CH_2COO^-$ and I is the ionic strength.

AUXILIARY INFORMATION

METHOD/APPARATUS/PROCEDURE:

Saturated column method as described in the preceding paper (1). The saturated solutions were passed several times through the saturating column.

The author stated that the concentration of $Ca(IO_3)_2$ in the saturated solution was determined by the ordinary volumetric method. The method is probably similar to that reported in the preceding paper (1), which has been compiled elsewhere in this volume.

SOURCE AND PURITY OF MATERIALS:

The source of calcium iodate was not given, but probably same as in the preceding paper (1): see compilation of the paper for details. Sodium β-hydroxybutyrate was recrystallized from absolute alcohol and dried in a vacuum.

ESTIMATED ERROR:

Soly: the mean of two titrations agreed within at least 0.2%

Temp: ± 0.01°C (author)

REFERENCES:

1. Wise, W. C. A.; Davies, C. A. *J. Chem. Soc.* 1938, 273.

COMPONENTS:

(1) Calcium iodate; $Ca(IO_3)_2$; [7789-80-2]

(2) Sodium glycyl glycinate; $C_4H_7N_2O_3Na$; [1070-67-3]

(3) Water; H_2O; [7732-18-5]

ORIGINAL MEASUREMENTS:

Davies, C. W.; Waind, G. M.

J. Chem. Soc. 1950, 301-3.

VARIABLES:

$T/K = 298$

$10^3 c_2/\text{mol dm}^{-3} = 0$ to 48.52

PREPARED BY:

Hiroshi Miyamoto

EXPERIMENTAL VALUES:

t/°C	Sodium Glycyl Glycinate $10^3 c_2/\text{mol dm}^{-3}$	Calcium Iodate $10^3 c_1/\text{mol dm}^{-3}$
25	0	7.84
	11.55	8.51
	13.06	8.56
	23.10	9.04
	34.85	9.55
	48.52	10.05

NaA was assumed to be completely dissociated.

$K_D^\circ = 0.057$

AUXILIARY INFORMATION

METHOD/APPARATUS/PROCEDURE:

The saturating column method was employed (see ref (1) for details). Where only small amounts of saturating salt were available, a smaller saturator was used. The saturation was ensured by passing a portion of the solution through the saturating column a second time. The analyses were effected by withdrawing 10 cm^3 of the saturating solution in a calibrated pipette, and running it into an acidic KI solution. The liberated iodine was titrated by weight against $Na_2S_2O_3$ solution, iodine solution being used for the back titration.

SOURCE AND PURITY OF MATERIALS:

$Ca(IO_3)_2 \cdot 6H_2O$ was prepared by dropwise addition of solutions of KIO_3 and $CaCl_2$ in equivalent amounts to a large volume of conductivity water. The hexahydrate separated, and was washed. (see also compilation of ref (1)). The purity of the dipeptide was checked by potentiometric titration of the solution against standard NaOH. The sodium salt was then made up in accordance with the observed end-point.

ESTIMATED ERROR:

Soly: nothing specified

Temp: ± 0.01°C (authors)

REFERENCES:

1. Wise, W. C. A.; Davies, C. W. *J. Chem. Soc.* 1938, 273.

COMPONENTS:

(1) Calcium iodate; $Ca(IO_3)_2$; [7789-80-2]

(2) *D,L*-Glutamic acid, disodium salt (disodium glutamate); $C_5H_7NO_4Na_2$; [149-65-5]

(3) Water; H_2O; [7732-18-5]

ORIGINAL MEASUREMENTS:

Davies, C. W.; Waind, G. M.

J. Chem. Soc. 1950, 301-3.

VARIABLES:

$T/K = 298$

$10^3 c_2/\text{mol dm}^{-3} = 0$ to 22.49

PREPARED BY:

Hiroshi Miyamoto

EXPERIMENTAL VALUES:

t/°C	Disodium Glutamate $10^3 c_2/\text{mol dm}^{-3}$	Calcium Iodate $10^3 c_1/\text{mol dm}^{-3}$
25	0	7.84
	5.62	8.68
	11.42	9.34
	21.43	10.42
	22.49	10.44

Disodium glutamate solutions are extensively hydrolysed, and authors had to account for OH^-, $CaOH^+$, HA^-, $CaHA^+$ ($A = (O_2CCH_2CH_2CH(NH_2)CO_2)2-$), but details not given.

$K^{\circ}_D(CaNaA^+) = 0.0088$

AUXILIARY INFORMATION

METHOD/APPARATUS/PROCEDURE:

The saturating column method was employed (see ref (1) for details). The saturation was ensured by passing a portion of the solution through the saturating column a second time. The analyses were effected by withdrawing 10 cm^3 of the saturating solution in a calibrated pipette, and running it into an acidic KI solution. The liberated iodine was titrated by weight against $Na_2S_2O_3$ solution, iodine solution being used for the back titration.

SOURCE AND PURITY OF MATERIALS:

$Ca(IO_3)_2 \cdot 6H_2O$ was prepared by dropwise addition of solutions of KIO_3 and $CaCl_2$ in equivalent amounts to a large volume of conductivity water. The hexahydrate separated, and was washed (see also compilation of ref (1)). The purity of the acid was checked by potentiometric titration of the solution against standard NaOH. The sodium salt was then made up in accordance with the observed end-point.

ESTIMATED ERROR:

Soly: nothing specified

Temp: ± 0.01°C (authors)

REFERENCES:

1. Wise, W. C. A.; Davies, C. W. *J. Chem. Soc.* 1938, 273.

COMPONENTS:

(1) Calcium iodate; $Ca(IO_3)_2$; [7789-80-2]

(2) D,L-Glutamic acid, monosodium salt (monosodium glutamate); $C_5H_8NO_4Na$; [32221-81-1]

(3) Water; H_2O; [7732-18-5]

ORIGINAL MEASUREMENTS:

Davies, C. W.; Waind, G. M.

J. Chem. Soc. 1950, 301-3.

VARIABLES:

$T/K = 298$

$10^3c_2/mol\ dm^{-3} = 0$ to 92.85

PREPARED BY:

Hiroshi Miyamoto

EXPERIMENTAL VALUES:

t/°C	Monosodium Glutamate $10^3c_2/mol\ dm^{-3}$	Calcium Iodate $10^3c_1/mol\ dm^{-3}$
25	0	7.84
	22.22	8.87
	23.00	9.06
	44.50	9.86
	46.47	9.92
	90.26	11.05
	92.85	11.14

NaA assumed to be completely dissociated.

$K_D^\circ(CaA^+) = 0.067$

AUXILIARY INFORMATION

METHOD/APPARATUS/PROCEDURE:

The saturating column method was employed (see ref (1) for details). The saturation was ensured by passing a portion of the solution through the saturating column a second time. The analyses were effected by withdrawing 10 cm^3 of the saturating solution in a calibrated pipette, and running it into an acidic KI solution. The liberated iodine was titrated by weight against $Na_2S_2O_2$ solution, iodine solution being used for the back titration.

SOURCE AND PURITY OF MATERIALS:

$Ca(IO_3)_2 \cdot 6H_2O$ was prepared by dropwise addition of solutions of KIO_3 and $CaCl_2$ in equivalent amounts to a large volume of conductivity water. The hexahydrate separated, and was washed (see also compilation of ref (1)). The purity of the acid was checked by potentiometric titration of the solution against standard NaOH. The sodium salt was then made up in accordance with the observed end-point.

ESTIMATED ERROR:

Soly: nothing specified
Temp: ± 0.01°C (authors)

REFERENCES:

1. Wise, W. C. A.; Davies, C. W. *J. Chem. Soc.* 1938, 273.

COMPONENTS:

(1) Calcium iodate; $Ca(IO_3)_2$; [7789-80-2]

(2) Alanyl glycine, monosodium salt (monosodium alanyl glycinate); $C_5H_9N_2O_3Na$; [82808-62-6][a]

(3) Water; H_2O; [7732-18-5]

ORIGINAL MEASUREMENTS:

Davies, C. W.; Waind, G. M.

J. Chem. Soc. 1950, 301-3.

VARIABLES:

$T/K = 298$

$10^3 c_2/\text{mol dm}^{-3} = 0$ and 61.22

PREPARED BY:

Hiroshi Miyamoto

EXPERIMENTAL VALUES:

t/°C	Monosodium Alanyl Glycinate $10^3 c_2/\text{mol dm}^{-3}$	Calcium Iodate $10^3 c_1/\text{mol dm}^{-3}$
25	0	7.84
	61.22	9.83

NaA was assumed to be completely dissociated.

$K_D^\circ(CaA^+) = 0.22$

AUXILIARY INFORMATION

METHOD/APPARATUS/PROCEDURE:

The saturating column method was employed (see ref (1) for details). Where only small amounts of saturating salt were available, a smaller saturator was used. The saturation was ensured by passing a portion of the solution through the saturating column a second time. The analyses were effected by withdrawing 10 cm^3 of the saturating solution in a calibrated pipette, and running it into an acidic KI solution. The liberated iodine was titrated by weight against $Na_2S_2O_3$ solution, iodine solution being used for the back titration.

[a] *N*-L-Alanylglycine, monosodium salt

SOURCE AND PURITY OF MATERIALS:

$Ca(IO_3)_2 \cdot 6H_2O$ was prepared by drop-wise addition of solutions of KIO_3 and $CaCl_2$ in equivalent amounts to a large volume of conductivity water. The hexahydrate separated, and was washed. (see also compilation of ref (1)). The purity of the dipeptide was checked by potentiometric titration of the solution against standard NaOH. The sodium salt was then made up in accordance with the observed end-point.

ESTIMATED ERROR:

Soly: nothing specified

Temp: ± 0.01°C (authors)

REFERENCES:

1. Wise, W. C. A.; Davies, C. W. *J. Chem. Soc.* 1938, 273.

COMPONENTS:

(1) Calcium iodate; $Ca(IO_3)_2$; [7789-80-2]

(2) 2-Hydroxybenzoic acid, monosodium salt (sodium salicylate); $C_7H_5O_3Na$; [54-21-7]

(3) Water; H_2O; [7732-18-5]

ORIGINAL MEASUREMENTS:

Davies, C. W.

J. Chem. Soc. 1938, 277-81.

VARIABLES:

$T/K = 298$

$10^3 c_2/\text{mol dm}^{-3} = 20$ and 40

PREPARED BY:

Hiroshi Miyamoto

EXPERIMENTAL VALUES:

t/°C	Sodium Salycilate $10^3 c_2/\text{mol dm}^{-3}$	Calcium Iodate $10^3 c_1/\text{mol dm}^{-3}$	Dissociation constant $K_D^\circ/\text{mol dm}^{-3}$
25	20.0	8.652	0.24
	40.0	9.241	0.31

K_D° was calculated from the equation

$$\log K_D^\circ = \log[Ca^{2+}][X^-]/[CaX^+] - 2.02I^{1/2} + 2.8I$$

where $X^- = HOC_6H_4COO^-$ and I is the ionic strength.

AUXILIARY INFORMATION

METHOD/APPARATUS/PROCEDURE:

Saturated column method as described in the preceding paper (1). The saturated solutions were passed several times through the saturating column.

The author stated that the concentration of $Ca(IO_3)_2$ in the saturated solution was determined by the ordinary volumetric method. The method is probably similar to that reported in the preceding paper (1), which has been compiled elsewhere in this volume.

SOURCE AND PURITY OF MATERIALS:

The source of calcium iodate was not given, but probably same as in the preceding paper (1): see compilation of the paper for details. Salicylic acid was a commercial sample, the acid equivalent determined to be within 0.1% of theoretical. The acid and AnalaR NaOH used to prepare sodium salicylate.

ESTIMATED ERROR:

Soly: the mean of two titrations agreed within at least 0.2%.
Temp: ± 0.01°C (author)

REFERENCES:

1. Wise, W. C. A.; Davies, C. A. *J. Chem. Soc.* 1938, 273.

COMPONENTS:

(1) Calcium iodate; $Ca(IO_3)_2$; [7789-80-2]

(2) Sodium mandelate; $C_8H_7O_3Na$; [19944-52-6]

(3) Water; H_2O; [7732-18-5]

ORIGINAL MEASUREMENTS:

Wise, W. C. A.; Davies, C. W.

J. Chem. Soc. 1938, 273-7.

VARIABLES:

$T/K = 298$

$10^3 c_2/\text{mol dm}^{-3} = 0$ to 100.0

PREPARED BY:

Hiroshi Miyamoto
Mark Salomon

EXPERIMENTAL VALUES:

t/°C	Sodium Mandelate $10^3 c_2/\text{mol dm}^{-3}$	Calcium Iodate $10^3 c_1/\text{mol dm}^{-3}$	Density $\rho/\text{g cm}^{-3}$
25	0	7.840	0.9998
	20.0	9.177	1.0002
	50.0	10.69	1.0047
	100.0	12.83	1.0090

COMMENTS AND/OR ADDITIONAL DATA:

The conductivity data at 18°C were used to evaluate the thermodynamic ion pair dissociation constant, K_D°; it was found that $K_D^\circ = 0.13$. Using this value for the ion pair dissociation constant, the concentration of the ion pair $CaIO_3^+$ was calculated from

$$\log [CaIO_3^+] = \log [Ca^{2+}][IO_3^-] - \log K_D^\circ - 2.02 I^{1/2} + 2.0 I$$

where I is the ionic strength. Utilizing this relation to compute the ionic concentrations of Ca^{2+} and IO_3^-, the authors plotted (1/3) log $[Ca^{2+}] \times [IO_3^-]$ against the ionic strength and extrapolated to zero ionic strength to obtain the thermodynamic solubility product constant. The result of this extrapolation is $K_{s0}^\circ = 6.953 \times 10^{-7}$.

AUXILIARY INFORMATION

METHOD/APPARATUS/PROCEDURE:

Saturating column method as in (1) and modified as in (2). A bulb containing the solvent solution is attached to a column containing the slightly soluble salt, and the solvent is allowed to flow through the column at a rate sufficient to insure saturation (1). The modification (2) consisted of connecting the column by capillary tubing to a second parallel arm in which the saturated solution collected. The entire apparatus was placed in a thermostat. Weighed samples of the satd slns were taken for analysis by method described in (3): i.e. the satd slns added to acidified KI sln and the liberated I_2 titrd by weight against an approx 0.15N thiosulfate sln, 0.01N iodine sln being used for the back titrn. The densities of the satd slns were measured at 25°C, and the molar conductivities at 18°C for the binary $Ca(IO_3)_2$-H_2O system are also reported.

SOURCE AND PURITY OF MATERIALS:

$Ca(IO_3)_2 \cdot 6H_2O$ was prepared by dropwise addition of solutions of KIO_3 and $CaCl_2$ in equivalent amounts to a large volume of conductivity water. The precipitate was washed first by decantation and then in the solubility columns until a constant solubility was obtained.

ESTIMATED ERROR:

Soly: not specified, but reproducibility probably around ± 0.3% as in ref. (3).

Temp: ± 0.01°K (authors)

REFERENCES:

1. Brönsted, J. N.; La Mer, V. K. *J. Am. Chem. Soc.* 1924, *46*, 555.
2. Money, R. W.; Davies, C. W. *J. Chem. Soc.* 1934, 400.
3. Macdougall, G.; Davies, C. W. *J. Chem. Soc.* 1935, 1416.

COMPONENTS:

(1) Calcium iodate; $Ca(IO_3)_2$; [7789-80-2]

(2) Sodium mandelate; $C_8H_7O_3Na$; [114-21-6]

(3) Water; H_2O; [7732-18-5]

ORIGINAL MEASUREMENTS:

Das, A. R.; Nair, V. S. K.

J. Inorg. Nucl. Chem. 1975, *37*, 991-3.

EXPERIMENTAL VALUES:

t/°C	Sodium Mandelate 10^3c_2/mol dm^{-3}	Calcium Iodate 10^3c_1/mol dm^{-3}	log K_A[a]
25	10.970	8.527	1.45
	27.380	9.362	1.41
	29.858	9.622	1.49
	39.950	10.095	1.49
	51.226	10.529	1.47
	56.148	10.952	1.48
			mean 1.46 ± 0.05
30	27.069	11.403	1.56
	40.580	12.285	1.59
	49.310	12.781	1.59
	56.350	13.284	1.62
	72.340	14.052	1.62
			mean 1.58 ± 0.06
35	7.042	12.41	1.62
	14.084	13.096	1.65
	28.162	14.158	1.60
	35.210	14.721	1.62
	56.336	16.095	1.62
			mean 1.62 ± 0.02
40	30.340	17.227	1.79
	37.925	17.741	1.74
	53.095	19.053	1.75
	60.680	19.698	1.76
			mean 1.75 ± 0.05
45	37.921	21.312	1.80
	45.510	22.265	1.80
	53.095	23.299	1.84
	60.680	24.665	1.88
			mean 1.83 ± 0.03

[a] K_A is the association constant for $CaC_6H_5CH(OH)CO_2^+$.

COMMENTS AND/OR ADDITIONAL DATA:

The authors state that a plot of log K_A against $(T/K)^{-1}$ is linear and have calculated free energies, enthalpies, and entropies for the association reaction. Individual ionic entropies have also been calculated, and the original paper should be consulted for these details.

The authors do not discuss the nature of the solid phases in equilibrium with the saturated solutions. At 40° and 45°C, the solid phase is probably the monohydrate, $Ca(IO_3)_2 \cdot H_2O$, and below these temperatures, the solid phase is probably the hexahydrate $Ca(IO_3)_2 \cdot 6H_2O$ (compilers).

The results of the calculations for the thermodynamic solubility product constant for $Ca(IO_3)_2$ as a function of temperature are:

t/°C	25	30	35	40	45
$-\log K^0_{s0}$	6.15	5.95	5.72	5.54	5.34

COMPONENTS:

(1) Calcium iodate; $Ca(IO_3)_2$; [7789-80-2]

(2) Sodium mandelate; $C_8H_7O_3Na$; [114-21-6]

(3) Water; H_2O; [7732-18-5]

ORIGINAL MEASUREMENTS:

Das, A. R.; Nair, V. S. K.

J. Inorg. Nucl. Chem. 1975, *37*, 991-3.

VARIABLES:

T/K = 298-318

$10^3 c_2$/mol dm^{-3} = 7 - 72

PREPARED BY:

Hiroshi Miyamoto

Mark Salomon

METHOD/APPARATUS/PROCEDURE:

The saturating column method was used as described by Davies (1) (see the compilation for this reference for details). All solutions were made by dilution of the stock sln, and the pH was measured before and after percolation through the column. The pH of all slns were in the range of 3.6 - 3.8 (pK_A of mandelic acid = 3.68 at 25°C). Other details same as in ref. (2): i.e. the slns were passed through the column at least four times (twice was sufficient for saturation), and satd slns were analysed for iodate by iodometric titrn in a nitrogen atmosphere.

The thermodynamic soly product constant was calcd from

$$K^\circ_{s0} = [Ca^{2+}][IO_3^-]^2 y_\pm(Ca^{2+}) y_\pm(IO_3^-)^2$$

In these calculations the ion pairing constant for $NaIO_3$ was taken as 3.0 at 25°C (3) and the association constant for $CaIO_3^+$ was estimated for 30-45°C by the electrostatic theory of Bjerrum (4): values for these association constants are not given in the paper. The association constant for $CaMan^+$ (K_A in the tables, Man^- = $C_6H_5CH(OH)CO_2^-$) was calcd from

$$K_A(CaMan^+) = [CaMan^+]/[Ca^{2+}][Man^-]y_\pm(Ca^{2+})$$

All activity coefficients were calculated from the Davies equation.

AUXILIARY INFORMATION

SOURCE AND PURITY OF MATERIALS:

$Ca(IO_3)_2$ prepd by stoichiometric addn of $CaCl_2$ and KIO_3 slns over a 2 h period to a large volume of water at 65°C (authors do not state which hydrate was formed). The ppt was washed 6 times by decantation and digested in double dist water for 4h at 80°C. B.D.H. mandelic acid was "repeatedly" recrystallized from water and dried in vacuo before use. Stock solutions of sodium mandelate were prepd from the acid and CO_2-free NaOH sln so as to obtain a 1:1 buffer ratio. Conductivity water prepared by mixed-bed ionization (6) was used in all preparations and solubility experiments.

ESTIMATED ERROR:

Soly: nothing specified

Temp: precision probably ± 0.1% as in (2).

REFERENCES:

1. Davies, C. W. *J. Chem. Soc.* 1930, 2471.
2. Ghosh, R.; Nair, V. S. K. *J. Inorg. Nucl. Chem.* 1970, *32*, 3025.
3. Wise, W. C. A.; Davies, C. W. *J. Chem. Soc.* 1938, 273.
4. Bjerrum, N. *K. Danske Vidensk. Selsk. Math-Fys. Medd.* 1926, *7*, No. 9.
5. Davies, C. W. *Ion Association.* Butterworths. London. 1960.
6. Davies, C. W.; Nancollas, G. H. *Chem. Ind.* 1950, *7*, 129.

COMPONENTS:

(1) Calcium iodate; $Ca(IO_3)_2$; [7789-80-2]

(2) Leucyl glycine, monosodium salt (monosodium leucyl glycinate) $C_8H_{15}N_2O_3Na$; [82808-61-5][a]

(3) Water; H_2O; [7732-18-5]

ORIGINAL MEASUREMENTS:

Davies, C. W.; Waind, G. M.

J. Chem. Soc. 1950, 301-3.

VARIABLES:

$T/K = 298$

$10^3 c_2/\text{mol dm}^{-3} = 0$ and 52.55

PREPARED BY:

Hiroshi Miyamoto

EXPERIMENTAL VALUES:

t/°C	Monosodium Leucyl Glycinate $10^3 c_2/\text{mol dm}^{-3}$	Calcium Iodate $10^3 c_1/\text{mol dm}^{-3}$
25	0	7.84
	52.55	9.61

NaA was assumed to be completely dissociated.

$K_D^\circ(CaA^+) = 0.20$

AUXILIARY INFORMATION

METHOD/APPARATUS/PROCEDURE:

The saturating column method was employed (see ref (1) for details). Where only small amounts of saturating salt were available, a smaller saturator was used. The saturation was ensured by passing a portion of the solution through the saturating column a second time. The analyses were effected by withdrawing 10 cm^3 of the saturating solution in a calibrated pipette, and running it into an acidic KI solution. The liberated iodine was titrated by weight against $Na_2S_2O_3$ solution, iodine solution being used for the back titration.

[a] *N*-L-Leucylglycine, monosodium salt

SOURCE AND PURITY OF MATERIALS:

$Ca(IO_3)_2\cdot 6H_2O$ was prepared by dropwise addition of solutions of KIO_3 and $CaCl_2$ in equivalent amounts to a large volume of conductivity water. The hexahydrate separated, and was washed. (see also compilation of ref (1)). The purity of the dipeptide was checked by potentiometric titration of the solution against standard NaOH. The sodium salt was then made up in accordance with the observed end-point.

ESTIMATED ERROR:

Soly: nothing specified
Temp: ± 0.01°C (authors)

REFERENCES:

1. Wise, W. C. A.; Davies, C. W. *J. Chem. Soc.* 1938, 273.

COMPONENTS:

(1) Calcium iodate; $Ca(IO_3)_2$; [7789-80-2]

(2) Tyrosine, 3,5-diiodo, monosodium salt; $C_9H_8I_2NO_3Na$; [76841-98-0]

(3) Water; H_2O; [7732-18-5]

ORIGINAL MEASUREMENTS:

Davies, C. W.; Waind, G. M.

J. Chem. Soc. 1950, 301-3.

VARIABLES:

$T/K = 298$

$10^3 c_2/\text{mol dm}^{-3} = 0$ to 41.89

PREPARED BY:

Hiroshi Miyamoto

EXPERIMENTAL VALUES:

t/°C	Tyrosine, 3,5-diiodo, monosodium salt $10^3 c_2/\text{mol dm}^{-3}$	Calcium Iodate $10^3 c_1/\text{mol dm}^{-3}$
25	0	7.84
	18.64	9.17
	41.89	10.44

NaA was assumed to be completely dissociated.

$K_D^\circ(CaA^+) = 0.029$

AUXILIARY INFORMATION

METHOD/APPARATUS/PROCEDURE:

The saturating column method was employed (see ref (1) for details). Where only small amounts of saturating salt were available, a smaller saturator was used. The saturation was ensured by passing a portion of the solution through the saturating column a second time. The analyses were effected by withdrawing 10 cm^3 of the saturating solution in a calibrated pipette, and running it into an acidic KI solution. The liberated iodine was titrated by weight against $Na_2S_2O_3$ solution, iodine solution being used for the back titration.

SOURCE AND PURITY OF MATERIALS:

$Ca(IO_3)_2 \cdot 6H_2O$ was prepared by dropwise addition of solutions of KIO_3 and $CaCl_2$ in equivalent amounts to a large volume of conductivity water. The hexahydrate separated, and was washed. (see also compilation of ref (1)). The purity of the acid was checked by potentiometric titration of the solution against standard NaOH. The sodium salt was then made up in accordance with the observed end-point.

ESTIMATED ERROR:

Soly: nothing specified

Temp: ± 0.01°C (authors)

REFERENCES:

1. Wise, W. C. A.; Davies, C. W. *J. Chem. Soc.* 1938, 273.

COMPONENTS:

(1) Calcium iodate; $Ca(IO_3)_2$; [7789-80-2]

(2) Sodium hippurate; $C_9H_8NO_3Na$; [532-94-5]

(3) Water; H_2O; [7732-18-5]

ORIGINAL MEASUREMENTS:

Davies, C. W.; Waind, G. M.

J. Chem. Soc. 1950, 301-3.

VARIABLES:

$T/K = 298$

$10^3c_2/mol\ dm^{-3} = 0$ to 50.22

PREPARED BY:

Hiroshi Miyamoto

EXPERIMENTAL VALUES:

t/°C	Sodium Hippurate $10^3c_2/mol\ dm^{-3}$	Calcium Iodate $10^3c_1/mol\ dm^{-3}$
25	0	7.84
	24.15	8.76
	25.14	8.81
	48.36	9.45
	50.22	9.48

NaA was assumed to be completely dissociated.

$K_D^\circ(CaA^+) = 0.37$

AUXILIARY INFORMATION

METHOD/APPARATUS/PROCEDURE:

The saturating column method was employed (see ref (1) for details). The saturation was ensured by passing a portion of the solution through the saturating column a second time.
The analyses were effected by withdrawing 10 cm^3 of the saturating solution in a calibrated pipette, and running it into an acidic KI solution. The liberated iodine was titrated by weight against $Na_2S_2O_3$ solution, iodine solution being used for the back titration.

SOURCE AND PURITY OF MATERIALS:

$Ca(IO_3)_2 \cdot 6H_2O$ was prepared by dropwise addition of solutions of KIO_3 and $CaCl_2$ in equivalent amounts to a large volume of conductivity water. The hexahydrate separated, and was washed (see also compilation of ref (1)). The purity of the acid was checked by potentiometric titration of the solution against standard NaOH. The sodium salt was then made up in accordance with the observed end-point.

ESTIMATED ERROR:

Soly: nothing specified
Temp: ± 0.01°C (authors)

REFERENCES:

1. Wise, W. C. A.; Davies, C. W. *J. Chem. Soc.* 1938, 273.

COMPONENTS:

(1) Calcium iodate; $Ca(IO_3)_2$; [7789-80-2]

(2) Tyrosine, monosodium salt; $C_9H_{10}NO_3Na$; [16655-52-0]

(3) Water; H_2O; [7732-18-5]

ORIGINAL MEASUREMENTS:

Davies, C. W.; Waind, G. M.

J. Chem. Soc. 1950, 301-3.

VARIABLES:

T/K = 298

$10^3 c_2$/mol dm^{-3} = 0 to 35.89

PREPARED BY:

Hiroshi Miyamoto

EXPERIMENTAL VALUES:

t/°C	Tyrosine, Monosodium salt $10^3 c_2$/mol dm^{-3}	Calcium Iodate $10^3 c_1$/mol dm^{-3}
25	0	7.84
	19.09	9.02
	32.20	9.69
	35.89	9.98

NaA was assumed to be completely dissociated.

$K^{\circ}_{D}(CaA^+)$ = 0.033

AUXILIARY INFORMATION

METHOD/APPARATUS/PROCEDURE:

The saturating column method was employed (see ref (1) for details). The saturation was ensured by passing a portion of the solution through the saturating column a second time. The analyses were effected by withdrawing 10 cm^3 of the saturating solution in a calibrated pipette, and running it into an acidic KI solution. The liberated iodine was titrated by weight against $Na_2S_2O_3$ solution, iodine solution being used for the back titration.

SOURCE AND PURITY OF MATERIALS:

$Ca(IO_3)_2 \cdot 6H_2O$ was prepared by dropwise addition of solutions of KIO_3 and $CaCl_2$ in equivalent amounts to a large volume of conductivity water. The hexahydrate separated, and was washed (see also compilation of ref (1)). The purity of the acid was checked by potentiometric titration of the solution against standard NaOH. The sodium salt was then made up in accordance with the observed end-point.

ESTIMATED ERROR:

Soly: nothing specified

Temp: ± 0.01°C (authors)

REFERENCES:

1. Wise, W. C. A.; Davies, C. W. *J. Chem. Soc.* 1938, 273.

COMPONENTS:

(1) Calcium iodate; $Ca(IO_3)_2$; [7789-80-2]

(2) Potassium hydroxide; KOH; [1310-58-3]

(3) Water; H_2O; [7732-18-5]

ORIGINAL MEASUREMENTS:

Bell, R. P.; George, J. H. B.

Trans. Faraday Soc. 1953, *49*, 619-27.

VARIABLES:

T/K=273; 10^3c_2/mol dm^{-3}=0 to 68.27
T/K=298; 10^3c_2/mol dm^{-3}=0 to 54.29
T/K=313; 10^3c_2/mol dm^{-3}=0 to 65.25

PREPARED BY:

Hiroshi Miyamoto
Mark Salomon

EXPERIMENTAL VALUES:

t/°C	0		25		40	
	Potassium Hydroxide 10^3c_2/mol dm^{-3}	Calcium Iodate 10^3c_1/mol dm^{-3}	Potassium Hydroxide 10^3c_2/mol dm^{-3}	Calcium Iodate 10^3c_1/mol dm^{-3}	Potassium Hydroxide 10^3c_2/mol dm^{-3}	Calcium Iodate 10^3c_1/mol dm^{-3}
	0	2.315	0	7.838	0	13.06
	13.80	2.716	5.73	8.236	15.77	14.48
	22.92	2.919	6.95	8.318	27.17	15.46
	29.32	3.058	8.07	8.389	32.34	15.70
	38.36	3.183	10.70	8.565	38.77	16.18
	44.24	3.309	14.64	8.765	49.75	17.02
	54.24	3.436	17.84	8.943	65.25	17.84
	68.27	3.709	19.59	9.059		
			22.38	9.245		
			27.08	9.504		
			35.86	9.907		
			54.29	10.718		

Solid phase is the hexahydrate at 0° and 25°C and the monohydrate at 40°C.

Ion dissociation constant K_D° for $CaOH^+$ detd to be 0.043, 0.040, and 0.033 at 0°, 25° and 40°C, respectively. In these calculations, the authors used the 25°C data for $K_D^\circ(CaIO_3^+)$ = 0.13 (1), and $K_D^\circ(KIO_3)$ = 2.0 (2); to estimate these latter K_D° values at 0° and 40°C, authors assumed their temp dependence is given by the electrostatic theory of Bjerrum (3). Combining these equilibrium constants with the experimental soly data, the authors calculated the concns of Ca^{2+} and IO_3^- and computed the thermodynamic solubility products at the three temperatures:

$K_{s0}^\circ = 2.859 \times 10^{-8}$ at 0°C; $K_{s0}^\circ = 7.119 \times 10^{-7}$ at 25°C;

$K_{s0}^\circ = 2.437 \times 10^{-6}$ at 40°C.

AUXILIARY INFORMATION

METHOD/APPARATUS/PROCEDURE:

The solubilities were detd by passing about 50 ml of solution through a column of solid iodate (8 cm deep and 0.5 cm^2 cross-section) similar to that described in (4). The time of passage required to reach equil for the 50 ml of sln varied from 6 h at 0°C to 1.5 h at 25° and 40°C. The slns were brought to the thermostat temp before passage through the column, and repeated tests showed that complete saturation was reached after one passage. The iodate concns were detd iodometrically. The titrn was carried out by weight except that the final addition of around 0.3 ml of thiosulfate sln was made volumetrically with a microburet. Four independent analyses were made with each sln, and the average spread was about 0.3% at 0°C and 0.1% at 25° and 40°C.

SOURCE AND PURITY OF MATERIALS:

Calcium iodate was prepared by dropping solutions of A.R. grade KIO_3 and $CaCl_2$ into a large volume of water at 40°C. At this temperature, the solid is the monohydrate, $Ca(IO_3)_3 \cdot H_2O$. Conductivity water was used in all preparations and solubility measurements.

ESTIMATED ERROR: Reproducibility in soly probably equal to or slightly poorer than the reproducibility in the analysis of IO_3^-. Error in temp not specified but could add ~0.1% to above error.

REFERENCES:

1. Davies, C. W. *J. Chem. Soc.* 1930, 2410.
2. Davies, C. W. *Trans. Faraday Soc.* 1927, *23*, 592.
3. Bjerrum, J. *Kgl. Danske vid. Selsk. Math-fys Medd.* 1926, *7*, 9.
4. Davies, C. W. *J. Chem. Soc.* 1938, 277.

COMPONENTS:

(1) Calcium iodate; $Ca(IO_3)_2$; [7789-80-2]

(2) Potassium chloride; KCl; [7447-40-7]

(3) Water; H_2O; [7732-18-5]

ORIGINAL MEASUREMENTS:

Gross, P.; Klinghoffer, St. S.

Monatsh. Chem. 1930, *55*, 338-41.

VARIABLES:

$T/K = 298$

$10^3c_2/\text{mol dm}^{-3} = 0$ to 1946

PREPARED BY:

J. W. Lorimer and H. Miyamoto

EXPERIMENTAL VALUES:

t/°C	Potassium Chloride[a] $10^3c_2/\text{mol dm}^{-3}$	Calcium Iodate[b] $10^3c_1/\text{mol dm}^{-3}$
25	0	7.976
	50.14	9.551
	99.74	10.60
	149.8	11.39
	298.7	13.26
	500.0	15.16
	747.4	17.51
	898.0	18.78
	1497	22.75
	1946	25.59

[a] Concentrations of KCl appear to be initial values.

[b] Solid phase is $Ca(IO_3)_2 \cdot 6H_2O$.

COMMENTS:

See compilation of the author's work on the system $Ca(IO_3)_2-LiCl-H_2O$.

AUXILIARY INFORMATION

METHOD/APPARATUS/PROCEDURE:

Salts and water were shaken in sealed flasks in a thermostat for 24-48 h. One sample (two for determinations in pure water) was heated above 25°C and shaken for a short time to produce a solution which was supersaturated at 25°C. After settling, the solutions were filtered through cotton wool filters which were found to be inert to calcium iodate. Two samples were removed for analysis, presumably (compiler) by titration with $AgNO_3$.

SOURCE AND PURITY OF MATERIALS:

Calcium iodate hexahydrate was made from solutions of Merck p.a. $CaCl_2$ and KIO_3. KCl was Merck p.a.

ESTIMATED ERROR:

Temperature: control to ± 0.005 K, accuracy within ± 0.05 K.

Analyses for IO_3: precision within ± 0.3 %.

REFERENCES:

COMPONENTS:

(1) Calcium iodate; $Ca(IO_3)_2$; [7789-80-2]

(2) Potassium chloride; KCl; [7447-40-7]

(3) Water; H_2O; [7732-18-5]

ORIGINAL MEASUREMENTS:

Wise, W. C. A.; Davies, C. W.

J. Chem. Soc. 1938, 273-7.

VARIABLES:

$T/K = 298$

$10^3 c_2/\text{mol dm}^{-3} = 0$ to 100.0

PREPARED BY:

Hiroshi Miyamoto
Mark Salomon

EXPERIMENTAL VALUES:

t/°C	Potassium Chloride $10^3 c_2/\text{mol dm}^{-3}$	Calcium Iodate $10^3 c_1/\text{mol dm}^{-3}$	Density $\rho/\text{g cm}^{-3}$
25	0	7.840	0.9998
	12.5	8.312	1.0008
	25.0	8.730	1.0017
	50.0	9.387	1.0032
	100.0	10.42	1.0057

COMMENTS AND/OR ADDITIONAL DATA:

The conductivity data at 18°C were used to evaluate the thermodynamic ion pair dissociation constant, K_D°; it was found that $K_D^\circ = 0.13$. Using this value for the ion pair dissociation constant, the concentration of the ion pair $CaIO_3^+$ was calculated from

$$\log [CaIO_3^+] = \log [Ca^{2+}][IO_3^-] - \log K_D^\circ - 2.02I^{1/2} + 2.0I$$

where I is the ionic strength. Utilizing this relation to compute the ionic concentrations of Ca^{2+} and IO_3^-, the authors plotted (1/3) log $[Ca^{2+}] \times [IO_3^-]$ against the ionic strength and extrapolated to zero ionic strength to obtain the thermodynamic solubility product constant. The result of this extrapolation is $K_{s0}^\circ = 6.953 \times 10^{-7}$.

AUXILIARY INFORMATION

METHOD/APPARATUS/PROCEDURE:

Saturating column method as in (1) and modified as in (2). A bulb containing the solvent solution is attached to a column containing the slightly soluble salt, and the solvent is allowed to flow through the column at a rate sufficient to insure saturation (1). The modification (2) consisted of connecting the column by capillary tubing to a second parallel arm in which the saturated solution collected. The entire apparatus was placed in a thermostat. Weighed samples of the satd slns were taken for analysis by method described in (3): i.e. the satd slns added to acidified KI sln and the liberated I_2 titrd by weight against an approx 0.15N thiosulfate sln, 0.01N iodine sln being used for the back titrn. The densities of the satd slns were measured at 25°C, and the molar conductivities at 18°C for the binary $Ca(IO_3)_2$-H_2O system are also reported.

SOURCE AND PURITY OF MATERIALS:

$Ca(IO_3)_2 \cdot 6H_2O$ was prepared by dropwise addition of solutions of KIO_3 and $CaCl_2$ in equivalent amounts to a large volume of conductivity water. The precipitate was washed first by decantation and then in the solubility columns until a constant solubility was obtained.

ESTIMATED ERROR:

Soly: not specified, but reproducibility probably around ± 0.3% as in ref. (3).
Temp: ± 0.01°K (authors)

REFERENCES:

1. Brönsted, J. N.; La Mer, V. K. *J. Am. Chem. Soc.* 1924, *46*, 555.
2. Money, R. W.; Davies, C. W. *J. Chem. Soc.* 1934, 400.
3. Macdougall, G.; Davies, C. W. *J. Chem. Soc.* 1935, 1416.

COMPONENTS:

(1) Calcium iodate; $Ca(IO_3)_2$; [7789-80-2]

(2) Potassium chloride; KCl; [7447-40-7]

(3) Water; H_2O; [7732-18-5]

ORIGINAL MEASUREMENTS:

Keefer, R. M.; Reiber, H. G.; Bisson, C. S.

J. Am. Chem. Soc. 1940, *62*, 2951-5.

VARIABLES:

$T/K = 298$

$m_2/mol\ kg^{-1} = 0$ to 0.1008

PREPARED BY:

Hiroshi Miyamoto

EXPERIMENTAL VALUES:

t/°C	Potassium Chloride $m_2/mol\ kg^{-1}$	Calcium Iodate $10^3 m_1/mol\ kg^{-1}$
25	0	7.86
	0.02514	8.85
	0.05036	9.70
	0.1008	10.53

AUXILIARY INFORMATION

METHOD/APPARATUS/PROCEDURE:

KCl solutions were prepared from distilled water using calibrated volumetric equipment. An excess of air-dried calcium iodate was placed in a glass-stoppered Pyrex flask and 200 ml of KCl solution added. The flasks were rotated in a thermostat for at least 12 hours. Equilibrium was obtained in 4-5 hours.

The saturated solutions were analyzed iodometrically. Analyses and solubility measurements were done in duplicate. Densities of all solutions were determined, but the data were not given in the original paper.

SOURCE AND PURITY OF MATERIALS:

$Ca(IO_3)_2$ was prepared by dropwise addition of 1.0 mol dm^{-3} $CaCl_2$ solution to 2 dm^3 of 0.38 mol dm^{-3} KIO_3 solution. The mixture was stirred, the precipitate filtered, washed, and then dried at room temperature. The number of hydrated waters was not given. C.p. grade KCl was crystallized from water, and air dried at 180°C.

ESTIMATED ERROR:

Soly: nothing specified
Temp: ± 0.02°C (authors)

REFERENCES:

COMPONENTS:

(1) Calcium iodate; $Ca(IO_3)_2$; [7789-80-2]

(2) Potassium chloride; KCl; [7447-40-7]

(3) Water; H_2O; [7732-18-5]

ORIGINAL MEASUREMENTS:

Rens, G.

Sucr. Belge 1958, *77*, 193-208.

VARIABLES:

$T/K = 293$

$10^3 c_2/\text{mol dm}^{-3} = 0.000$ to 190.02

PREPARED BY:

Hiroshi Miyamoto

EXPERIMENTAL VALUES:

t/°C	Potassium Chloride $10^3 c_2/\text{mol dm}^{-3}$	1/2 Iodate Ion $10^3 (c/2)/\text{mol dm}^{-3}$	Calcium Ion $10^3 c/\text{mol dm}^{-3}$	Calcium Iodate $10^3 c_1/\text{mol dm}^{-3}$ [a]
20	0.000	6.231	6.230	6.231
	8.002	6.554	6.563	6.559
	50.02	7.635	7.630	7.633
	120.00	8.782	8.785	8.784
	190.02	9.651	9.652	9.652

[a] Average value calculated by the compiler

pK°_{s0} was calculated from the equation:

1) $I < 0.07$ mol dm^{-3}, $pK_{s0} = pK^\circ_{s0} - 6AI^{1/2} + 3CI$

where I is the ionic strength, and $A = 0.5046$ mol$^{-1/2}$ dm$^{3/2}$ (ref 1)

2) $0.07 < I < 0.25$ mol dm^{-3}, $pK_{s0} = pK^\circ_{s0} - 6AI^{1/2}/(1 + aBI^{1/2})$

where a is distance of closest approach.

For $a = 3.4$A°, the value of K°_{s0} was 4.159×10^{-7} mol^3 dm^{-9}.

AUXILIARY INFORMATION

METHOD/APPARATUS/PROCEDURE:

Excess $Ca(IO_3)_2 \cdot 6H_2O$ and aqueous KCl solution were placed in sealed Erlenmeyer flasks. The flasks were rotated in a thermostat for 24 hours. Aliquots of saturated solution were filtered.
The concentration of iodate was determined iodometrically. The calcium content was determined by chelatometric titration using Eriochrome black T as an indicator.

SOURCE AND PURITY OF MATERIALS:

$Ca(IO_3)_2 \cdot 6H_2O$ was prepared by slowly adding the solution of KIO_3 (about 50g dm^{-3}) to an equivalent solution of $CaCl_2 \cdot 2H_2O$ at 20°C. The precipitate was washed by decantation, and was air-dried at room temperature. An analysis of the product gave the following values: IO_3 99.64% and Ca 99.95% of theoretical.

ESTIMATED ERROR:

Soly: nothing specified
Temp: ± 0.1°K (author)

REFERENCES:

1. Harned, H.; Owen, B. *The Physical Chemistry of Electrolytic Solutions.* Reinhold, New York, 1950, 447.

COMPONENTS:

(1) Calcium iodate; $Ca(IO_3)_2$; [7789-80-2]

(2) Potassium chloride; KCl; [7447-40-7]

(3) Glycine; $C_2H_5NO_2$; [56-40-6]

(4) Water; H_2O; [7732-18-5]

ORIGINAL MEASUREMENTS:

Keefer, R. M.; Reiber, H. G.; Bisson, C. B.

J. Am. Chem. Soc. 1940, *62*, 2951-5.

VARIABLES: $T/K = 298$

$m_2/mol\ kg^{-1} = 0.02516$ to 0.1011

$m_3/mol\ kg^{-1} = 0.02511$ to 0.07570

PREPARED BY:

Hiroshi Miyamoto

EXPERIMENTAL VALUES:

t/°C	Potassium Chloride $m_2/mol\ kg^{-1}$	Glycine $m_3/mol\ kg^{-1}$	Calcium Iodate $10^3m_1/mol\ kg^{-1}$
25	0.02516	0.02511	8.99
	0.02519	0.05030	9.20
	0.02522	0.07552	9.39
	0.02525	0.1008	9.59
	0.05036	0.02513	9.70
	0.05041	0.05032	9.90
	0.05047	0.07558	10.06
	0.05053	0.1009	10.27
	0.1009	0.02517	10.73
	0.1010	0.05041	10.93
	0.1011	0.07570	11.15

AUXILIARY INFORMATION

METHOD/APPARATUS/PROCEDURE:

KCl and glycine solutions were prepared from distilled water using calibrated volumetric equipment. An excess of air-dried calcium iodate was placed in a glass-stoppered Pyrex flask and 200 ml of glycine solution containing KCl added. The flasks were rotated in a thermostat for at least 12 hours. Equilibrium was obtained in 4-5 hours.

The saturated solutions were analyzed iodometrically. Analyses and solubility measurements were done in duplicate. Densities of all solutions were determined, but the data were not given in the original paper.

SOURCE AND PURITY OF MATERIALS:

$Ca(IO_3)_2$ was prepared by dropwise addition of 1.0 mol dm^{-3} $CaCl_2$ solution in 2 dm^3 of 0.38 mol dm^{-3} KIO_3 solution. The mixture was stirred, the precipitate filtered, washed, and then dried at room temperature. The number of hydrated waters was not given. C.p. grade KCl was recrystallized from water, and air-dried at 180°C. C.p. grade glycine was recrystallized twice from water by addition of EtOH. The product was dried in a vacuum oven at about 35°C.

ESTIMATED ERROR:

Soly: nothing specified
Temp: ± 0.02°C (authors)

COMPONENTS:

(1) Calcium iodate; $Ca(IO_3)_2$; [7789-80-2]

(2) Potassium bromide; KBr; [7758-02-3]

(3) Water; H_2O; [7732-18-5]

ORIGINAL MEASUREMENTS:

Rens, G.

Sucr. Belge 1958, *77*, 193-208.

VARIABLES: $T/K = 293$

$10^3c_2/\text{mol dm}^{-3} = 0.000$ to 184.94

PREPARED BY:

Hiroshi Miyamoto

EXPERIMENTAL VALUES:

t/°C	Potassium Bromide $10^3c_2/\text{mol dm}^{-3}$	1/2 Iodate Ion $10^3(c/2)/\text{mol dm}^{-3}$	Calcium Ion $10^3c/\text{mol dm}^{-3}$	Calcium Iodate $10^3c_1/\text{mol dm}^{-3}$ [a]
20	0.000	6.231	6.230	6.231
	6.999	6.456	6.470	6.463
	45.01	7.472	7.477	7.475
	114.91	8.623	8.620	8.622
	184.94	9.496	9.510	9.503

[a] Average value calculated by the compiler

pK°_{s0} was calculated from the equation:

1) $I < 0.07$ mol dm^{-3}, $pK_{s0} = pK^\circ_{s0} - 6AI^{1/2} + 3CI$

where I is the ionic strength, and $A = 0.5046$ mol$^{-1/2}$ dm$^{3/2}$ (ref 1)

2) $0.07 < I < 0.25$ mol dm^{-3}, $pK_{s0} = pK^\circ_{s0} - 6AI^{1/2}/(1 + aBI^{1/2})$

where a is distance of closest approach.

For $a = 3.5$, the value of K°_{s0} was 4.159×10^{-7} mol^3 dm^{-9}.

AUXILIARY INFORMATION

METHOD/APPARATUS/PROCEDURE:

Excess $Ca(IO_3)_2 \cdot 6H_2O$ and aqueous KBr solution were placed in sealed Erlenmeyer flasks. The flasks were rotated in a thermostat for 24 hours. Aliquots of saturated solution were filtered.
The concentration of iodate was determined iodometrically. The calcium content was determined by chelatometric titration using Eriochrome black T as an indicator.

SOURCE AND PURITY OF MATERIALS:

$Ca(IO_3)_2 \cdot 6H_2O$ was prepared by slowly adding the solution of KIO_3 (about 50g dm^{-3}) to an equivalent solution of $CaCl_2 \cdot 2H_2O$ at 20°C. The precipitate was washed by decantation, and was air-dried at room temperature. An analysis of the product gave the following values: IO_3 99.64% and Ca 99.95% of theoretical.

ESTIMATED ERROR:

Soly: nothing specified
Temp: ± 0.1°K (author)

REFERENCES:

1. Harned, H.; Owen, B. *The Physical Chemistry of Electrolytic Solutions.* Reinhold, New York, 1950, 447.

COMPONENTS:

(1) Calcium iodate; $Ca(IO_3)_2$; [7789-80-2]

(2) Potassium iodide; KI; [7681-11-0]

(3) Water; H_2O; [7732-18-5]

ORIGINAL MEASUREMENTS:

Rens, G.

Sucr. Belge 1958, *77*, 193-208.

VARIABLES:

$T/K = 293$

$10^3 c_2/mol\ dm^{-3} = 0.000$ to 190.17

PREPARED BY:

Hiroshi Miyamoto

EXPERIMENTAL VALUES:

t/°C	Potassium Iodide $10^3 c_2/mol\ dm^{-3}$	1/2 Iodate Ion $10^3 (c/2)/mol\ dm^{-3}$	Calcium Ion $10^3 c/mol\ dm^{-3}$	Calcium Iodate $10^3 c_1/mol\ dm^{-3}$ [a]
20	0.000	6.231	6.230	6.231
	9.002	6.538	6.546	6.542
	47.00	7.474	7.479	7.477
	119.92	8.623	8.621	8.622
	190.17	9.654	9.679	9.667

[a] Average value calculated by the compiler

pK°_{s0} was calculated from the equation:

1) $I < 0.07\ mol\ dm^{-3}$, $pK_{s0} = pK^\circ_{s0} - 6AI^{1/2} + 3CI$

where I is the ionic strength, and $A = 0.5046\ mol^{-1/2}\ dm^{3/2}$ (ref 1)

2) $0.07 < I < 0.25\ mol\ dm^{-3}$, $pK_{s0} = pK^\circ_{s0} - 6AI^{1/2}/(1 + aBI^{1/2})$

where a is distance of closest approach.

For a = 3.5A°, the value of K°_{s0} was $4.159 \times 10^{-7}\ mol^3\ dm^{-9}$.

AUXILIARY INFORMATION

METHOD/APPARATUS/PROCEDURE:

Excess $Ca(IO_3)_2 \cdot 6H_2O$ and aqueous KI solution were placed in sealed Erlemeyer flasks. The flasks were rotated in a thermostat at 24 hours. Aliquots of saturated solution were filtered.
The concentration of iodate was determined iodometrically. The calcium content was determined by chelatometric titration using Eriochrome black T as an indicator.

SOURCE AND PURITY OF MATERIALS:

$Ca(IO_3)_2 \cdot 6H_2O$ was prepared by slowly adding the solution of KIO_3 (about $50g\ dm^{-3}$) to an equivalent solution of $CaCl_2 \cdot 6H_2O$ at 20°C. The precipitate was washed by decantation, and was air-dried at room temperature. An analysis of the product gave the following values: IO_3 99.64% and Ca 99.95% of theoretical.

ESTIMATED ERROR:

Soly: nothing specified
Temp: ± 0.1°K (author)

REFERENCES:

1. Harned, H.; Owen, B. *The Physical Chemistry of Electrolytic Solutions.* Reinhold, New York, 1950, 447.

COMPONENTS:

(1) Calcium iodate; $Ca(IO_3)_2$; [7789-80-2]

(2) Potassium iodate; KIO_3; [7758-05-6]

(3) Water; H_2O; [7732-18-5]

ORIGINAL MEASUREMENTS:

Kilde, G.

Z. Anorg. Allg. Chem. 1934, *218*, 113-28.

VARIABLES:

T/K = 291, 298, and 303

c_2/mol dm^{-3} = 0 to 0.0167

PREPARED BY:

Hiroshi Miyamoto

EXPERIMENTAL VALUES:

t/°C	Potassium iodate c_2/mol dm^{-3}	Calcium Iodate $10^3 c_1$/mol dm^{-3}	$10^6 K_{s0}$/mol^3 dm^{-9}
18	0	5.69	0.737
	0.0167	2.03	0.875
25	0	7.84	1.93
	0.0167	3.75	2.20
30	0	9.91	3.89
	0.0167	5.37	4.04

Concentration solubility product $K_{s0} = [Ca^{+}][IO_3^-]^2$

K°_{s0} was calculated from the equation

$$1/3 \log K^\circ_{s0} = 1/3 \log K_{s0} - Z_1 Z_2 A I^{1/2} + BI$$

where $Z_1 Z_2 A$ = 0.998 at 18°C, 1.008 at 25°C and 1.018 at 30°C, and I is the ionic strength.

The values obtained were the following: $K^\circ_{s0} = 0.329 \times 10^{-6}$ at 18°C, 0.736×10^{-6} at 25°C and 1.35×10^{-6} at 30°C.

AUXILIARY INFORMATION

METHOD/APPARATUS/PROCEDURE:

An excess of calcium iodate hexahydrate was shaken with aqueous KIO_3 solutions for at least 24 hours in a thermostat at the desired temperature. Aliquots of saturated solutions were filtered through cotton wool, and the iodate content was determined iodometrically.

SOURCE AND PURITY OF MATERIALS:

Calcium iodate hexahydrate was prepared by mixing calcium chloride solution and KIO_3 solution. The precipitate was washed and dried at room temperature. Reagent grade KIO_3 was used.

ESTIMATED ERROR:

Soly: precision within 1 %

Temp: nothing specified

REFERENCES:

COMPONENTS:

(1) Calcium iodate; $Ca(IO_3)_2$; [7789-80-2]

(2) Potassium iodate; KIO_3; [7758-05-6]

(3) Glycine; $C_2H_5NO_2$; [56-40-6]

(4) Water; H_2O; [7732-18-5]

ORIGINAL MEASUREMENTS:

Monk, C. B.

Trans. Faraday Soc. 1951, *47*, 1233-40.

VARIABLES:

$T/K = 298$

$10^3 c_3/\text{mol dm}^{-3} = 0$ to 200.9

PREPARED BY:

Hiroshi Miyamoto

EXPERIMENTAL VALUES:

t/°C	Glycine $10^3 c_3/\text{mol dm}^{-3}$	Potassium Iodate $10^3 c_2/\text{mol dm}^{-3}$	Calcium Iodate $10^3 c_1/\text{mol dm}^{-3}$	$-\log K^\circ_{s0}$
25	0	0	7.84	6.148
	50.3	0	8.20	6.090
	50.3	1.42	7.74	6.090
	50.3	2.83	7.29	6.089
	100.9	0	8.58	6.032
	100.9	4.38	7.20	6.032
	100.9	5.66	6.84	6.030
	200.9	0	9.32	5.927
	200.9	2.19	8.60	5.928
	200.9	5.47	7.59	5.928

COMMENTS AND/OR ADDITIONAL DATA:

Thermodynamic solubility product constant, K°_{s0}, was calculated from

$$\log K^\circ_{s0} = \log [Ca^{2+}][IO_3^-]^2 - 3F$$

where $F = (78.54/\varepsilon)^{3/2}[I^{1/2}/(1 + I^{1/2}) - 0.2I]$, ε = dielectric constant, I = ionic strength.
In solving for $[Ca^{2+}]$ and $[IO_3^-]$, the author made allowance for ion pair formation, i.e. $K^\circ_D(CaIO_3^+) = 0.13(1)$ and $K^\circ_D(KIO_3) = 2.0(2)$.

AUXILIARY INFORMATION

METHOD/APPARATUS/PROCEDURE:

The saturating column method was employed similar to that reported in ref (1).
The analyses were effected by withdrawing $100\ cm^3$ of the saturated solution in a calibrated pipet, and running the solution into an acidic KI solution. The liberated iodine was titrated by weight against 0.15 mol dm^{-3} $Na_2S_2O_3$ solution, 0.005 mol dm^{-3} iodine solution being used for the back titration.

SOURCE AND PURITY OF MATERIALS:

$Ca(IO_3)_2 \cdot 6H_2O$ was prepared by dropwise addition of solution of KIO_3 and $CaCl_2$ in equivalent amounts to conductivity water as in ref (1). Glycine (A.R.) was dried to constant weight in a vacuum oven at 80°C.

ESTIMATED ERROR:

Soly: nothing specified
Temp: ± 0.03°C (author)

REFERENCES:

1. Macdougall, G.; Davies, C. W. *J. Chem. Soc.* 1935, 1416.

2. Davies, C. W. *J. Chem. Soc.* 1930, 2410.

COMPONENTS:

(1) Calcium iodate; $Ca(IO_3)_2$; [7789-80-2]
(2) Potassium iodate; KIO_3; [7758-05-6]
(3) Alanine; $C_3H_7NO_2$; [302-72-7]
(4) Water; H_2O; [7732-18-5]

ORIGINAL MEASUREMENTS:

Monk, C. B.

Trans. Faraday Soc. 1951, *47*, 1233-40.

VARIABLES:

$T/K = 298$
$10^3c_3/\text{mol dm}^{-3} = 0$ to 196.3

PREPARED BY:

Hiroshi Miyamoto

EXPERIMENTAL VALUES:

t/°C	Alanine $10^3c_3/\text{mol dm}^{-3}$	Potassium Iodate $10^3c_2/\text{mol dm}^{-3}$	Calcium Iodate $10^3c_1/\text{mol dm}^{-3}$	$-\log K^\circ_{s0}$
25	0	--	7.84	6.148
	68.3	--	8.21	6.087
	99.9	--	8.38	6.059
	196.3	--	8.86	5.982
	196.3	2.14	8.21	5.977

COMMENTS AND/OR ADDITIONAL DATA:

Thermodynamic solubility product constant, K°_{s0}, was calculated from

$$\log K^\circ_{s0} = \log [Ca^{2+}][IO_3^-]^2 - 3F$$

where $F = (78.54/\varepsilon)^{3/2}[I^{1/2}/(1 + I^{1/2}) - 0.2I]$, ε = dielectric constant, I = ionic strength.
In solving for $[Ca^{2+}]$ and $[IO_3^-]$, the author made allowance for ion pair formation, i.e. $K^\circ_D(CaIO_3^+) = 0.13(1)$ and $K^\circ_D(KIO_3) = 2.0(2)$.

AUXILIARY INFORMATION

METHOD/APPARATUS/PROCEDURE:

The saturating column method was employed similar to that reported in ref (1).
The analyses were effected by withdrawing $100 cm^3$ of the saturated solution in a calibrated pipet, and running the solution into an acidic KI solution. The liberated iodine was titrated by weight against 0.15 mol dm^{-3} $Na_2S_2O_3$ solution, 0.005 mol dm^{-3} iodine solution being used for the back titration.

SOURCE AND PURITY OF MATERIALS:

$Ca(IO_3)_2 \cdot 6H_2O$ was prepared by dropwise addition of solutions of KIO_3 and $CaCl_2$ in equivalent amounts to conductivity water as in ref (1). Laboratory grade alanine was recrystallized from aqueous alcohol. The acid was dried to constant weight in a vacuum oven at 80°C.

ESTIMATED ERROR:

Soly: nothing specified
Temp: ± 0.03°C (author)

REFERENCES:

1. Macdougall, G.; Davies, C. W. *J. Chem. Soc.* 1935, 1416.

2. Davies, C. W. *J. Chem. Soc.* 1930, 2410.

COMPONENTS:

(1) Calcium iodate; $Ca(IO_3)_2$; [7789-80-2]

(2) Potassium iodate; KIO_3; [7758-05-6]

(3) Glycyl glycine; $C_4H_8N_2O_3$; [556-50-3]

(4) Water; H_2O; [7732-18-5]

ORIGINAL MEASUREMENTS:

Monk, C. B.

Trans. Faraday Soc. 1951, *47*, 1233-40.

VARIABLES:

T/K = 298

$10^3 c_3$/mol dm^{-3} = 0 to 91.26

PREPARED BY:

Hiroshi Miyamoto

EXPERIMENTAL VALUES:

t/°C	Glycyl Glycine $10^3 c_3$/mol dm^{-3}	Potassium Iodate $10^3 c_2$/mol dm^{-3}	Calcium Iodate $10^3 c_1$/mol dm^{-3}	$-\log K^\circ_{s0}$
25	0	--	7.84	6.148
	34.40	--	8.23	6.078
	65.17	--	8.62	6.013
	82.53	--	8.88	5.972
	91.26	--	8.98	5.957
	82.53	2.76	7.98	5.973

COMMENTS AND/OR ADDITIONAL DATA:

Thermodynamic solubility product constant, K°_{s0}, was calculated from

$$\log K^\circ_{s0} = \log [Ca^{2+}][IO_3^-]^2 - 3F$$

where $F = (78.54/\varepsilon)^{3/2}[I^{1/2}/(1 + I^{1/2}) - 0.2I]$, ε = dielectric constant, I = ionic strength.
In solving for $[Ca^{2+}]$ and $[IO_3^-]$, the author made allowance for ion pair formation, i.e. $K^\circ_D(CaIO_3^+) = 0.13(1)$ and $K^\circ_D(KIO_3) = 2.0(2)$.

AUXILIARY INFORMATION

METHOD/APPARATUS/PROCEDURE:

The saturating column method was employed similar to that reported in ref (1).
The analyses were effected by withdrawing 100cm^3 of the saturated solution in a calibrated pipet, and running the solution into an acidic KI solution. The liberated iodine was titrated by weight against 0.15 mol dm^{-3} $Na_2S_2O_3$ solution, 0.005 mol dm^{-3} iodine solution being used for the back titration.

SOURCE AND PURITY OF MATERIALS:

$Ca(IO_3)_2 \cdot 6H_2O$ was prepared by dropwise addition of solutions of KIO_3 and $CaCl_2$ in equivalent amounts to conductivity water as in ref (1).
Glycyl glycine (Roche Products Chemicals) was "certified pure," and dried to constant weight at 80°C in vacuum.

ESTIMATED ERROR:

Soly: nothing specified
Temp: ± 0.03°C (author)

REFERENCES:

1. Macdougall, G.; Davies, C. W. *J. Chem. Soc.* 1935, 1416.

2. Davies, C. W. *J. Chem. Soc.* 1930, 2410.

COMPONENTS:

(1) Calcium iodate; $Ca(IO_3)_2$; [7789-80-2]

(2) Potassium sulfate; K_2SO_4; [7778-80-5]

(3) Water; H_2O; [7732-18-5]

ORIGINAL MEASUREMENTS:

Bell, R. P.; George, J. H. B.

Trans. Faraday Soc. 1953, *49*, 619-27.

VARIABLES:

T/K=273; 10^3c_2/mol dm^{-3}= 0 to 23.71
T/K=298; 10^3c_2/mol dm^{-3}=0 to 20.12
T/K=313; 10^3c_2/mol dm^{-3}=0 to 14.19

PREPARED BY:

Hiroshi Miyamoto
Mark Salomon

EXPERIMENTAL VALUES:

t/°C	0		25		40	
	Potassium Sulfate 10^3c_2/mol dm^{-3}	Calcium Iodate 10^3c_1/mol dm^{-3}	Potassium Sulfate 10^3c_2/mol dm^{-3}	Calcium Iodate 10^3c_1/mol dm^{-3}	Potassium Sulfate 10^3c_2/mol dm^{-3}	Calcium Iodate 10^3c_1/mol dm^{-3}
	0	2.315	0	7.838	0	13.06
	3.33	2.620	3.41	8.439	3.05	13.74
	5.85	2.821	5.14	8.744	5.96	14.38
	9.67	3.026	8.16	9.184	8.88	14.91
	12.78	3.192	10.70	9.552	11.83	15.48
	15.75	3.328	12.59	9.772	14.09	15.81
	20.53	3.471	15.08	10.077		
	23.71	3.557	20.12	10.580		

Solid phase is the hexahydrate at 0°C and 25°C and the monohydrate at 40°C.

Dissociation constant K_D^o for $CaSO_4$ detd to be 0.0060, 0.0049, and 0.0041 at 0°, 25° and 40°C, respectively. In these calculations, the authors used the 25°C data of $K_D^o(KSO_4^-)$ = 0.15 (1), and K_D^o = 2.0 (2): to estimate the latter K_D^o values at 0° and 40°C, authors assumed their temp dependence is given by the electrostatic theory of Bjerrum (3). Combining these equilibrium constants with the experimental soly data, the authors calculated the concns of Ca^{2+} and IO_3^- and computed the thermodynamic solubility products at the three temperatures:

K_{s0}^o = 2.859 x 10^{-8} at 0°C; K_{s0}^o = 7.119 x 10^{-7} at 25°C;

K_{s0}^o = 2.437 x 10^{-6} at 40°C

AUXILIARY INFORMATION

METHOD/APPARATUS/PROCEDURE:

The solubilities were detd by passing about 50 ml of solution through a column of ground solid iodate (8 cm deep and 0.5 cm^2 cross-section) similar to that described in (4). The time of passage required to reach equil for the 50 ml of sln varied from 6 h at 0°C to 1.5 h at 25° and 40°C. The slns were brought to the thermostat temp before passage through the column, and repeated tests showed that complete saturation was reached after one passage. The iodate concns were detd iodometrically. The titrn was carried out by weight except that the final addition of around 0.3 ml of thiosulfate sln was made volumetrically with a microburet. Four independent analyses were made with each sln, and the average spread was about 0.3% at 0°C and 0.1% at 25° and 40°C.

SOURCE AND PURITY OF MATERIALS:

Calcium iodate was prepared by dropping slns of A.R. grade KIO_3 and $CaCl_2$ into a large volume of water at 40°C. At this temperature, the solid is the monohydrate, $Ca(IO_3)_2 \cdot H_2O$. Conductivity water was used in all preparations and solubility measurements.

ESTIMATED ERROR:

Reproducibility in soly probably equal to or slightly poorer than the reproducibility in the analysis of IO_3^-. Error in temp not specified but could add ∿0.1% to above error.

REFERENCES:

1. Righellato, D. C.; Davies, C. W. *Trans. Faraday Soc.* 1930, *26*, 292.
2. Davies, C. W. *Trans. Faradau Soc.* 1927, *23*, 592.
3. Bjerrum, J. *Kgl. Danske vid. Selsk. Math-fys. Medd.* 1926, *7*, 9.
4. Davies, C. W. *J. Chem. Soc.* 1938, 277.

COMPONENTS:

(1) Calcium iodate; $Ca(IO_3)_2$; [7789-80-2]

(2) Potassium sulfate; K_2SO_4; [7778-80-5]

(3) Water; H_2O; [7732-18-5]

ORIGINAL MEASUREMENTS:

Chloupek, J. B.; Daneš, VL. Z.; Danešova, B. A.

Collect. Czech. Chem. Commun. 1933, *5*, 339-42.

VARIABLES:

T/K = 298

m_2/mol kg^{-1} = 0 to 0.1

PREPARED BY:

Hiroshi Miyamoto

EXPERIMENTAL VALUES:

Potassium Sulfate	Calcium Iodate	
m_2/mol kg^{-1}	g kg^{-1} (H_2O)	$10^3 m_1$/mol kg^{-1} [a]
0	3.031	7.774
0.002	3.196	8.197
0.005	3.389	8.692
0.01	3.666	9.403
0.02	4.108	10.54
0.05	5.077	13.02
0.1	6.290	16.13

[a] Compiler calculations using 1977 IUPAC recommended atomic weights.

AUXILIARY INFORMATION

METHOD/APPARATUS/PROCEDURE:

The solutions with the solid calcium iodate were rotated in a thermostat for at least 12 hours. Aliquots of saturated solutions were withdrawn with a filtering pipet equipped with a sintered glass disc. The analysis of the solution was carried out by iodometric titration.

SOURCE AND PURITY OF MATERIALS:

$Ca(IO_3)_2$ was prepared by adding a cold solution of KIO_3 to a solution of $CaCl_2$. $CaCl_2$ was obtained by dissolving $CaCO_3$ (Merk pro an) in HCl (c.p. grade). The precipitate was separated from the mother liquor by suction filtering, suspended 3 times in distilled cold water and left 48 hours. Then it was decanted, washed and dried. Calcd for $Ca(IO_3)_2$: CaO 14.38; I_2O_5 85.62. Found: CaO 14.28; I_2O_5 85.04.

ESTIMATED ERROR:

Soly: the mean deviation is ± 0.303%.
Temp: ± 0.005°C (authors)

COMPONENTS:

(1) Calcium iodate; $Ca(IO_3)_2$; [7789-80-2]

(2) Potassium nitrate; KNO_3; [7757-79-1]

(3) Water; H_2O; [7732-18-5]

ORIGINAL MEASUREMENTS:

Chloupek, J. B.; Daneš, VL. Z.; Danešova, B. A.

Collect. Czech. Chem. Commun. 1933, *5*, 339-42.

VARIABLES:

$T/K = 298$

$m_2/\text{mol kg}^{-1} = 0$ to 0.5

PREPARED BY:

Hiroshi Miyamoto

EXPERIMENTAL VALUES:

Potassium Nitrate $m_2/\text{mol kg}^{-1}$	Calcium Iodate g $\text{kg}^{-1}(H_2O)$	Calcium Iodate $10^3m_1/\text{mol kg}^{-1}$ [a]
0	3.031	7.774
0.005	3.125	8.015
0.01	3.207	8.225
0.02	3.393	8.703
0.05	3.695	9.477
0.1	4.102	10.52
0.2	4.677	12.00
0.5	5.861	15.03

[a] Compiler calculations using 1977 IUPAC recommended atomic weight.

AUXILIARY INFORMATION

METHOD/APPARATUS/PROCEDURE:

The solutions with the solid calcium iodate were rotated in a thermostat for at least 12 hours. Aliquots of saturated solutions were withdrawn with a filtering pipet equipped with a sintered glass disc. The analysis of the solution was carried out by iodometric titration.

SOURCE AND PURITY OF MATERIALS:

$Ca(IO_3)_2$ was prepared by adding a cold solution of KIO_3 to a solution of $CaCl_2$. $CaCl_2$ was obtained by dissolving $CaCO_3$ (Merk pro an) in HCl (c.p. grade). The precipitate was separated from the mother liquor by suction filtering, suspended 3 times in distilled cold water and left 48 hours. Then it was decanted, washed and dried. Calcd for $Ca(IO_3)_2$: CaO 14.38; I_2O_5 85.62. Found: CaO 14.28; I_2O_5 85.04.

ESTIMATED ERROR:

Soly: the mean deviation is ± 0.303%.
Temp: ± 0.005°C (authors)

COMPONENTS:

(1) Calcium iodate; $Ca(IO_3)_2$; [7789-80-2]

(2) Potassium hexacyanoferrate (II); $K_4[Fe(CN)_6]$; [13943-58-3]

(3) Water; H_2O; [7732-18-5]

ORIGINAL MEASUREMENTS:

Wise, W. C. A.; Davies, C. W.

J. Chem. Soc. 1938, 273-7.

VARIABLES:

T/K = 298

$10^3 c_2$/mol dm^{-3} = 0 to 5.0

PREPARED BY:

Hiroshi Miyamoto
Mark Salomon

EXPERIMENTAL VALUES:

t/°C	Potassium Hexacyanoferrate $10^3 c_2$/mol dm^{-3}	Calcium Iodate $10^3 c_1$/mol dm^{-3}	Density ρ/g cm^{-3}
25	0	7.840	0.9998
	1.25	8.377	1.0007
	2.5	8.839	1.0013
	3.75	9.430	1.0018
	5.0	9.804	1.0023

COMMENTS AND/OR ADDITIONAL DATA:

The conductivity data at 18°C were used to evaluate the thermodynamic ion pair dissociation constant, K_D°; it was found that K_D° = 0.13. Using this value for the ion pair dissociation constant, the concentration of the ion pair $CaIO_3^+$ was calculated from

$$\log [CaIO_3^+] = \log [Ca^{2+}][IO_3^-] - \log K_D^\circ - 2.02I^{1/2} + 2.0I$$

where I is the ionic strength. Utilizing this relation to compute the ionic concentrations of Ca^{2+} and IO_3^-, the authors plotted (1/3) log $[Ca^{2+}]$x $[IO_3^-]$ against the ionic strength and extrapolated to zero ionic strength to obtain the thermodynamic solubility product constant. The result of this extrapolation is $K_{s0}^\circ = 6.953 \times 10^{-7}$.

AUXILIARY INFORMATION

METHOD/APPARATUS/PROCEDURE:

Saturating column method as in (1) and modified as in (2). A bulb containing the solvent solution is attached to a column containing the slightly soluble salt, and the solvent is allowed to flow through the column at a rate sufficient to insure saturation (1). The modification (2) consisted of connecting the column by capillary tubing to a second parallel arm in which the saturated solution collected. The entire apparatus was placed in a thermostat. Weighed samples of the satd slns were taken for analysis by method described in (3): i.e. the satd slns added to acidified KI sln and the liberated I_2 titrd by weight against an approx 0.15N thiosulfate sln, 0.01N iodine sln being used for the back titrn. The densities of the satd slns were measured at 25°C, and the molar conductivities at 18°C for the binary $Ca(IO_3)_2$-H_2O system are also reported.

SOURCE AND PURITY OF MATERIALS:

$Ca(IO_3)_2 \cdot 6H_2O$ was prepared by dropwise addition of solutions of KIO_3 and $CaCl_2$ in equivalent amounts to a large volume of conductivity water. The precipitate was washed first by decantation and then in the solubility columns until a constant solubility was obtained.

ESTIMATED ERROR:

Soly: not specified, but reproducibility probably around ± 0.3% as in ref. (3).
Temp: ± 0.01°K (authors)

REFERENCES:

1. Brönsted, J. N.; La Mer, V. K. *J. Am. Chem. Soc.* 1924, *46*, 555.
2. Money, R. W.; Davies, C. W. *J. Chem. Soc.* 1934, 400.
3. Macdougall, G.; Davies, C. W. *J. Chem. Soc.* 1935, 1416.

COMPONENTS:

(1) Calcium iodate; $Ca(IO_3)_2$; [7789-80-2]

(2) Magnesium chloride; $MgCl_2$; [7786-30-3]

(3) Water; H_2O; [7732-18-5]

ORIGINAL MEASUREMENTS:

Chloupek, J. B.; Daneš, VL. Z.; Danešova, B. A.

Collect. Czech. Chem. Commun. 1933, *5*, 339-42.

VARIABLES:

T/K = 298

m_2/mol kg^{-1} = 0 to 0.1

PREPARED BY:

Hiroshi Miyamoto

EXPERIMENTAL VALUES:

Magnesium Chloride m_2/mol kg^{-1}	Calcium Iodate g kg^{-1} (H_2O)	$10^3 m_1$/mol kg^{-1} [a]
0	3.031	7.774
0.002	3.121	8.005
0.005	3.262	8.367
0.01	3.444	8.833
0.02	3.736	9.582
0.05	4.331	11.11
0.1	4.924	12.63

[a] Compiler calculations using 1977 IUPAC recommended atomic weights.

AUXILIARY INFORMATION

METHOD/APPARATUS/PROCEDURE:

The solutions with the solid calcium iodate were rotated in a thermostat for at least 12 hours. Aliquots of saturated solutions were withdrawn with a filtering pipet equipped with a sintered glass disc. The analysis of the solution was carried out by iodometric titration.

SOURCE AND PURITY OF MATERIALS:

$Ca(IO_3)_2$ was prepared by adding a cold solution of KIO_3 to a solution of $CaCl_2$. $CaCl_2$ was obtained by dissolving $CaCO_3$ (Merk pro an) in HCl (c.p. grade). The precipitate was separated from the mother liquor by suction filtering, suspended 3 times in distilled cold water and left 48 hours. Then it was decanted, washed and dried. Calcd for $Ca(IO_3)_2$: CaO 14.38; I_2O_5 85.62. Found: CaO 14.28; I_2O_5 85.04.

ESTIMATED ERROR:

Soly: the mean deviation is ± 0.303%.
Temp: ± 0.005°C (authors)

COMPONENTS:

(1) Calcium iodate; $Ca(IO_3)_2$; [7789-80-2]

(2) Magnesium chloride; $MgCl_2$; [7786-30-3]

(3) Water; H_2O; [7732-18-5]

ORIGINAL MEASUREMENTS:

Kilde, G.

Z. Anorg. Allg. Chem. 1934, *218*, 113-28.

VARIABLES:

T/K = 291, 298, and 303

$c_2/\text{mol dm}^{-3}$ = 0 to 0.505

PREPARED BY:

Hiroshi Miyamoto

EXPERIMENTAL VALUES:

t/°C	Magnesium Chloride $c_2/\text{mol dm}^{-3}$	Calcium Iodate $10^3 c_1/\text{mol dm}^{-3}$	$10^6 K_{s0}/\text{mol}^3\ \text{dm}^{-9}$
18	0	5.69	0.737
	0.0503	8.41	2.38
	0.505	15.3	14.3
25	0	7.84	1.93
	0.0503	11.2	5.62
	0.505	20.0	32.0
30	0	9.91	3.89
	0.0503	13.8	10.5
	0.505	24.3	57.4

Concentration solubility product $K_{s0} = [Ca^{2+}][IO_3^-]^2$

K°_{s0} was calculated from the equation

$$1/3 \log K^\circ_{s0} = 1/3 \log K_{s0} - Z_1 Z_2 A I^{1/2} + BI$$

where $Z_1 Z_2 A$ = 0.998 at 18°C, 1.008 at 25°C and 1.018 at 30°C, and I is the ionic strength.

The values obtained were the following: $K^\circ_{s0} = 0.329 \times 10^{-6}$ at 18°C, 0.736×10^{-6} at 25°C and 1.35×10^{-6} at 30°C.

AUXILIARY INFORMATION

METHOD/APPARATUS/PROCEDURE:

An excess of $Ca(IO_3)_2 \cdot 6H_2O$ was shaken with aqueous $MgCl_2$ solutions for at least 24 hours in a thermostat at the desired temperature. Aliquots of saturated solutions were filtered through cotton wool, and the iodate content was determined iodometrically.

SOURCE AND PURITY OF MATERIALS:

Calcium iodate hexahydrate was prepared by mixing calcium chloride solution and KIO_3 solution. The precipitate was washed and dried at room temperature. Reagent grade $MgCl_2$ was used.

ESTIMATED ERROR:

Soly: precision within 1 %

Temp: nothing specified

REFERENCES:

COMPONENTS:

(1) Calcium iodate; $Ca(IO_3)_2$; [7789-80-2]

(2) Magnesium chloride; $MgCl_2$; [7786-30-3]

(3) Water; H_2O; [7732-18-5]

ORIGINAL MEASUREMENTS:

Rens, G.

Sucr. Belge 1958, *77*, 193-208.

VARIABLES:

$T/K = 293$

$10^3 c_2/\text{mol dm}^{-3} = 0.000$ to 62.327

PREPARED BY:

Hiroshi Miyamoto

EXPERIMENTAL VALUES:

t/°C	Magnesium Chloride $10^3 c_2/\text{mol dm}^{-3}$	1/2 Iodate Ion $10^3(c/2)/\text{mol dm}^{-3}$	Calcium Ion $10^3 c/\text{mol dm}^{-3}$	Calcium Iodate $10^3 c_1/\text{mol dm}^{-3}$ [a]
20	0.000	6.231	6.230	6.231
	4.942	6.758	6.742	6.750
	9.887	7.175	7.130	7.153
	19.706	7.779	7.723	7.751
	34.519	8.498	8.482	8.490
	49.206	9.048	9.055	9.052
	52.729	9.213	9.145	9.179
	62.327	9.509	9.425	9.467

[a] Average value calculated by the compiler

pK°_{s0} was calculated from the equation:

1) $I < 0.07$ mol dm^{-3}, $pK_{s0} = pK^\circ_{s0} - 6AI^{1/2} + 3CI$

where I is the ionic strength, and $A = 0.5046$ mol$^{-1/2}$ dm$^{3/2}$ (ref 1)

2) $0.07 < I < 0.25$ mol dm^{-3}, $pK_{s0} = pK^\circ_{s0} - 6AI^{1/2}/(1 + aBI^{1/2})$

where a is distance of closest approach.

For $a = 3.5$A°, the value of K°_{s0} was 4.159×10^{-7} mol^3 dm^{-9}

AUXILIARY INFORMATION

METHOD/APPARATUS/PROCEDURE:

Excess $Ca(IO_3)_2 \cdot 6H_2O$ and aqueous $MgCl_2$ solution were placed in sealed Erlemeyer flasks. The flasks were rotated in a thermostat for 24 hours. Aliquots of saturated solution were filtered.
The concentration of iodate was determined iodometrically. The calcium content was determined by chlatometric titration using Eriochrome black T as an indicator.

SOURCE AND PURITY OF MATERIALS:

$Ca(IO_3)_2 \cdot 6H_2O$ was prepared by slowly adding the solution of KIO_3 (about 50g dm^{-3}) to an equivalent solution of $CaCl_2 \cdot 2H_2O$ at 20°C. The precipitate was washed by decantation, and was air-dried at room temperature. An analysis of the product gave the following values: IO_3 99.64% and Ca 99.95% of theoretical.

ESTIMATED ERROR:

Soly: nothing specified
Temp: ± 0.1°K (author)

REFERENCES:

1. Harned, H.; Owen, B. *The Physical Chemistry of Electrolytic Solutions.* Reinhold, New York, 1950, 447.

COMPONENTS:

(1) Calcium iodate; $Ca(IO_3)_2$; [7789-80-2]

(2) Magnesium iodate; $Mg(IO_3)_2$; [7790-32-1]

(3) Water; H_2O; [7732-18-5]

ORIGINAL MEASUREMENTS:

Pedersen, K. J.

K. Dan. Vidensk. Selsk. Mat-Fys. Medd. 1941, *18*, 21-4.

VARIABLES:

$T/K = 291.1$

$10^3 c_2/\text{mol dm}^{-3} = 0.00$ to 10.00

PREPARED BY:

Hiroshi Miyamoto

EXPERIMENTAL VALUES:

t/°C	Magnesium Iodate $10^3 c_2/\text{mol dm}^{-3}$	Calcium Iodate $10^3 c_1/\text{mol dm}^{-3}$
17.9	0.00	5.686
	1.009	5.087
	2.011	4.528
	3.012	4.029
	5.000	3.174
	10.00	1.771

AUXILIARY INFORMATION

METHOD/APPARATUS/PROCEDURE:

The equilibrium procedure was not given.
The analyses of the saturated solutions were carried out by iodometry.
The solubility of $Ca(IO_3)_2$ was found by subtracting the concentration of $Mg(IO_3)_2$ from the total iodate concentration.
The details of equilibrium procedure and analyses of saturated solutions were similar to those described in the compilation of $Ca(IO_3)_2$-1,4-dioxane-H_2O system.

SOURCE AND PURITY OF MATERIALS:

$Mg(IO_3)_2 \cdot 6H_2O$ was prepared by adding 4.6g of basic magnesium carbonate to a solution of 18g of iodic acid at 50°C. The solution was evaporated slowly at 50°C.
The large crystals formed were separated and analyzed by iodometry.
The molecular weight found was 445.6, but the calculated one for $Mg(IO_3)_2 \cdot 4H_2O$ is 446.1712.

ESTIMATED ERROR:

Nothing specified

REFERENCES:

The compiler calculated to be 446.1712 for $Mg(IO_3)_2 \cdot 4H_2O$ using 1977 IUPAC recommended atomic weights.

COMPONENTS:

(1) Calcium iodate; $Ca(IO_3)_2$; [7789-80-2]

(2) Magnesium sulfate; $MgSO_4$; [7487-88-9]

(3) Water; H_2O; [7732-18-5]

ORIGINAL MEASUREMENTS:

Chloupek, J. B.; Daneš, VL. Z.; Danešova, B. A.

Collect. Czech. Chem. Commun. 1933, *5*, 339-42.

VARIABLES:

T/K = 298

m_2/mol kg^{-1} = 0 to 0.1

PREPARED BY:

Hiroshi Miyamoto

EXPERIMENTAL VALUES:

Magnesium Sulfate	Calcium Iodate	
m_2/mol kg^{-1}	g kg^{-1} (H_2O)	$10^3 m_1$/mol kg^{-1} [a]
0	3.031	7.774
0.002	3.219	8.256
0.005	3.397	8.713
0.01	3.663	9.395
0.02	4.090	10.49
0.05	4.952	12.70
0.1	5.768	14.79

[a] Compiler calculations using 1977 IUPAC recommended atomic weights.

AUXILIARY INFORMATION

METHOD/APPARATUS/PROCEDURE:

The solutions with the solid calcium iodate were rotated in a thermostat for at least 12 hours. Aliquots of saturated solutions were withdrawn with a filtering pipet equipped with a sintered glass disc. The analysis of the solution was carried out by iodometric titration.

SOURCE AND PURITY OF MATERIALS:

$Ca(IO_3)_2$ was prepared by adding a cold solution of KIO_3 to a solution of $CaCl_2$. $CaCl_2$ was obtained by dissolving $CaCO_3$ (Merk pro an) in HCl (c.p. grade). The precipitate was separated from the mother liquid by suction filtering, suspended 3 times in distilled cold water and left 48 hours. Then it was decanted, washed and dried. Calcd for $Ca(IO_3)_2$: CaO 14.38; I_2O_5 85.62. Found: CaO 14.28; I_2O_5 85.04.

ESTIMATED ERROR:

Soly: the mean deviation is ± 0.303%.
Temp: ± 0.005°C (authors)

COMPONENTS:

(1) Calcium iodate; $Ca(IO_3)_2$; [7789-80-2]

(2) Magnesium sulfate; $MgSO_4$; [7487-88-9]

(3) Water; H_2O; [7732-18-5]

ORIGINAL MEASUREMENTS:

Wise, W. C. A.; Davies, C. W.

J. Chem. Soc. 1938, 273-7.

VARIABLES:

T/K = 298

10^3c_2/mol dm^{-3} = 0 to 25.0

PREPARED BY:

Hiroshi Miyamoto
Mark Salomon

EXPERIMENTAL VALUES:

t/°C	Magnesium Sulfate 10^3c_2/mol dm^{-3}	Calcium Iodate 10^3c_1/mol dm^{-3}	Density ρ/g cm^{-3}
25	0	7.840	0.9998
	6.25	9.038	1.001
	12.5	9.788	1.002
	18.75	10.42	1.004
	25.0	10.95	1.004

COMMENTS AND/OR ADDITIONAL DATA:

The conductivity data at 18°C were used to evaluate the thermodynamic ion pair dissociation constant, K_D°; it was found that K_D° = 0.13. Using this value for the ion pair dissociation constant, the concentration of the ion pair $CaIO_3^+$ was calculated from

$$\log [CaIO_3^+] = \log [Ca^{2+}][IO_3^-] - \log K_D^\circ - 2.02I^{1/2} + 2.0I$$

where I is the ionic strength. Utilizing this relation to compute the ionic concentrations of Ca^{2+} and IO_3^-, the authors plotted (1/3) log $[Ca^{2+}]$x$[IO_3^-]$ against the ionic strength and extrapolated to zero ionic strength to obtain the thermodynamic solubility product constant. The result of this extrapolation is K_{s0}° = 6.953 x 10^{-7}.

AUXILIARY INFORMATION

METHOD/APPARATUS/PROCEDURE:

Saturating column method as in (1) and modified as in (2). A bulb containing the solvent solution is attached to a column containing the slightly soluble salt, and the solvent is allowed to flow through the column at a rate sufficient to insure saturation (1). The modification (2) consisted of connecting the column by capillary tubing to a second parallel arm in which the saturated solution collected. The entire apparatus was placed in a thermostat. Weighed samples of the satd slns were taken for analysis by method described in (3): i.e. the satd slns added to acidified KI sln and the liberated I_2 titrd by weight against an approx 0.15N thiosulfate sln, 0.01N iodine sln being used for the back titrn. The densities of the satd slns were measured at 25°C, and the molar conductivities at 18°C for the binary $Ca(IO_3)_2$-H_2O system are also reported.

SOURCE AND PURITY OF MATERIALS:

$Ca(IO_3)_2 \cdot 6H_2O$ was prepared by dropwise addition of solutions of KIO_3 and $CaCl_2$ in equivalent amounts to a large volume of conductivity water. The precipitate was washed first by decantation and then in the solubility columns until a constant solubility was obtained.

ESTIMATED ERROR:

Soly: not specified, but reproducibility probably around ± 0.3% as in ref. (3).
Temp: ± 0.01°K (authors)

REFERENCES:

1. Brönsted, J. N.; La Mer, V. K. *J. Am. Chem. Soc.* 1924, *46*, 555.
2. Money, R. W.; Davies, C. W. *J. Chem. Soc.* 1934, 400.
3. Macdougall, G.; Davies, C. W. *J. Chem. Soc.* 1935, 1416.

COMPONENTS:

(1) Calcium iodate; $Ca(IO_3)_2$; [7789-80-2]

(2) Calcium hydroxide; $Ca(OH)_2$; [1305-62-0]

(3) Water; H_2O; [7732-18-5]

ORIGINAL MEASUREMENTS:

Davies, C. W.; Hoyle, B. E.

J. Chem. Soc. 1951, 233-4.

VARIABLES:

$T/K = 298$

$10^3 c_2/\text{mol dm}^{-3} = 0$ to 20.93

PREPARED BY:

Hiroshi Miyamoto

EXPERIMENTAL VALUES:

t/°C	Calcium Hydroxide $10^3 c_2/\text{mol dm}^{-3}$	Calcium Iodate $10^3 c_1/\text{mol dm}^{-3}$	Dissociation Constant ($CaOH^+$) $K^\circ_D/\text{mol dm}^{-3}$
25	0	7.84	--
	3.63	7.27	0.050
	4.29	7.19	0.049
	10.28	6.65	0.050
	12.86	6.48	0.051
	14.26	6.42	0.049
	15.13	6.37	0.050
	18.60	6.23	0.048
	20.93	6.15	0.048

COMMENTS AND/OR ADDITIONAL DATA:

Activity coefficients were estimated from the equation (ref 1):

$$-\log y_\pm = 0.52\, z^2\{I^{1/2}/(1 + I^{1/2} - 0.20\, I\}$$

and by successive approximation the equation:

$$\log [Ca^{2+}][IO_3^-]/[CaIO_3^+] = \log 0.13 + 2F(I)$$

and

$$\log [Ca^{2+}][IO_3^-] = \log c_2 + 3F(I)$$ were solved to $[CaIO_3^+]$, $[Ca^{2+}]$

and by difference, $[CaOH^+]$.

The dissociation constants of $[CaOH^+]$ were obtained from

$\log K^\circ_D(CaOH^+) = \log [Ca^{2+}][OH^-]/[CaOH^+] - 2F(I)$. $K^\circ_D(CaOH^+) = 0.050$.

AUXILIARY INFORMATION

METHOD/APPARATUS/PROCEDURE:

Saturated column method as described in (2 and 3): see compilation of these papers for details.
The carbon dioxide free-$Ca(OH)_2$ solutions was introduced into the saturator after the saturator had been swept out with CO_2-free air. When saturation had been attained, samples were withdrawn and rapidly titrated with 0.05 mol dm^{-3} HCl after which the iodate was determined in the usual manner by iodometry.

SOURCE AND PURITY OF MATERIALS:

$Ca(IO_3)_2 \cdot 6H_2O$ was prepared by dropwise addition of solutions of KIO_3 and $CaCl_2$ in equivalent amounts to a large volume of conductivity water as in (2).
$Ca(OH)_2$ was prepared from AnalaR $CaCO_3$ which was heated in a platinum crucible, and after cooling added to CO_2 free water.

ESTIMATED ERROR:

Soly: nothing specified, but reproducibility probably around ± 0.3% as in ref 1 and 2.
Temp: ± 0.01°C (authors)

REFERENCES:

1. Davies, C. W. *J. Chem. Soc.* 1938, 2093.
2. Wise, W. C. A.; Davies, C. W. *J. Chem. Soc.* 1938, 273.
3. Davies, C. W. *J. Chem. Soc.* 1938, 277.

COMPONENTS:

(1) Calcium iodate; $Ca(IO_3)_2$; [7789-80-2]

(2) Calcium chloride; $CaCl_2$; [10043-52-4]

(3) Water; H_2O; [7732-18-5]

ORIGINAL MEASUREMENTS:

Kilde, G.

Z. Anorg. Allg. Chem. 1934, *218*, 113-28.

VARIABLES:

T/K = 291, 298, and 303

$c_2/\text{mol dm}^{-3}$ = 0 to 0.0500

PREPARED BY:

Hiroshi Miyamoto

EXPERIMENTAL VALUES:

t/°C	Calcium Chloride $c_2/\text{mol dm}^{-3}$	Calcium Iodate $10^3c_1/\text{mol dm}^{-3}$	$10^6K_{s0}/\text{mol}^3\ \text{dm}^{-9}$
18	0	5.69	0.737
	0.0050	4.78	0.896
	0.0100	4.33	1.08
	0.0250	3.75	1.62
	0.0500	3.30	2.32
25	0	7.84	1.93
	0.0050	6.81	2.19
	0.0100	6.44	2.71
	0.0250	5.57	3.80
	0.0500	4.91	5.30
30	0	9.91	3.89
	0.0050	8.65	4.09
	0.0100	8.22	4.90
	0.0250	7.24	6.09
	0.0500	6.58	9.80

AUXILIARY INFORMATION

METHOD/APPARATUS/PROCEDURE:

An excess of $Ca(IO_3)_2\cdot 6H_2O$ was shaken with aqueous $CaCl_2$ solutions for at least 24 hours in a thermostat at the desired temperature. Aliquots of saturated solutions were filtered through cotton wool, and the iodate content was determined iodometrically.

Concentration solubility product
$K_{s0} = [Ca^{2+}][IO_3^-]^2$

K°_{s0} was calculated from the equation

$$1/3 \log K^\circ_{s0} = 1/3 \log K_{s0} - Z_1Z_2AI^{1/2} + BI$$

where Z_1Z_2A = 0.998 at 18°C, 1.008 at 25°C and 1.018 at 30°C, and I is the ionic strength.

The values obtained were the following: $K^\circ_{s0} = 0.329 \times 10^{-6}$ at 18°C, 0.736×10^{-6} at 25°C and 1.35×10^{-6} at 30°C.

SOURCE AND PURITY OF MATERIALS:

Calcium iodate hexahydrate was prepared by mixing calcium chloride solution and KIO_3 solution. The precipitate was washed and dried at room temperature. Reagent grade $CaCl_2$ was used.

ESTIMATED ERROR:

Soly: precision within 1 %
Temp: nothing specified

COMPONENTS:

(1) Calcium iodate; $Ca(IO_3)_2$; [7789-80-2]

(2) Calcium chloride; $CaCl_2$; [10043-52-4]

(3) Water; H_2O; [7732-18-5]

ORIGINAL MEASUREMENTS:

Wise, W. C. A.; Davies, C. W.

J. Chem. Soc. 1938, 273-7.

VARIABLES:

$T/K = 298$

$10^3 c_2/\text{mol dm}^{-3} = 0$ to 50.0

PREPARED BY:

Hiroshi Miyamoto
Mark Salomon

EXPERIMENTAL VALUES:

t/°C	Calcium Chloride $10^3 c_2/\text{mol dm}^{-3}$	Calcium Iodate $10^3 c_1/\text{mol dm}^{-3}$	Density $\rho/\text{g cm}^{-3}$
25	0	7.840	0.9998
	6.25	6.692	1.0001
	25.0	5.444	1.0016
	50.0	4.900	1.0036

COMMENTS AND/OR ADDITIONAL DATA:

The conductivity data at 18°C were used to evaluate the thermodynamic ion pair dissociation constant, K_D°; it was found that $K_D^\circ = 0.13$. Using this value for the ion pair dissociation constant, the concentration of the ion pair $CaIO_3^+$ was calculated from

$$\log [CaIO_3^+] = \log [Ca^{2+}][IO_3^-] - \log K_D^\circ - 2.02I^{1/2} + 2.0I$$

where I is the ionic strength. Utilizing this relation to compute the ionic concentrations of Ca^{2+} and IO_3^-, the authors plotted (1/3) log $[Ca^{2+}] \times [IO_3^-]$ against the ionic strength and extrapolated to zero ionic strength to obtain the thermodynamic solubility product constant. The result of this extrapolation is $K_{s0}^\circ = 6.953 \times 10^{-7}$.

AUXILIARY INFORMATION

METHOD/APPARATUS/PROCEDURE:

Saturating column method as in (1) and modified as in (2). A bulb containing the solvent solution is attached to a column containing the slightly soluble salt, and the solvent is allowed to flow through the column at a rate sufficient to insure saturation (1). The modification (2) consisted of connecting the column by capillary tubing to a second parallel arm in which the saturated solution collected. The entire apparatus was placed in a thermostat. Weighed samples of the satd slns were taken for analysis by method described in (3): i.e. the satd slns added to acidified KI sln and the liberated I_2 titrd by weight against an approx 0.15N thiosulfate sln, 0.01N iodine sln being used for the back titrn. The densities of the satd slns were measured at 25°C, and the molar conductivities at 18°C for the binary $Ca(IO_3)_2-H_2O$ system are also reported.

SOURCE AND PURITY OF MATERIALS:

$Ca(IO_3)_2 \cdot 6H_2O$ was prepared by dropwise addition of solutions of KIO_3 and $CaCl_2$ in equivalent amounts to a large volume of conductivity water. The precipitate was washed first by decantation and then in the solubility columns until a constant solubility was obtained.

ESTIMATED ERROR:

Soly: not specified, but reproducibility probably around ± 0.3% as in ref. (3).
Temp: ± 0.01°K (authors)

REFERENCES:

1. Brönsted, J. N.; La Mer, V. K. *J. Am. Chem. Soc.* 1924, *46*, 555.
2. Money, R. W.; Davies, C. W. *J. Chem. Soc.* 1934, 400.
3. Macdougall, G.; Davies, C. W. *J. Chem. Soc.* 1935, 1416.

COMPONENTS:

(1) Calcium iodate; $Ca(IO_3)_2$; [7789-80-2]

(2) Calcium chloride; $CaCl_2$; [10043-52-4]

(3) Water; H_2O; [7732-18-5]

ORIGINAL MEASUREMENTS:

Rens, G.

Sucr. Belge 1958, *77*, 193-208.

VARIABLES: $T/K = 293$

$10^3 c_2/\text{mol dm}^{-3}$ = 0.000 to 64.134

PREPARED BY:

Hiroshi Miyamoto

EXPERIMENTAL VALUES:

t/°C	Calcium Chloride $10^3 c_2/\text{mol dm}^{-3}$	1/2 Iodate Ion $10^3 (c/2)/\text{mol dm}^{-3}$	Calcium Ion $10^3 c/\text{mol dm}^{-3}$	Calcium Iodate $10^3 c_1/\text{mol dm}^{-3}$ [a]
20	0.000	6.231	6.230	6.231
	4.983	5.322	5.329	5.326
	9.971	4.816	4.836	4.826
	19.730	4.271	4.296	4.284
	24.934	4.106	4.106	4.106
	34.541	3.861	3.907	4.884
	49.828	3.675	3.627	3.651
	54.284	3.622	3.618	3.620
	64.134	3.546	3.460	3.503

[a] Average value calculated by the compiler

pK°_{s0} was calculated from the equation:

1) $I < 0.07 \text{ mol dm}^{-3}$, $pK_{s0} = pK^\circ_{s0} - 6AI^{1/2} + 3CI$

where I is the ionic strength, and $A = 0.5046 \text{ mol}^{-1/2} \text{ dm}^{3/2}$ (ref 1)

2) $0.07 < I < 0.25 \text{ mol dm}^{-3}$, $pK_{s0} = pK^\circ_{s0} - 6AI^{1/2}/(1 + aBI^{1/2})$

where a is distance of closest approach.

For a = 3.4A°, the value of K°_{s0} was $4.159 \times 10^{-7} \text{ mol}^3 \text{ dm}^{-9}$

AUXILIARY INFORMATION

METHOD/APPARATUS/PROCEDURE:

Excess $Ca(IO_3)_2 \cdot 6H_2O$ and aqueous $CaCl_2$ solution were placed in sealed Erlenmeyer flasks. The flasks were rotated in a thermostat for 24 hours. Aliquots of saturated solution were filtered.
The concentration of iodate was determined iodometrically. The calcium content was determined by chlatometric titration using Eriochrome black T as an indicator.

SOURCE AND PURITY OF MATERIALS:

$Ca(IO_3)_2 \cdot 6H_2O$ was prepared by slowly adding the solution of KIO_3 (about 50g dm^{-3}) to an equivalent solution of $CaCl_2 \cdot 2H_2O$ at 20°C. The precipitate was washed by decantation, and was air-dried at room temperature. An analysis of the product gave the following values IO_3 99.64% and Ca 99.95% of theoretical.

ESTIMATED ERROR:

Soly: nothing specified
Temp: ± 0.1°K (author)

REFERENCES:

1. Harned, H.; Owen, B. *The Physical Chemistry of Electrolytic Solutions.* Reinhold, New York, 1950, 447.

COMPONENTS:

(1) Calcium iodate; $Ca(IO_3)_2$; [7789-80-2]
(2) Calcium chloride; $CaCl_2$; [10043-52-4]
(3) Glycine; $C_2H_5NO_2$; [56-40-6]
(4) Water; H_2O; [7732-18-5]

ORIGINAL MEASUREMENTS:

Monk, C. B.

Trans. Faraday Soc. 1951, *47*, 1233-40.

VARIABLES:

$T/K = 298$
$10^3 c_3/\text{mol dm}^{-3} = 0$ to 200.9

PREPARED BY:

Hiroshi Miyamoto

EXPERIMENTAL VALUES:

t/°C	Glycine $10^3 c_3/\text{mol dm}^{-3}$	Calcium Chloride $10^3 c_2/\text{mol dm}^{-3}$	Calcium Iodate $10^3 c_1/\text{mol dm}^{-3}$	$-\log K^\circ_{s0}$
25	0	0	7.84	6.148
	50.3	5.71	7.17	6.093
	100.9	5.71	7.52	6.035
	200.9	5.71	8.23	5.931

COMMENTS AND/OR ADDITIONAL DATA:

Thermodynamic solubility product constant, K°_{s0}, was calculated from

$$\log K^\circ_{s0} = \log [Ca^{2+}][IO_3^-]^2 - 3F$$

where $F = (78.54/\varepsilon)^{3/2}[I^{1/2}/(1 + I^{1/2}) - 0.2I]$, ε = dielectric constant, I = ionic strength.

AUXILIARY INFORMATION

METHOD/APPARATUS/PROCEDURE:

The saturating column method was employed similar to that reported in ref (1).
The analyses were effected by withdrawing 100 cm^3 of the saturated solution in a calibrated pipet, and running the solution into an acidic KI solution. The liberated iodine was titrated by weight against 0.15 mol dm^{-3} $Na_2S_2O_3$ solution, 0.005 mol dm^{-3} iodine solution being used for the back titration.

SOURCE AND PURITY OF MATERIALS:

$Ca(IO_3)_2 \cdot 6H_2O$ was prepared by dropwise addition of solutions of KIO_3 and $CaCl_2$ in equivalent amounts to conductivity water as in (1).
Analytical reagent grade glycine was dried to constant weight in a vacuum oven at 80°C.

ESTIMATED ERROR:

Soly: nothing specified
Temp: ± 0.03°C (author)

REFERENCES:

1. Macdougall, G.; Davies, C. W. *J. Chem. Soc.* 1935, 1416.

2. Davies, C. W. *J. Chem. Soc.* 1930, 2410.

COMPONENTS:

(1) Calcium iodate; $Ca(IO_3)_2$; [7789-80-2]

(2) Calcium hydroxide; $Ca(OH)_2$; [1305-62-0]

(3) Sucrose; $C_{12}H_{22}O_{11}$; [57-50-1]

(4) Water; H_2O; [7732-18-5]

ORIGINAL MEASUREMENTS:

Kilde, G.

Z. Anorg. Allg. Chem. 1934, *218*, 113-28.

VARIABLES:

T/K = ?

c_2/mol dm^{-3} = 0.0576 and 0.0896

PREPARED BY:

Hiroshi Miyamoto

EXPERIMENTAL VALUES:

Sucrose	Calcium Hydroxide	Calcium Iodate		Calcium Ion[a]
c_3/mol dm^{-3}	c_2/mol dm^{-3}	$10^3 c_1$/mol dm^{-3}	$10^6 K_{s0}$/mol^3 dm^{-9}	$10^3 c_{Ca^{2+}}$/mol dm^{-3}
0.100	0.0576	8.13	1.35	5.2
0.100	0.0896	12.2	1.35	2.3

[a] $c_{Ca^{2+}}$ was calculated from $K_{s0}/4c_1^2$.

AUXILIARY INFORMATION

METHOD/APPARATUS/PROCEDURE:

The compiler assumes that the method of the solubility determination of this system was similar to that adopted in the case of neutral salt solutions.

An excess of $Ca(IO_3)_2 \cdot 6H_2O$ was shaken with aqueous calcium hydroxide solution containing sucrose for at least 24 hours in a thermostat at the desired temperature. Aliquots of saturated solutions were filtered through cotton wool, and the iodate content was determined iodometrically.

SOURCE AND PURITY OF MATERIALS:

Calcium iodate hexahydrate was prepared by mixing calcium chloride solution and potassium iodate solution. The precipitate was washed and dried at room temperature. The source of sucrose and calcium hydroxide was not given.

ESTIMATED ERROR:

Soly: precision within 1 %

Temp: nothing specified

REFERENCES:

COMPONENTS:

(1) Calcium iodate; $Ca(IO_3)_2$; [7789-80-2]

(2) Propanoic acid, 2-hydroxy-calcium salt(2:1)(calcium lactate); $C_6H_{10}O_6Ca$; [814-80-2]

(3) Water; H_2O; [7732-18-5]

ORIGINAL MEASUREMENTS:

Kilde, G.

Z. Anorg. Allg. Chem. 1936, *229*, 321-36.

VARIABLES:

T/K = 291, 298, 303

c_2/mol dm^{-3} = 0 to 0.0992

PREPARED BY:

Hiroshi Miyamoto

EXPERIMENTAL VALUES:

t/°C	Calcium Lactate	Calcium Iodate		
	c_2/mol dm^{-3}	c_1/mol dm^{-3}	$10^6\ K_{s0}$/mol^{3}dm^{-9} [a]	K_D/mol dm^{-3} [b]
18	0	5.68	--	--
	0.00496	4.99	0.891	0.079
	0.00992	4.64	1.02	0.075
	0.0198	4.29	1.26	0.080
	0.0496	3.98	1.88	0.093
	0.0992	3.92	2.48	0.088
25	0	7.81	--	--
	0.00496	7.15	2.26	0.099
	0.00992	6.77	2.60	0.098
	0.0196	6.33	3.16	0.105
	0.0496	5.93	4.32	0.092
	0.0992	5.85	5.69	0.088
30	0	9.78	--	--
	0.00496	9.14	4.17	0.065
	0.00992	8.74	4.57	0.066
	0.0196	8.30	5.41	0.072
	0.0496	7.84	7.78	0.090
	0.0992	7.75	10.6	0.094

K_D° = 0.040

[a] $K_{s0} = [Ca^{2+}][IO_3^-]^2$

[b] $K_D = [Ca^{2+}][C_3H_5O_3^-]/[CaC_3H_5O_3^+]$

AUXILIARY INFORMATION

METHOD/APPARATUS/PROCEDURE:

An excess of calcium iodate hexahydrate and calcium lactate solution were placed into bottles. The bottles were shaken in a thermostat for at least 24 hours. Samples of saturated solutions were withdrawn through a filter fitted with cotton wool, and the iodate estimation was made by addition of KI to about 25 cm^3 of the saturated solution, followed by sulfuric acid, and titrated with thiosulfate solution.

SOURCE AND PURITY OF MATERIALS:

$Ca(IO_3)_2 \cdot 6H_2O$ was prepared by dropwise addition of $CaCl_2$ solution to KIO_3 solution.
Calcium lactate solution was prepared by dissolving an excess of calcium carbonate in lactic acid solution. The solution was boiled to remove lactylacetic acid.
Calcium lactate of Pharmacopeia Dan. 33 was also used.

ESTIMATED ERROR:

Soly: precision within 1 %
Temp: nothing specified

REFERENCES:

COMPONENTS:

(1) Lithium iodate; $LiIO_3$; [13765-03-2]

(2) Calcium iodate; $Ca(IO_3)_2$; [7789-80-2]

(3) Water; H_2O; [7732-18-5]

ORIGINAL MEASUREMENTS:

Azarova, L. A.; Vinogradov, E. E.

Zh. Neorg. Khim. 1977, *22*, 273-5; *Russ. J. Inorg. Chem. (Engl. Transl.)* 1977, *22*, 153-4.

VARIABLES:

T/K = 323

$LiIO_3$/mass % = 6 - 43

PREPARED BY:

Hiroshi Miyamoto

EXPERIMENTAL VALUES:

Composition of Saturated Solutions

t/°C	Lithium Iodate mass %	Lithium Iodate mol %[a]	Calcium Iodate mass %	Calcium Iodate mol %[a]	Nature of the Solid Phase[b]
50	--	--	0.544[c]	0.0253	C
	6.06	0.635	0.058	0.0028	C
	8.53	0.916	0.020	0.0010	C
	15.56	1.794	0.062	0.0033	C
	28.50	3.802	0.062	0.0039	C
	38.14	5.774	0.204	0.0144	C
	43.67	7.137	0.038	0.0029	C + L
	42.68	6.879	0.082	0.0062	C + L
	43.57	7.108	0.023	0.0018	L
	43.28	7.028	--	--	L

[a] Mol % values and molalities calculated by compiler.

[b] C = $Ca(IO_3)_2 \cdot H_2O$; L = $LiIO_3$

[c] For the binary systems at 50°C:

Soly of $LiIO_3$ in H_2O = 4.196 mol kg^{-1}

Soly of $Ca(IO_3)_2$ in H_2O = 0.0140 mol kg^{-1}

AUXILIARY INFORMATION

METHOD/APPARATUS/PROCEDURE:

The starting materials plus a known amount of water were placed into solubility vessels and stirred continually in a thermostat at 50°C for 10-14 hours.
The iodate in liquid and solid phases was determined iodometrically, and calcium complexometrically. The lithium ion was determined by difference. The composition and nature of the solid phase were determined by Schreinemakers' method of residues, and crystal-optically.

SOURCE AND PURITY OF MATERIALS:

$LiIO_3$ was made from $LiCO_3$ and HIO_3. Calcium iodate was precipitated from calcium nitrate solution with iodic acid. The quality of the products obtained was controlled by chemical and X-ray diffraction analysis. The analysis showed that the products obtained were α-$LiIO_3$ and $Ca(IO_3)_2 \cdot 6H_2O$.

ESTIMATED ERROR:

Soly: nothing specified
Temp: ± 0.1°C (authors)

REFERENCES:

COMPONENTS:	ORIGINAL MEASUREMENTS:
(1) Lithium iodate; $LiIO_3$; [13765-03-2] (2) Calcium iodate; $Ca(IO_3)_2$; [7789-80-2] (3) Water; H_2O; [7732-18-5]	Azarova, L. A.; Vinogradov, E. E. *Zh. Neorg. Khim.* 1977, *22*, 273-5; *Russ. J. Inorg. Chem. (Engl. Transl.)* 1977, *22*, 153-4.

COMMENTS AND/OR ADDITIONAL DATA:

The phase diagram is given below

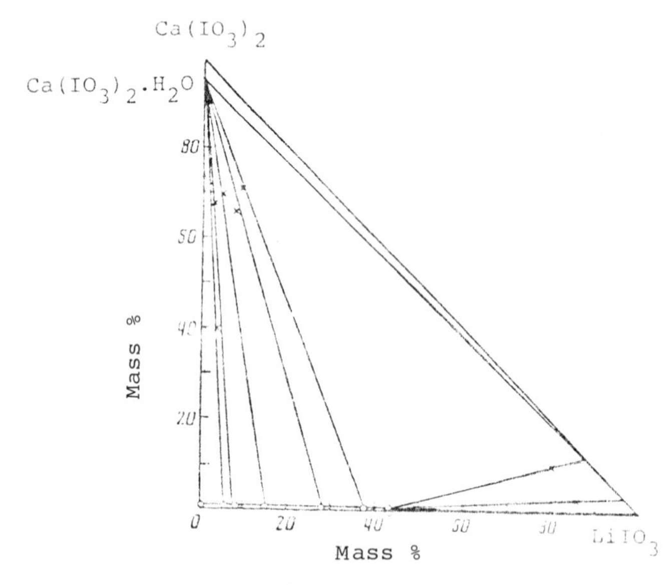

50 °C Solubility Isotherm

ACKNOWLEDGEMENT:

The figure reprinted from *Zh. Neorg. Khim.* by permission of the copyright owners, VAAP, The Copyright Agency of the USSR.

COMPONENTS:

(1) Lithium iodate; $LiIO_3$; [13765-03-2]

(2) Calcium iodate; $Ca(IO_3)_2$; [7789-80-2]

(3) Water; H_2O; [7732-18-5]

ORIGINAL MEASUREMENTS:

Arkhipov, S. M.; Kashina, N. I.; Kidyarov, B. I.

Zh. Neorg. Khim. 1978, *23*, 1422-3; *Russ. J. Inorg. Chem. (Engl. Transl.)* 1978, *23*, 784-5.

VARIABLES:

T/K = 298

$LiIO_3$/mass % = 0 - 44

PREPARED BY:

Hiroshi Miyamoto

EXPERIMENTAL VALUES:

t/°C	Composition of Saturated Solutions				Nature of the Solid Phase[a]
	Lithium Iodate		Calcium Iodate		
	mass %	mol % (compiler)	mass %	mol % (compiler)	
25	--	--	0.304	0.0141	A
	3.67	0.376	0.014	0.00067	A
	9.94	1.082	0.010	0.00051	A
	17.91	2.116	0.015	0.00083	A
	24.52	3.119	0.020	0.0012	A
	32.58	4.570	0.018	0.0012	A
	43.38[b]	7.056	0.015	0.0011	A + B
	43.60	7.115	0.012	0.00091	B
	43.79	7.165	--	--	B

[a] A = $Ca(IO_3)_2 \cdot 6H_2O$; B = $LiIO_3$

[b] Eutonic point

[c] For the binary systems at 25°C the compiler computes the following:

Soly of $LiIO_3$ = 4.284 mol kg^{-1}

Soly of $Ca(IO_3)_2$ = 0.00782 mol kg^{-1}

AUXILIARY INFORMATION

METHOD/APPARATUS/PROCEDURE:

$Ca(IO_3)_2$-$LiIO_3$-H_2O system was studied by the isothermal method. Equilibrium was established in 15 days. The iodate in liquid and solid phases was determined volumetrically (presumably by iodometric titration: compiler). The Ca content was determined by titration with EDTA at high concentrations and by flame photometry at low concentrations. Lithium was calculated by difference. The compositions of the solid phases were determined by the method of residues and verified by X-ray diffraction. The detail of the "residues" method is not given in the paper.

SOURCE AND PURITY OF MATERIALS:

Highly pure grade $LiIO_3$ was used. Calcium iodate was prepared from highly pure garde iodic acid and calcium nitrate.

ESTIMATED ERROR:

Soly: nothing specified

Temp: ± 0.1°K (authors)

REFERENCES:

COMPONENTS:

(1) Sodium iodate; $NaIO_3$; [7681-55-2]

(2) Calcium iodate; $Ca(IO_3)_2$; [7789-80-2]

(3) Water; H_2O; [7732-18-5]

ORIGINAL MEASUREMENTS:

Hill, A. E.; Brown, S. F.

J. Am. Chem. Soc. 1931, *53*, 4316-20.

VARIABLES:

$T/K = 298$

$NaIO_3$/mass % = 0 to 8.58

PREPARED BY:

Hiroshi Miyamoto

EXPERIMENTAL VALUES:

Composition of the Saturated Solutions					Solution Density	Nature of the Solid Phase[b]
Sodium iodate		Calcium iodate				
mass %	mol %[a]	mass %	mol %[a]	$10^3 m_1$/mol kg^{-1}[a]	ρ/g cm^{-3}	
0.000	0.000	0.306	0.0146	7.87	--	A
0.522	0.0478	0.084	0.0039	2.16	1.00	A
5.29	0.505	0.000	0.00	0	1.041	A
8.58	0.847	0.000	0.00	0	1.074	A + B
8.58	0.847	0.000	0.00	0	--	B

[a] Molalities and mol % values calculated by compiler.

[b] A = $Ca(IO_3)_2 \cdot 6H_2O$; B = $NaIO_3 \cdot H_2O$

COMMENTS AND/OR ADDITIONAL DATA:

By extrapolation of the tie-lines, it becomes apparent that the hexahydrate of calcium iodate is the only stable solid phase of that salt, and that the common ion from sodium iodate reduces its solubility to an amount too small to be determined for the greater part of the isotherm. The author said that 10g samples failed to give a weighable calcium precipitate on treatment with ammonium oxalate.

AUXILIARY INFORMATION

METHOD/APPARATUS/PROCEDURE:

The samples were rotated in a thermostat for about 2 weeks. Pipetted samples were analyzed for total iodate by the method of iodometry, and for calcium by precipitation with ammonium oxalate.

SOURCE AND PURITY OF MATERIALS:

Calcium iodate was prepared by addition of solutions of $Ca(NO_3)_2$ and KIO_3 to a large volume of water. The precipitate was washed and recrystallized from water. The hexahydrate was obtained by slow cooling within the temperature range below 30°C.

ESTIMATED ERROR:

nothing specified

REFERENCES:

COMPONENTS:

(1) Calcium iodate; $Ca(IO_3)_2$; [7789-80-2]

(2) Methanol; CH_4O; [67-56-1]

(3) Water; H_2O; [7732-18-5]

ORIGINAL MEASUREMENTS:

Monk, C. B.

J. Chem. Soc. 1951, 2723-6.

VARIABLES:

$T/K = 298$

Methanol/mass % = 0 to 14.43

PREPARED BY:

Hiroshi Miyamoto

EXPERIMENTAL VALUES:

t/°C	Methanol		Calcium Iodate	
	mass %	mol % (compiler)	$10^3 c_1$/mol dm^{-3}	$10^7 K^\circ_{s0}$/mol^{3}dm^{-9}
25	0	0	7.84	7.94
	4.72	2.71	4.88	3.16
	9.53	5.59	4.02	1.26
	14.43	8.66	2.94	0.52

COMMENTS AND/OR ADDITIONAL DATA:

Thermodynamic solubility product constant, K°_{s0}, was calculated from

$$\log K^\circ_{s0} = \log [Ca^{2+}][IO_3^-]^2 - 3F$$

where $F = (78.54/\varepsilon)^{3/2}[I^{1/2}/(1 + I^{1/2}) - 0.2I]$, ε = dielectric constant, I = ionic strength.

AUXILIARY INFORMATION

METHOD/APPARATUS/PROCEDURE:

Saturating column method used as in (1): see compilation of ref 2 for details.

The iodate concentrations in the saturated solutions were determined by titration with $Na_2S_2O_3$ standardized with KIO_3. Prior to the titration, excess KI was added and the solution acidified with dilute acetic acid.

SOURCE AND PURITY OF MATERIALS:

Calcium iodate was prepared by dropwise addition of solutions of KIO_3 and $CaCl_2$ in an equivalent amount to a large volume of conductivity water. The crystalline hexahydrate was separated and washed.

Methanol used was of laboratory grade.

ESTIMATED ERROR:

Soly: nothing specified

Temp: ± 0.03°C (author)

REFERENCES:

1. Money, R. W.; Davies, C. W. *J. Chem. Soc.* 1934, 400.

2. Macdougall, G.; Davies, C. W. *J. Chem. Soc.* 1935, 1416.

COMPONENTS:

(1) Calcium iodate; $Ca(IO_3)_2$; [7789-80-2]

(2) Ethanol; C_2H_6O; [64-17-5]

(3) Water; H_2O; [7732-18-5]

ORIGINAL MEASUREMENTS:

Monk, C. B.

J. Chem. Soc. 1951, 2723-6.

VARIABLES:

$T/K = 298$
Ethanol/mass % = 0 to 11.59

PREPARED BY:

Hiroshi Miyamoto

EXPERIMENTAL VALUES:

t/°C	Ethanol		Calcium Iodate	
	mass %	mol % (compiler)	$10^3 c_1$/mol dm^{-3}	$10^7 K^\circ_{s0}$/mol^{3}dm^{-9}
25	0	0	7.84	7.94
	3.82	1.53	5.83	3.44
	7.67	3.15	4.33	1.52
	11.59	4.88	3.23	0.68

COMMENTS AND/OR ADDITIONAL DATA:

Thermodynamic solubility product constant, K°_{s0}, was calculated from

$$\log K^\circ_{s0} = \log [Ca^{2+}][IO_3^-]^2 - 3F$$

where $F = (78.54/\varepsilon)^{3/2}[I^{1/2}/(1 + I^{1/2}) - 0.2I]$, ε = dielectric constant, I = ionic strength.

AUXILIARY INFORMATION

METHOD/APPARATUS/PROCEDURE:

Saturating column method used as in (1): see compilation of ref 2 for details.
The iodate concentrations in the saturated solutions were determined by titration with $Na_2S_2O_3$ standardized with KIO_3. Prior to the titration, excess KI was added and the solution acidified with dilute acetic acid.

SOURCE AND PURITY OF MATERIALS:

Calcium iodate was prepared by dropwise addition of solutions of KIO_3 and $CaCl_2$ in an equivalent amount to a large volume of conductivity water. The crystalline hexahydrate was separated and washed. Ethanol used was of laboratory grade.

ESTIMATED ERROR:

Soly: nothing specified
Temp: ± 0.03°C (author)

REFERENCES:

1. Money, R. W.; Davies, C. W. *J. Chem. Soc.* 1934, 400.

2. Macdougall, G.; Davies, C. W. *J. Chem. Soc.* 1935, 1416.

COMPONENTS:

(1) Calcium iodate; $Ca(IO_3)_2$; [7789-80-2]

(2) 1-Propanol; C_3H_8O; [71-23-8]

(3) Water; H_2O; [7732-18-5]

ORIGINAL MEASUREMENTS:

Monk, C. B.

J. Chem. Soc. 1951, 2723-6.

VARIABLES:

T/K = 298

1-Propanol/mass % = 0 to 12.71

PREPARED BY:

Hiroshi Miyamoto

EXPERIMENTAL VALUES:

t/°C	1-Propanol		Calcium Iodate	
	mass %	mol % (compiler)	$10^3 c_1$/mol dm^{-3}	$10^7 K^\circ_{s0}$/mol^{3}dm^{-9}
25	0	0	7.84	7.94
	4.16	1.28	5.72	3.30
	8.40	2.68	4.20	1.37
	12.71	4.18	3.07	0.57

COMMENTS AND/OR ADDITIONAL DATA:

Thermodynamic solubility product constant, K°_{s0}, was calculated from

$$\log K^\circ_{s0} = \log [Ca^{2+}][IO_3^-]^2 - 3F$$

where $F = (78.54/\varepsilon)^{3/2}[I^{1/2}/(1 + I^{1/2}) - 0.2I]$, ε = dielectric constant, I = ionic strength.

AUXILIARY INFORMATION

METHOD/APPARATUS/PROCEDURE:

Saturating column method used as in (1): see compilation of ref 2 for details.
The iodate concentrations in the saturated solutions were determined by titration with $Na_2S_2O_3$ standardized with KIO_3. Prior to the titration, excess KI was added and the solution acidified with dilute acetic acid.

SOURCE AND PURITY OF MATERIALS:

Calcium iodate was prepared by dropwise addition of solutions of KIO_3 and $CaCl_2$ in an equivalent amount to a large volume of conductivity water. The crystalline hexahydrate was separated and washed.
1-Propanol used was of laboratory grade.

ESTIMATED ERROR:

Soly: nothing specified
Temp: ± 0.03°C (author)

REFERENCES:

1. Money, R. W.; Davies, C. W. *J. Chem. Soc.* 1934, 400.

2. Macdougall, G.; Davies, C. W. *J. Chem. Soc.* 1935, 1416.

COMPONENTS:

(1) Calcium iodate; $Ca(IO_3)_2$; [7789-80-2]

(2) 1,2-Ethandiol (ethylene glycol); $C_2H_6O_2$; [107-21-1]

(3) Water; H_2O; [7732-18-5]

ORIGINAL MEASUREMENTS:

Monk, C. B.

J. Chem. Soc. 1951, 2723-6.

VARIABLES:

T/K = 298

Ethylene glycol/mass % = 0 to 16.85

PREPARED BY:

Hiroshi Miyamoto

EXPERIMENTAL VALUES:

t/°C	Ethylene Glycol		Calcium Iodate	
	mass %	mol % (compiler)	$10^3 c_1$/mol dm^{-3}	$10^7 K^\circ_{s0}$/mol^{3}dm^{-9}
25	0	0	7.84	7.94
	5.62	1.70	7.16	6.08
	11.24	3.55	6.60	4.78
	16.85	5.56	6.12	3.80

COMMENTS AND/OR ADDITIONAL DATA:

Thermodynamic solubility product constant, K°_{s0}, was calculated from

$$\log K^\circ_{s0} = \log [Ca^{2+}][IO_3^-]^2 - 3F$$

where $F = (78.54/\varepsilon)^{3/2}[I^{1/2}/(1 + I^{1/2}) - 0.2I]$, ε = dielectric constant, I = ionic strength.

AUXILIARY INFORMATION

METHOD/APPARATUS/PROCEDURE:

Saturating column method used as in (1): see compilation of ref 2 for details.
The iodate concentrations in the saturated solutions were determined by titration with $Na_2S_2O_3$ standardized with KIO_3. Prior to the titration, excess KI was added and the solution acidified with diluted acetic acid.

SOURCE AND PURITY OF MATERIALS:

Calcium iodate was prepared by dropwise addition of solutions of KIO_3 and $CaCl_2$ in an equivalent amount to a large volume of conductivity water. The crystalline hexahydrate was separated and washed. Ethylene glycol used was of laboratory grade.

ESTIMATED ERROR:

Soly: nothing specified
Temp: ± 0.03°C (author)

REFERENCES:

1. Money, R. W.; Davies, C. W. *J. Chem. Soc.* 1934, 400.

2. Macdougall, G.; Davies, C. W. *J. Chem. Soc.* 1935, 1416.

COMPONENTS:

(1) Calcium iodate; $Ca(IO_3)_2$; [7789-80-2]

(2) 1,2,3-Propanetriol(glycerol); $C_3H_8O_3$; [56-81-5]

(3) Water; H_2O; [7732-18-5]

ORIGINAL MEASUREMENTS:

Monk, C. B.

J. Chem. Soc. 1951, 2723-6.

VARIABLES:

$T/K = 298$

Glycerol/mass % = 0 to 18.43

PREPARED BY:

Hiroshi Miyamoto

EXPERIMENTAL VALUES:

t/°C	Glycerol		Calcium Iodate	
	mass %	mol % (compiler)	$10^3 c_1$/mol dm^{-3}	$10^7 K^{\circ}_{s0}$/mol^{3}dm^{-9}
25	0	0	7.84	7.94
	6.31	1.30	7.75	7.62
	12.44	2.71	7.70	7.13
	18.43	4.23	7.70	6.90

COMMENTS AND/OR ADDITIONAL DATA:

Thermodynamic solubility product constant, K°_{s0}, was calculated from

$$\log K^{\circ}_{s0} = \log [Ca^{2+}][IO_3^-]^2 - 3F$$

where $F = (78.54/\varepsilon)^{3/2}[I^{1/2}/(1 + I^{1/2}) - 0.2I]$, ε = dielectric constant, I = ionic strength.

AUXILIARY INFORMATION

METHOD/APPARATUS/PROCEDURE:

Saturating column method used as in (1): see compilation of ref 2 for details.
The iodate concentrations in the saturated solutions were determined by titration with $Na_2S_2O_3$ standardized with KIO_3. Prior to the titration, excess KI was added and the solution acidified with dilute acetic acid.

SOURCE AND PURITY OF MATERIALS:

Calcium iodate was prepared by dropwise addition of solutions of KIO_3 and $CaCl_2$ in an equivalent amount to a large volume of conductivity water. The crystalline hexahydrate was separated and washed.
Glycerol used was of laboratory grade.

ESTIMATED ERROR:

Soly: nothing specified
Temp: ± 0.03°C (author)

REFERENCES:

1. Money, R. W.; Davies, C. W. *J. Chem. Soc.* 1934, 400.

2. Macdougall, G.; Davies, C. W. *J. Chem. Soc.* 1935, 1416.

COMPONENTS:

(1) Calcium iodate; $Ca(IO_3)_2$; [7789-80-2]

(2) 1,4-Dioxane; $C_4H_8O_2$; [123-91-1]

(3) Water; H_2O; [7732-18-5]

ORIGINAL MEASUREMENTS:

Petersen, K. J.

K. Dan. Vidensk. Selsk. Nat.-Fys. Medd. 1941, *18*, 21-4.

VARIABLES:

$T/K = 291.15$

$c_2/mol\ dm^{-3} = 0 - 1.0$

PREPARED BY:

Hiroshi Miyamoto

EXPERIMENTAL VALUES:

t/°C	Dioxane $c_2/mol\ dm^{-3}$	Barium Iodate $10^3c_1/mol\ dm^{-3}$
18.00	0.000	5.702
	0.125	5.423
	0.250	5.159
	0.375	4.904
	0.500	4.652
	0.750	4.194
	1.000	3.771

AUXILIARY INFORMATION

METHOD/APPARATUS/PROCEDURE:

Excess $Ca(IO_3)_2 \cdot 6H_2O$ and aqueous dioxane solution were placed in glass stoppered-bottles. The bottles were rotated in an electrically regulated water thermostat. Samples of the saturated solutions were analyzed after different times of rotation in order to make sure that saturation was attained.
The samples were sucked from the bottle through a porous glass filter into a pipet.
The iodate contents were determined by iodometry. Analyses and solubility measurements were done in duplicate.

SOURCE AND PURITY OF MATERIALS:

$Ca(IO_3)_2 \cdot 6H_2O$ was prepared from calcium chlcride and iodic acid. Dioxane (Haardt u. Co., "Exluan 05") was left for two days over solid sodium hydroxide, refluxed with sodium for several hours, and then distilled in an all glass apparatus. The main fraction had a freezing point of 11.65°C, compared with 11.80°C for pure dioxane (1).

ESTIMATED ERROR:

Soly: within the limit of accuracy of the analytical method (author)
Temp: nothing specified

REFERENCES:

1. Hess, K.; Frahm, H. *Ber. Dtsch. Chem. Ges.* 1938, *71*, 2627.

COMPONENTS:

(1) Calcium iodate; $Ca(IO_3)_2$; [7789-80-2]

(2) 1,4-Dioxane; $C_4H_8O_2$; [123-91-1]

(3) Water; H_2O; [7732-18-5]

ORIGINAL MEASUREMENTS:

Monk, C. B.

J. Chem. Soc. 1951, 2723-6.

VARIABLES:

T/K = 298
1,4-Dioxane/mass % = 0 to 9.4

PREPARED BY:

Hiroshi Miyamoto

EXPERIMENTAL VALUES:

t/°C	1,4-Dioxane		Calcium Iodate	
	mass %	mol % (compiler)	$10^3 c_1$/mol dm^{-3}	$10^7 K^\circ_{s0}$/mol^{3}dm^{-9}
25	0	0	7.84	7.94
	2.2	0.46	7.04	5.77
	4.7	1.00	6.26	4.09
	9.4	2.08	4.93	2.02

COMMENTS AND/OR ADDITIONAL DATA:

Thermodynamic solubility product constant, K°_{s0}, was calculated from

$$\log K^\circ_{s0} = \log[Ca^{2+}][IO_3^-]^2 - 3F$$

where $F = (78.54/\varepsilon)^{3/2}[I^{1/2}/(1 + I^{1/2}) - 0.2I]$, ε = dielectric constant, I = ionic strength.

AUXILIARY INFORMATION

METHOD/APPARATUS/PROCEDURE:

Saturating column method used as in (1): see compilation of ref 2 for details.
The iodate concentrations in the saturated solutions were determined by titration with $Na_2S_2O_3$ standardized with KIO_3. Prior to the titration, excess KI was added and the solution acidified with dilute acetic acid.

SOURCE AND PURITY OF MATERIALS:

Calcium iodate was prepared by dropwise addition of solutions of KIO_3 and $CaCl_2$ in an equivalent amount to a large volume of conductivity water. The crystalline hexahydrate was separated and washed.
Dioxane used was of AnalaR reagent.

ESTIMATED ERROR:

Soly: nothing specified
Temp: ± 0.03°C (author)

REFERENCES:

1. Money, R. W.; Davies, C. W. *J. Chem. Soc.* 1934, 400.

2. Macdougall, G.; Davies, C. W. *J. Chem. Soc.* 1935, 1416.

COMPONENTS:

(1) Calcium iodate; $Ca(IO_3)_2$; [7789-80-2]

(2) 2-Propanone(acetone); C_3H_6O; [67-64-1]

(3) Water; H_2O; [7732-18-5]

ORIGINAL MEASUREMENTS:

Monk, C. B.

J. Chem. Soc. 1951, 2723-6.

VARIABLES:

$T/K = 298$
Acetone/mass % = 0 to 12.46

PREPARED BY:

Hiroshi Miyamoto

EXPERIMENTAL VALUES:

t/°C	Acetone mass %	Acetone mol % (compiler)	Calcium Iodate $10^3 c_1$/mol dm^{-3}	Calcium Iodate $10^7 K^\circ_{s0}$/mol^{3}dm^{-9}
25	0	0	7.84	7.94
	4.09	1.31	5.91	3.58
	8.25	2.71	4.46	1.63
	12.46	4.23	3.32	0.72

COMMENTS AND/OR ADDITIONAL DATA:

Thermodynamic solubility product constant, K°_{s0}, was calculated from

$$\log K^\circ_{s0} = \log [Ca^{2+}][IO_3^-]^2 - 3F$$

where $F = (78.54/\varepsilon)^{3/2}[I^{1/2}/(1 + I^{1/2}) - 0.2I]$, ε = dielectric constant, I = ionic strength.

AUXILIARY INFORMATION

METHOD/APPARATUS/PROCEDURE:

Saturating column method used as in (1): see compilation of ref 2 for details.
The iodate concentrations in the saturated solutions were determined by titration with $Na_2S_2O_3$ standardized with KIO_3. Prior to the titration, excess KI was added and the solution acidified with dilute acetic acid.

SOURCE AND PURITY OF MATERIALS:

Calcium iodate was prepared by dropwise addition of solutions of KIO_3 and $CaCl_2$ in an equivalent amount to a large volume of conductivity water. The crystalline hexahydrate was separated and washed.
Acetone used was of AnalaR reagent.

ESTIMATED ERROR:

Soly: nothing specified
Temp: ± 0.03°C (author)

REFERENCES:

1. Money, R. W.; Davies, C. W. *J. Chem. Soc.* 1934, 400.

2. Macdougall, G.; Davies, C. W. *J. Chem. Soc.* 1935, 1416.

COMPONENTS:

(1) Calcium iodate; $Ca(IO_3)_2$; [7789-80-2]

(2) Tetrahydrofuran; C_4H_8O; [109-99-9]

(3) Water; H_2O; [7732-18-5]

ORIGINAL MEASUREMENTS:

Miyamoto, H.

Nippon Kagaku Kaishi 1972, 659-61.

VARIABLES:

T/K = 298
Tetrahydrofuran/mass % = 0 to 40

PREPARED BY:

Hiroshi Miyamoto

EXPERIMENTAL VALUES:

t/°C	Tetrahydrofuran mass %	Tetrahydrofuran mol % (compiler)	Calcium Iodate 10^3c_1/mol dm^{-3}
25	0	0	7.84
	5	1.3	5.67
	10	2.7	4.10
	15	4.2	2.88
	20	5.9	2.09
	25	7.7	1.47
	30	9.7	1.06
	40	14.3	0.51

AUXILIARY INFORMATION

METHOD/APPARATUS/PROCEDURE:

Excess $Ca(IO_3)_2 \cdot 6H_2O$ and solvent mixtures were placed in glass-stoppered bottles. The bottles were rotated in a thermostat at 25°C for 48 hours.
After the saturated solution settled, the solution was withdrawn through a siphon tube equipped with a glass-sintered filter.
The iodate content was determined iodometrically.

SOURCE AND PURITY OF MATERIALS:

$Ca(IO_3)_2 \cdot 6H_2O$ was prepared by adding solutions of $CaCl_2$ (Wako Co G.R.) and KIO_3 (Wako Co G.R.) to a large volume of water containing KNO_3. The precipitate was filtered off, washed and dried under reduced pressure.
Tetrahydrofuran was distilled from NaOH and then redistilled from sodium metal.

ESTIMATED ERROR:

Soly: nothing specified
Temp: ± 0.02°C (author)

REFERENCES:

COMPONENTS:

(1) Calcium iodate; $Ca(IO_3)_2$; [7789-80-2]

(2) Ethyl acetate; $C_4H_8O_2$; [141-78-6]

(3) Water; H_2O; [7732-18-5]

ORIGINAL MEASUREMENTS:

Monk, C. B.

J. Chem. Soc. 1951, 2723-6.

VARIABLES:

$T/K = 298$
Ethyl acetate/mass % = 0 to 6.1

PREPARED BY:

Hiroshi Miyamoto

EXPERIMENTAL VALUES:

t/°C	Ethyl Acetate		Calcium Iodate	
	mass %	mol % (compiler)	$10^3 c_1$/mol dm^{-3}	$10^7 K°_{s0}$/mol^{3}dm^{-9}
25	0	0	7.84	7.94
	3.8	0.80	6.26	4.19
	6.1	1.31	5.50	2.88

COMMENTS AND/OR ADDITIONAL DATA:

Thermodynamic solubility product constant, $K°_{s0}$, was calculated from

$$\log K°_{s0} = \log [Ca^{2+}][IO_3^-]^2 - 3F$$

where $F = (78.54/\varepsilon)^{3/2}[I^{1/2}/(1 + I^{1/2}) - 0.2I]$, ε = dielectric constant, I = ionic strength.

AUXILIARY INFORMATION

METHOD/APPARATUS/PROCEDURE:

Saturating column method used as in (1): see compilation of ref 2 for details.
The iodate concentrations in the saturated solutions were determined by titration with $Na_2S_2O_3$ standardized with KIO_3. Prior to the titration, excess KI was added and the solution acidified with dilute acetic acid.

SOURCE AND PURITY OF MATERIALS:

Calcium iodate was prepared by dropwise addition of solutions of KIO_3 and $CaCl_2$ in an equivalent amount to a large volume of conductivity water. The crystalline hexahydrate was separated and washed.
Ethyl acetate used was of laboratory grade.

ESTIMATED ERROR:

Soly: nothing specified
Temp: ± 0.03°C (author)

REFERENCES:

1. Money, R. W.; Davies, C. W. *J. Chem. Soc.* 1934, 400.

2. Macdougall, G.; Davies, C. W. *J. Chem. Soc.* 1935, 1416.

COMPONENTS:

(1) Calcium iodate; $Ca(IO_3)_2$; [7789-80-2]

(2) Urea; CH_4N_2O; [57-13-6]

(3) Water; H_2O; [7732-18-5]

ORIGINAL MEASUREMENTS:

Petersen, K. J.

K. Dan. Vidensk. Selsk. Nat-Fys. Medd. 1941, *18*, 21-4.

VARIABLES:

$T/K = 291.1$

$c_2/\text{mol dm}^{-3} = 0.000$ to 8.000

PREPARED BY:

Hiroshi Miyamoto

EXPERIMENTAL VALUES:

t/°C	Urea $c_2/\text{mol dm}^{-3}$	Calcium Iodate $10^3c_1/\text{mol dm}^{-3}$
17.9	0.000	5.686
	0.100	5.821
	0.200	5.957
	0.400	6.233
	0.600	6.512
	0.800	6.805
	1.000	7.103
	2.000	8.689
	4.000	12.58
	6.000	17.67
	8.000	24.96

AUXILIARY INFORMATION

METHOD/APPARATUS/PROCEDURE:

Excess $Ca(IO_3)_2 \cdot 6H_2O$ and aqueous urea solution were placed in glass stoppered-bottles. The bottles were rotated in an electrically regulated water thermostat. Samples of the saturated solutions were analyzed after different times of rotation in order to make sure that saturation was attained.
The samples were sucked from the bottle through a porous glass filter into a pipet.
The iodate contents were determined by iodometry. Analyses and solubility measurements were done in duplicate.

SOURCE AND PURITY OF MATERIALS:

$Ca(IO_3)_2 \cdot 6H_2O$ was prepared from calcium chloride and iodic acid.
Urea (Kahlbaum, "für wissenschaftliche Zweeke") was used without further purification. It contained traces of calcium which could not be removed by recrystallization from alcohol.
1 to 3 mg of ash and 1 to 2 x 10^{-5} moles of calcium were found per mole of urea.

ESTIMATED ERROR:

Soly: within the limit of accuracy of the analytical method (author)
Temp: nothing specified

REFERENCES:

COMPONENTS:

(1) Calcium iodate; $Ca(IO_3)_2$; [7789-80-2]

(2) N,N,-Dimethylformamide; C_3H_7NO; [68-12-2]

(3) Water; H_2O; [7732-18-5]

ORIGINAL MEASUREMENTS:

Miyamoto, H.; Suzuki, K.; Yanai, K.

Nippon Kagaku Kaishi 1978, 1150-2.

VARIABLES:

T/K = 293, 298 and 303
Dimethylformamide/mass % = 0 - 41

PREPARED BY:

Hiroshi Miyamoto

EXPERIMENTAL VALUES:

t/°C	Dimethylformamide mass %	mol % (compiler)	Calcium Iodate $10^3 c_1$/mol dm^{-3}
20	0	0	6.31
	4.98	1.275	4.98
	10.01	2.668	3.94
	15.47	4.316	3.00
	20.07	5.828	2.39
	24.94	7.569	1.87
	29.67	9.418	1.44
	40.81	14.525	0.78
25	0	0	7.84
	5.07	1.299	6.18
	10.18	2.717	4.89
	15.00	4.168	3.90
	19.95	5.787	3.10
	24.88	7.547	2.39
	31.43	10.150	1.74
	40.33	14.279	1.08
30	0	0	9.90
	4.88	1.249	7.83
	9.98	2.660	6.16
	15.27	4.253	4.74
	19.97	5.794	3.79
	24.97	7.581	2.89
	30.12	9.603	2.21
	39.92	14.072	1.27

AUXILIARY INFORMATION

METHOD/APPARATUS/PROCEDURE:

$Ca(IO_3)_2 \cdot 6H_2O$ crystals and solvent mixtures were loaded into glass-stoppered bottles. The bottles were placed in a thermostat at a given temperature, and rotated for 72 hours.
After the saturated solutions were obtained, the solutions were separated from the solid phase using a sintered glass filter.
After the saturated solutions were diluted with water, the concentration of iodate was determined iodometrically. The solubility of $Ca(IO_3)_2$ was calculated from the observed values.

SOURCE AND PURITY OF MATERIALS:

Calcium iodate was prepared by adding dilute solutions of $CaCl_2$ and KIO_3 to a boiled water. The product was washed and dried at room temperature. $Ca(IO_3)_2 \cdot 6H_2O$ crystals was obtained.
Dimethylformamide (from Mitsubishi Gas Co.) was distilled under reduced pressure. After the product was dried over Na_2CO_3, the distillation of the solvent was repeated 3 times.

ESTIMATED ERROR:

Soly: the probable errors of the observed mean value were within $\pm$ 0.2 x 10^{-6} mol dm^{-3}.
Temp: $\pm$ 0.02°C (authors)

REFERENCES:

COMPONENTS:

(1) Calcium iodate; $Ca(IO_3)_2$; [7789-80-2]

(2) Glycine; $C_2H_5NO_2$; [56-40-6]

(3) Water; H_2O; [7732-18-5]

ORIGINAL MEASUREMENTS:

Keefer, R. M.; Reiber, H. G.; Bisson, C. S.

J. Am. Chem. Soc. 1940, *62*, 2951-5.

VARIABLES:

$T/K = 298$

$m_2/mol\ kg^{-1} = 0.0251$ to 0.8261

PREPARED BY:

Hiroshi Miyamoto

EXPERIMENTAL VALUES:

t/°C	Glycine $m_2/mol\ kg^{-1}$	Calcium Iodate $10^3m_1/mol\ kg^{-1}$
25	0.0251	8.06
	0.0503	8.23
	0.0755	8.49
	0.1008	8.65
	0.2009	9.51
	0.4055	11.11
	0.6140	12.97
	0.8261	14.95

AUXILIARY INFORMATION

METHOD/APPARATUS/PROCEDURE:

Glycine solutions were prepared from boiled distilled water using calibrated volumetric equipment. An excess of air-dried calcium iodate was placed in a glass-stoppered Pyrex flask and 200 ml of glycine solution added. The flasks were rotated in a thermostat for at least 12 hours. Equilibrium was established in 4-5 hours.

The saturated solutions were analyzed iodometrically. Analyses and solubility measurements were done in duplicate. Densities of all solutions were determined, but the data were not given in the original paper.

SOURCE AND PURITY OF MATERIALS:

$Ca(IO_3)_2$ was prepared by dropwise addition of 1.0 mol dm^{-3} $CaCl_2$ solution to 2 dm^3 of 0.38 mol dm^{-3} KIO_3 solution. The mixture was stirred, the precipitate filtered, washed, and then dried at room temperature. The number of hydrated waters was not given. C.p. grade glycine was recrystallized twice from water by addition of EtOH. The product was dried in a vacuum oven at about 35°C.

ESTIMATED ERROR:

Soly: nothing specified

Temp: ± 0.02°C (authors)

REFERENCES:

COMPONENTS:

(1) Calcium iodate; $Ca(IO_3)_2$; [7789-80-2]

(2) Alanine; $C_3H_7NO_2$; [302-72-7]

(3) Water; H_2O; [7732-18-5]

ORIGINAL MEASUREMENTS:

Keefer, R. M.; Reilber, H. G.; Bisson, C. S.

J. Am. Chem. Soc. 1940, *62*, 2951-5.

VARIABLES:

T/K = 298

m_2/mol kg^{-1} = 0.0251 to 0.1008

PREPARED BY:

Hiroshi Miyamoto

EXPERIMENTAL VALUES:

t/°C	Alanine m_2/mol kg^{-1}	Calcium Iodate $10^3 m_1$/mol kg^{-1}
25	0.0251	8.00
	0.0503	8.14
	0.0755	8.29
	0.1008	8.45

AUXILIARY INFORMATION

METHOD/APPARATUS/PROCEDURE:

Alanine solutions were prepared from distilled water using calibrated volumetric equipment. An excess of air-dried calcium iodate was placed in a glass-stoppered Pyrex flask and 200 ml of alanine solution added. The flasks were rotated in a thermostat for at least 12 hours. Equilibrium was obtained in 4-5 hours.

The saturated solutions were analyzed iodometrically. Analyses and solubility measurements were done in duplicate. Densities of all solutions were determined, but the data were not given in original paper.

SOURCE AND PURITY OF MATERIALS:

$Ca(IO_3)_2$ was prepared by dropwise addition of 1.0 mol dm^{-3} $CaCl_2$ solution to 2 dm^3 of 0.38 mol dm^{-3} KIO_3 solution. The mixture was stirred, the precipitate filtered, washed, and then dried at room temperature. The number of hydrated waters was not given. C.p. grade alanine was recrystallized twice from water by addition of EtOH. The product was dried in vacuum oven at about 35°C.

ESTIMATED ERROR:

Soly: nothing specified

Temp: ± 0.02°C (authors)

REFERENCES:

COMPONENTS:

(1) Strontium chlorate; $Sr(ClO_3)_2$; [7791-10-8]

(2) Water; H_2O; [7732-18-5]

EVALUATOR:

Hiroshi Miyamoto
Department of Chemistry
Niigata University
Niigata, Japan

May, 1982

CRITICAL EVALUATION:

Solubility for the binary $Sr(ClO_3)_2$ - H_2O system

Solubility in the binary $Sr(ClO_3)_2$-H_2O system has been reported in 2 publications (1,2). In older work, Mylius and Funk (1) measured the solubility of strontium chlorate in water at only 291K. The strontium content was determined gravimetrically by evaporation of the saturated solution to dryness.

Linke (2) studied solubilities in the binary $Sr(ClO_3)_2$-H_2O system over the temperature range of 235.1 to 394K. He also studied the solubility in the ternary $Sr(ClO_3)_2$-$SrBr_2$-H_2O system, but the solubility data for strontium chlorate in water were not given.

The relation between temperature and composition of solid phases in equilibrium with the saturated solutions was discussed by Linke for the binary system. Linke's results are reproduced in Fig. 1. The following solid phases have been identified:

$Sr(ClO_3)_2 \cdot 3H_2O$ [82150-37-6]

$Sr(ClO_3)_2$ [7791-10-8]

The eutectic of system $Sr(ClO_3)_2$-H_2O lies at 236K and 54.5 mass % $Sr(ClO_3)_2$, with $Sr(ClO_3)_2 \cdot 3H_2O$ and ice as solids. The transition from trihydrate to anhydrous salt occurs at 283K. The composition of the trihydrate was confirmed by the change in its solubility at 253K when $SrBr_2$ was added to the solution.

The interpolated result of Linke (2) at 291.2K is in good agreement with that of Mylius and Funk (1), and the data reported in these two publications are designated as tentative values. The tentative values are given in Table 1, and were fitted to the following smoothing equations.

$$\ln(S_i/\text{mol kg}^{-1}) = -11455.23 + 15652.53/(T/100\text{K}) + 12692.41 \ln (T/100\text{K}) - 2573.794\ T/100\text{K}:\ \sigma = 0.23$$

$$\ln(S_3/\text{mol kg}^{-1}) = -37.86877 + 50.43290/(T/100\text{K}) + 21.13040 \ln (T/100\text{K}) :\ \sigma = 0.049$$

$$\ln(S_0/\text{mol kg}^{-1}) = -4.901121 + 9.150474/(T/100\text{K}) + 3.447384 +3.447384 \ln (T/100\text{K}) :\ \sigma = 0.028$$

where S_i, S_3 and S_0 are the solubilities of strontium chlorate in equilibrium with ice, the trihydrate, and the anhydrous salt as solids, respectively.

The results calculated from the smoothing equation are also given in Table 1.

COMPONENTS:

(1) Strontium chlorate; $Sr(ClO_3)_2$; [7791-10-8]

(2) Water; H_2O; [7732-18-5]

EVALUATOR:

Hiroshi Miyamoto
Department of Chemistry
Niigata University
Niigata, JAPAN

May 1982

CRITICAL EVALUATION:

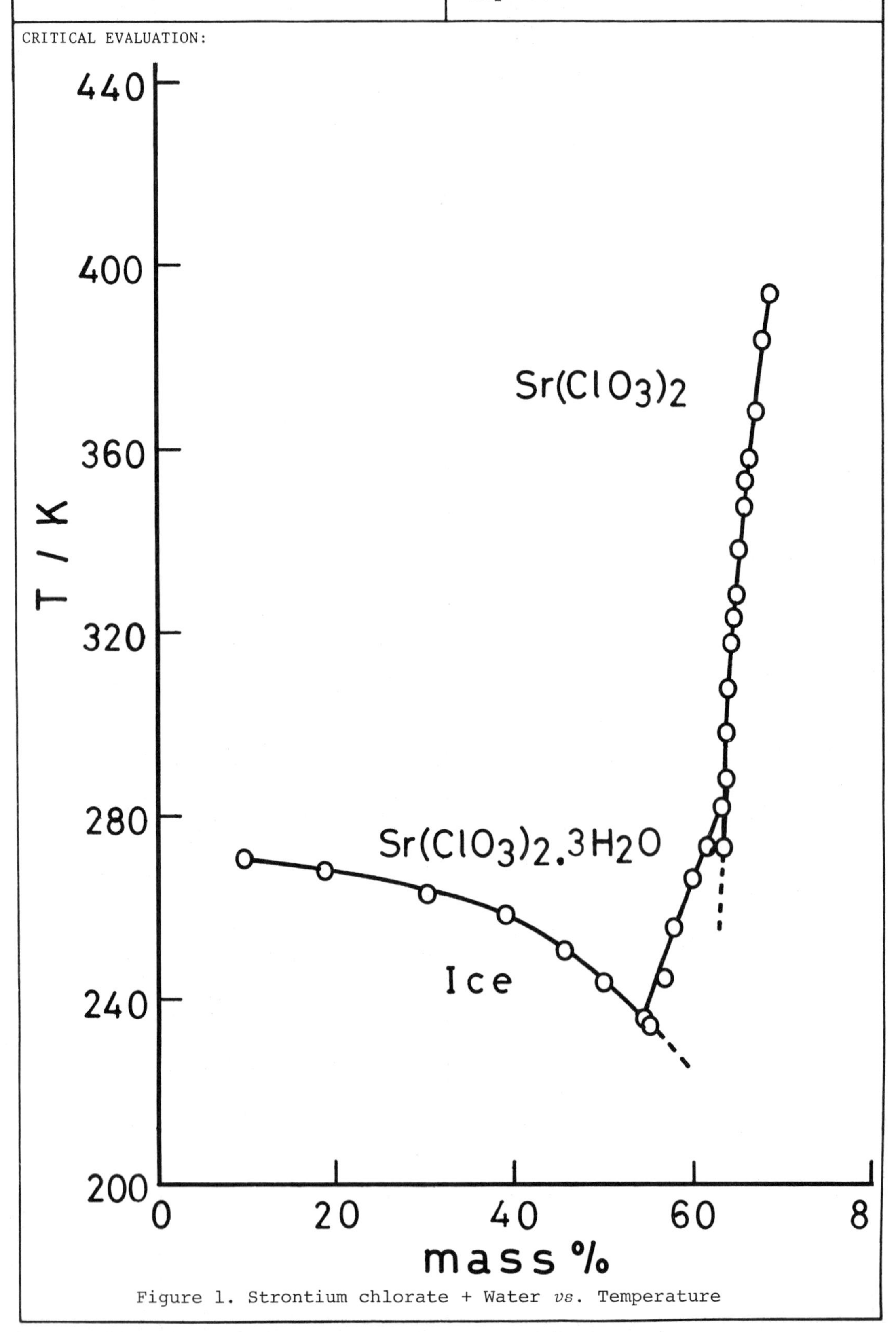

Figure 1. Strontium chlorate + Water *vs*. Temperature

COMPONENTS:

(1) Strontium chlorate; $Sr(ClO_3)_2$; [7791-10-8]

(2) Water; H_2O; [7732-18-5]

EVALUATOR:

Hiroshi Miyamoto
Department of Chemistry
Niigata University
Niigata, Japan

May, 1982

CRITICAL EVALUATION:

Table 1 Tentative values for the solubility of strontium chlorate in water

T/K	m_1/mol kg^{-1}	m_1'/mol kg^{-1}	Solid Phase
235.1	4.80	4.89	Ice
243.8	3.909	3.638	"
251.0	3.229	3.391	"
259.0	2.353	2.519	"
263.8	1.679	1.633	"
269.2	0.8578	0.7219	"
271.4	0.402	0.460	"
236.2	4.71		Ice + $Sr(ClO_3)_2 \cdot 3H_2O$
245.2	5.23	5.21	$Sr(ClO_3)_2 \cdot 3H_2O$
256.5	5.40	5.45	"
266.2	5.87	5.84	"
273.2	6.25	6.22	"
282.2	6.826	6.85	"
283.2	6.81		$Sr(ClO_3)_2 \cdot 3H_2O$ + $Sr(ClO_3)_2$
273.2	6.756	6.773	$Sr(ClO_3)_2$
288.2	6.850	6.841	"
291.2	6.865	6.861	"
298.2	6.918	6.917	"
308.2	7.037	7.016	"
318.2	7.154	7.135	"
323.2	7.185	7.201	"
328.2	7.281	7.272	"
338.2	7.400	7.426	"
348.2	7.627	7.597	"
353.2	7.651	7.689	"
358.2	7.768	7.784	"
368.2	8.006	7.985	"
383.2	8.27	8.314	"
394.2	8.62	8.575	"

m_1: experimental value

m_1': calculated value

REFERENCES:

1. Mylius, F., Funk, R. *Ber. Dtsch. Chem. Ges.* 1897, *30*, 1716.

2. Linke, W. *J. Am. Chem. Soc.* 1953, *75*, 5797.

COMPONENTS:

(1) Strontium chlorate; $Sr(ClO_3)_2$; [7791-10-8]

(2) Water; H_2O; [7732-18-5]

ORIGINAL MEASUREMENTS:

Mylius, F.; Funk, R.

Ber. Dtsch. Chem. Ges. 1897, *30*, 1716-25.

VARIABLES:

T/K = 291

PREPARED BY:

Hiroshi Miyamoto

EXPERIMENTAL VALUES:

The solubility of $Sr(ClO_3)_2$ in water at 18°C is given as below:

63.6 mass %	(authors)
174.9 g/100g[a] H_2O	(authors)
174.7 g/100g H_2O	(the compiler reculculated)
6.865 mol kg^{-1}	(compiler)

The density of the saturated solution at 18°C is also given:

1.839 g cm^{-3}

Based on this density, the compiler calculated the solubility in volume units as

4.597 mol dm^{-3}

[a] The compiler presumes that the first word in the fifth line from the end of page 1717 should read 100g.

AUXILIARY INFORMATION

METHOD/APPARATUS/PROCEDURE:

The salt and water were placed in a bottle and the bottle was shaken in a constant temperature bath for a long time. After the saturated solution settled, an aliquot of solution was removed with a pipet. Strontium chlorate was determined by evaporation of the solution by dryness. The density of the saturated solution was also determined.

SOURCE AND PURITY OF MATERIALS:

The salt used was purchased as a "pure" chemical and traces of impurities were not present. The purity sufficed for the solubility determination.

ESTIMATED ERROR:

Soly: precision within 1 %
Temp: nothing specified

REFERENCES:

COMPONENTS:

(1) Strontium chlorate; $Sr(ClO_3)_2$; [7791-10-8]

(2) Water; H_2O; [7732-18-5]

ORIGINAL MEASUREMENTS:

Linke, W. F.

J. Am. Chem. Soc. 1953, *75*, 5797-800.

EXPERIMENTAL VALUES:

t/°C	Strontium Chlorate			Density	Nature of the Solid Phase[a]
	mass %	mol % (compiler)	m_1/mol kg^{-1} (compiler)	ρ/g cm^{-3}	
- 38.1[b]	55.0[b]	7.96	4.80	--	Ice[c]
- 29.4	49.87	6.578	3.909	--	"
- 22.2	45.11	5.497	3.229	--	"
- 14.2	37.46	4.067	2.353	--	"
- 9.4	29.94	2.936	1.679	--	"
- 4.0	17.92	1.522	0.8578	--	"
- 1.8	9.29	0.720	0.402	--	"
- 37.0 ± 0.5	54.5[d]	7.82	4.71	--	Ice + A
- 28	57.1	8.61	5.23	--	A
- 16.7	57.9	8.87	5.40	--	"
- 7	59.9	9.56	5.87	--	"
0	61.4	10.1	6.25	--	"
9	63.47	10.95	6.826	--	"
10 ± 1	63.4[d]	10.9	6.81	1.829	A + B
0[a]	63.23[b]	10.85	6.756	1.828	B
15	63.55	10.98	6.850	1.830	"
25	63.78	11.08	6.918	1.831	"
35	64.17	11.25	7.037	1.833	"
45	64.55	11.42	7.154	1.835	B
50	64.65	11.46	7.185	1.837	"
55	64.95	11.60	7.281	1.838	"
65	65.32	11.76	7.400	1.842	"
75	66.0	12.08	7.627	1.845	"
80	66.07	12.11	7.651	1.847	"
85	66.41	12.28	7.768	1.849	"
95	67.08	12.60	8.006	1.853	"
110	67.8	13.0	8.27	1.861	"
121 ± 1	68.7[d]	13.4	8.62	1.867	"

[a] A = $Sr(ClO_3)_2 \cdot 3H_2O$; B = $Sr(ClO_3)_2$

[b] Metastable system

[c] Nature of solid phase not specified by author, but assumed by compiler based upon shape of the polytherm plotted in the source paper.

[d] Determined graphically

COMPONENTS:

(1) Strontium chlorate; $Sr(ClO_3)_2$; [7791-10-8]

(2) Water; H_2O; [7732-18-5]

ORIGINAL MEASUREMENTS:

Linke, W. F.

J. Am. Chem. Soc. 1953, *75*, 5797-800.

VARIABLES:

T/K = 235.1 to 394

PREPARED BY:

Hiroshi Miyamoto

EXPERIMENTAL VALUES:

AUXILIARY INFORMATION

METHOD/APPARATUS/PROCEDURE:

Temperatures between 0 and 100°C were maintained in water-baths, and at 110°C an acetic acid vapor bath was used (1). At 0°C and below, baths of melting ice and of melting mono-, di- and triethylene glycol were employed. Freezing points were determined from the cooling curves of known mixtures and reproducible to ± 0.05°C.
Equilibrium in saturated solutions was established by repeated analysis after several hours of stirring. Representative points were checked by approach from supersaturation. Each reported value is the average of at least two closely agreeing determinations. Filtered samples of the solution were withdrawn with preheated calibrated pipets, and approximate densities were calculated. Analysis for chlorate was made by reduction to chloride with nitrite and subsequent Volhard titration.

SOURCE AND PURITY OF MATERIALS:

Strontium chlorate anhydrate was prepared as follows: a chloric acid solution was prepared from roughly equivalent quantities of c.p. grade $Ba(ClO_3)_2$ and H_2SO_4, and small amounts of BaO and H_2SO_4 were then added until no significant tests for Ba^{2+} or SO_4^{2-} were obtained. Excess c.p. grade $SrCO_3$ (previously leached with a large volume of boiling water) was then added to the chloric acid solution. The mixture was filtered, and evaporation by boiling yielded pure anhydrous $Sr(ClO_3)_2$. The solid was recrystallized from water, air-dried, and stored at room temperature. A qualitative flame test showed that no sodium, and only traces of calcium present. Analysis by reduction to chloride and Volhard titration showed 99.6% $Sr(ClO_3)_2$. Loss in weight upon drying at 110°C was 0.26%.
$Sr(ClO_3)_2 \cdot 3H_2O$ was prepared by cooling a concentrated solution of the anhydrous salt in an acetone-dry-ice bath. When the solution had become very viscous, vigorous scratching produced the trihydrate. The excess solution was removed by suction, and the moist solid was stored at 7°C.

ESTIMATED ERROR:

Soly: nothing specified
Temp: the maximum variation between 0 and 100°C never exceed ± 0.1°C, and was usually much less.

REFERENCES:

1. Linke, W. F. *J. Chem. Educ.* 1952, *29*, 429.

COMPONENTS:

(1) Strontium chlorate; $Sr(ClO_3)_2$; [7791-10-8]

(2) Strontium bromide; $SrBr_2$; [10476-81-0]

(3) Water; H_2O; [7732-18-5]

ORIGINAL MEASUREMENTS:

Linke, W. F.

J. Am. Chem. Soc. 1953, *75*, 5797-800.

VARIABLES:

T/K = 249.7 to 255

c_2/mass % = 5.16 to 7.01

PREPARED BY:

Hiroshi Miyamoto

EXPERIMENTAL VALUES:

t/°C	Composition of Saturated Solutions				Nature of the Solid Phase
	Strontium Chlorate		Strontium Bromide		
	mass %	mol % (compiler)	mass %	mol % (compiler)	
-18	52.66	8.053	5.16	0.812	$Sr(ClO_3)\cdot 3H_2O$
-20	50.86	7.712	6.54	1.020	"
-20	50.14	7.566	7.01	1.088	"
-20	51.33	7.739	5.78	0.896	"
-23	49.97	7.485	6.81	1.049	"
-23.5	51.08	7.653	5.71	0.880	"

AUXILIARY INFORMATION

METHOD/APPARATUS/PROCEDURE:

Known complexes were prepared by weighing together anhydrous $Sr(ClO_3)_2$, $SrBr_2\cdot 6H_2O$ and water. Sufficient water was always present to dissolve the salts completely at about room temperature, and when they had dissolved the solutions were cooled with dry ice and seeded with $Sr(ClO_3)_2\cdot 3H_2O$. The mixtures were placed in a bath of melting ethylene glycol (about -20°C) and stirred for 1-2.5 hr. Separate samples of each solution were analyzed for (1) bromide, by Volhard titration, and (2) total halide, after reduction of ClO_3^- to Cl^- with NO_2^-.

SOURCE AND PURITY OF MATERIALS:

Strontium chlorate anhydrate was prepared as follows: A chloric acid solution was prepared from roughly equivalent quantities of c.p. grade $Ba(ClO_3)_2$ and H_2SO_4, and small amounts of BaO and H_2SO_4 were then added. Excess c.p. grade $SrCO_3$ (previously leached with a large volume of boiling water) was then added to the chloric acid solution. The mixture was filtered, and evaporation by boiling yielded pure anhydrous $Sr(ClO_3)_2$. The solid was recrystallized from water, air-dried, and stored at room temperature. A qualitative flame test showed that no sodium, and only traces of calcium were present. Analysis by reduction to chloride and Volhard titration showed 99.6% $Sr(ClO_3)_2$. Loss in weight upon drying at 110°C was 0.26%.

ESTIMATED ERROR:

Soly: accuracy 0.1 - 0.2%
Temp: ± 0.1°C (author)

COMPONENTS:

(1) Strontium bromate; $Sr(BrO_3)_2$; [14519-18-7]

(2) Water; H_2O; [7732-18-5]

EVALUATOR:

Hiroshi Miyamoto
Department of Chemistry
Niigata University
Niigata, Japan

May, 1982

CRITICAL EVALUATION:

Solubility in the binary $Sr(BrO_3)_2$-H_2O system

Data for the binary $Sr(BrO_3)_2$-H_2O system have been reported by Linke (1) only. He determined the bromate content iodometrically.

Depending upon temperature and composition, equilibrated solid phases of varying degrees of hydration have been reported by Linke (1). The following solid phases have been identified:

$Sr(BrO_3)_2 \cdot H_2O$	[10022-52-3]
$Sr(BrO_3)_2$	[14519-18-7]

The relation between the solubility and the temperature is given in Fig. 1.

The eutectic of the system $Sr(BrO_3)_2$-H_2O is at 270.97K and 17.50 mass % $Sr(BrO_3)_2$ with $Sr(BrO_3)_2 \cdot H_2O$ and ice as solids. The transition from monohydrate to anhydrous salt occurs at 348.7K, and a saturated solution boils at 377K and 41.0 mass % $Sr(BrO_3)_2$.

The data reported in (1) obtained by Linke are tentative values. The tentative values were fitted to the following smoothing equations.

$$\ln(S_i/\text{mol kg}^{-1}) = 37880.51 - 51198.15/(T/100\text{K}) - 19046.81 \ln(T/100\text{K}) \quad : \quad \sigma = 0.008$$

$$\ln(S_1/\text{mol kg}^{-1}) = 28.00873 - 47.35544/(T/100\text{K}) - 11.03203 \ln(T/100\text{K}) \quad : \quad \sigma = 0.014$$

$$\ln(S_0/\text{mol kg}^{-1}) = 46.49750 - 73.88421/(T/100\text{K}) - 19.73933 \ln(T/100\text{K}) \quad : \quad \sigma = 0.005$$

where S_i is the solubility of strontium bromate with ice as solid phase, S_1 and S_0 are the solubilities of the monohydrate and the anhydrous salt, respectively.

The tentative values with the values calculated from the smoothing equations are given in Table 1.

COMPONENTS:	EVALUATOR:
(1) Strontium bromate; $Sr(BrO_3)_2$; [145197-18-7] (2) Water; H_2O; [7732-18-5]	Hiroshi Miyamoto Department of Chemistry Niigata University Niigata, JAPAN May 1982

CRITICAL EVALUATION:

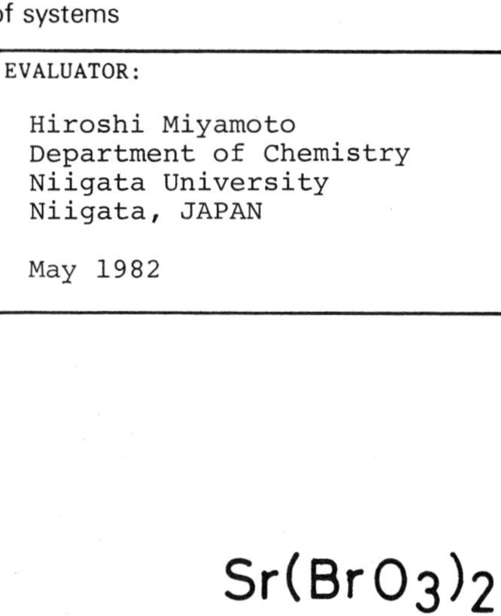

Figure 1. Strontium bromate + Water *vs.* Temperature

COMPONENTS:

(1) Strontium bromate; $Sr(BrO_3)_2$; [14519-18-7]

(2) Water; H_2O; [7732-18-5]

EVALUATOR:

Hiroshi Miyamoto
Department of Chemistry
Niigata University
Niigata, Japan

May, 1982

CRITICAL EVALUATION:

Table 1 Tentative values for the solubility of strontium bromate in water

T/K	m_1(exptl)/mol kg^{-1}	m_1(calcd)/mol kg^{-1}	Solid Phase
270.87	0.6601	0.6556	Ice
271.05	0.5901	0.5927	"
271.12	0.5608	0.5687	"
271.31	0.5114	0.5050	"
272.01	0.301	0.301	"
270.97	0.6177		Ice + $Sr(BrO_3)_2 \cdot H_2O$
273.2	0.6531	0.6620	$Sr(BrO_3)_2 \cdot H_2O$
277.6	0.7330	0.7305	"
288.2	0.9180	0.9049	"
298.2	1.091	1.078	"
308.2	1.250	1.254	"
318.2	1.414	1.429	"
328.2	1.578	1.598	"
338.2	1.752	1.758	"
345.2	1.873	1.863	"
348.2	1.921	1.906	"
348.7	1.933		$Sr(BrO_3)_2 \cdot H_2O$ + $Sr(BrO_3)_2$
349.2	1.93	1.92_6	$Sr(BrO_3)_2$
350.2	1.93	1.93_4	"
352.2	1.95	1.94_8	"
353.2	1.95	1.95_5	"
355.2	1.971	1.968	"
358.2	1.990	1.984	"
363.2	2.003	2.005	"
368.2	2.014	2.017	"
377.2	2.023	2.022	"

REFERENCES:

1. Linke, W. F. *J. Am. Chem. Soc.* 1953, *75*, 5797.

COMPONENTS:	ORIGINAL MEASUREMENTS:
(1) Strontium bromate; $Sr(BrO_3)_2$; [14519-18-7] (2) Water; H_2O; [7732-18-5]	Linke, W. F. *J. Am. Chem. Soc.* 1953, *75*, 5797-800.

EXPERIMENTAL VALUES:

t/°C	Strontium Bromate			Density	Nature of the Solid Phase[a]
	mass %	mol % (compiler)	m_1/mol kg^{-1} (compiler)	ρ/g cm^{-3}	
- 2.28[b]	18.48[b]	1.175	0.6601	--	Ice
- 2.10	16.85	1.052	0.5901	--	"
- 2.03	16.15	1.000	0.5608	--	"
- 1.84	14.94	0.9130	0.5114	--	"
- 1.14	9.38	0.540	0.301	--	"
- 2.18	17.50[c]	1.100	0.6177	1.165[c]	Ice + A
0	18.32	1.163	0.6531	1.177	A
4.4	20.11	1.303	0.7330	1.199	"
15	23.97	1.627	0.9180	1.241	"
25	27.25	1.927	1.091	1.285	"
35	30.03	2.202	1.250	1.320	"
45	32.69	2.484	1.414	1.356	"
55	35.15	2.765	1.578	1.384	"
65	37.57	3.060	1.752	1.422	"
72	39.15	3.265	1.873	--	"
75	39.75	3.345	1.921	1.458	"
75.5 ± 0.5	39.9 [c]	3.365	1.933	1.458[c]	A + B
76	39.9	3.37	1.93	--	B
77	39.9	3.37	1.93	1.457	"
79	40.1	3.39	1.95	--	"
80	40.1	3.39	1.95	1.461	"
82	40.37	3.430	1.971	--	"
85	40.60	3.461	1.990	1.462	"
90	40.75	3.482	2.003	1.465	"
95	40.89	3.502	2.014	1.465	"
104 ± 1	41.0[c]	3.517	2.023	1.470[c]	"

[a] A = $Sr(BrO_3)_2 \cdot H_2O$; B = $Sr(BrO_3)_2$

[b] metastable

[c] determined graphically

COMPONENTS:

(1) Strontium bromate; $Sr(BrO_3)_2$; [14519-18-7]

(2) Water; H_2O; [7732-18-5]

ORIGINAL MEASUREMENTS:

Linke, W. F.

J. Am. Chem. Soc. 1953, *75*, 5797-800.

VARIABLES:

T/K = 270.8 to 377

PREPARED BY:

Hiroshi Miyamoto

EXPERIMENTAL VALUES:

AUXILIARY INFORMATION

METHOD/APPARATUS/PROCEDURE:

Temperatures between 0 and 100°C were maintained in water-baths, and at 110°C an acetic acid vapor bath was used (1). At 0°C and below, baths of melting ice and of melting mono-, di- and triethylene glycol were employed. Freezing points were determined from the cooling curves of known mixtures.
Equilibrium in saturated solutions was established by repeated analysis after several hours of internal stirring. Representative points were checked by approach from super-saturation. Each reported value is the average of at least two closely agreeing determinations. Filtered samples of the solution were with-drawn with preheated calibrated pipets, and approximate densities were calculated.
The bromate content was determined iodometrically.

SOURCE AND PURITY OF MATERIALS:

Strontium bromate monohydrate was prepared as follows: A bromic acid was prepared from roughly equivalent quantities c.p. grade $Ba(BrO_3)_2$ and H_2SO_4. Excess c.p. grade $SrCO_3$ (previously leached with a large volume of boiling water) was then added to the bromic acid solution. The mixture was filtered, and then the solution of $Sr(BrO_3)_2$ was evapo-rated. The salt was recrystallized from water and air-dried. Iodometry showed 95.2% $Sr(BrO_3)_2$ (Calcd. for $Sr(BrO_3)_2 \cdot H_2O$ 95.0%). Loss in weight at 110°C was 4.89% (Calcd. 4.99%).

ESTIMATED ERROR:

Soly: accuracy 0.1 - 0.2%
Temp: ± 0.1°C (compiler assumes)

REFERENCES:

1. Linke, W. F. *J. Chem. Educ.* 1952, *29*, 492.

COMPONENTS:	EVALUATOR:
(1) Strontium iodate; $Sr(IO_3)_2$; [13470-01-4] (2) Water; H_2O; [7732-18-5]	Hiroshi Miyamoto Department of Chemistry Niigata University Niigata, Japan March, 1982

CRITICAL EVALUATION:

1. The binary system: $Sr(IO_3)_2$-H_2O

Solubilities in the binary $Sr(IO_3)_2$-H_2O system have been reported in 6 publications (1-6), and are summarized in Table 1.

Table 1 Solubility studies of strontium iodate in water

Reference	T/K	Solid Phase	Soly and/or Soly Product	Method of Analysis
Colman-Porter; Monk(1)	298	$Sr(IO_3)_2$	Soly, K°_{s0}	iodometric
Linke(2)	273-373	$Sr(IO_3)_2$	Soly	iodometric
	273-373	$Sr(IO_3)_2 \cdot H_2O$	"	"
	273-298	$Sr(IO_3)_2 \cdot 6H_2O$	"	"
Bousquet; Mathurin;	280.5-373	$Sr(IO_3)_2$	K°_{s0}	iodometric
Vermande(3)	283-313	$Sr(IO_3)_2 \cdot H_2O$	"	"
	273-298	$Sr(IO_3)_2 \cdot 6H_2O$	"	"
Miyamoto(4)	298	$Sr(IO_3)_2 \cdot H_2O$	Soly	iodometric
Miyamoto; Suzuki; Yanai(5)	293,298 303	$Sr(IO_3)_2$	Soly	iodometric
Vinogradov; Azarova; Pakhomov(6)	323	$Sr(IO_3)_2 \cdot H_2O$	Soly	complexometric(Sr^{2+}) iodometric(IO_3^-)

Linke (2) measured solubilities in the binary $Sr(IO_3)_2$-H_2O over a wide temperature range from 273.2 to 393.2K. Other investigations deal with ternary systems, and include the solubility in the binary system.

Solubilities of strontium iodate in aqueous NaOH solutions have been reported by Colman-Porter (1), and in aqueous NaCl solutions by Bousquet, Mathurin and Vermande (3) who also determined the solubility in the binary $Sr(IO_3)_2$-H_2O at 298.2K. Vinogradov, Azarova and Pakhomov (6) studied the solubility in the ternary $Sr(IO_3)_2$-HIO_3-H_2O by the iodometric method and the solubility in the binary system is given as one point on the phase diagram.

Kolosov (7) studied the solubility of strontium iodate in aqueous KIO_3 containing 1 mol dm^{-3} $HClO_4$, and Fedorov, Pobov, Shmyd'ko, Vorontsova and Mironov (8) measured solubilities in aqueous $LiNO_3$ solutions containing $LiClO_4$ to adjust the ionic strength.

Depending upon temperature and composition, equilibrated solid phases of varying degrees of hydration have been reported. The following solid phases have been identified:

$Sr(IO_3)_2 \cdot 6H_2O$	[7790-36-5]
$Sr(IO_3)_2 \cdot H_2O$	[19495-49-9]
$Sr(IO_3)_2$	[13470-01-4]

COMPONENTS:	EVALUATOR:
(1) Strontium iodate; $Sr(IO_3)_2$; [13470-01-4] (2) Water; H_2O; [7732-18-5]	Hiroshi Miyamoto Department of Chemistry Niigata University Niigata, Japan March, 1982

CRITICAL EVALUATION:

The temperature dependence of the solubility of strontium iodate in pure water has been reported by Linke (2). The system $Sr(IO_3)_2$-H_2O was studied from 273.2 to 368.2K. The relation between the solubility (based on mass %) of strontium iodate in pure water and the temperature studied is given in Fig. 1 which is based on Linke's data (2).

Below 279.2K the stable phase is the hexahydrate, and above 279.2 the solid phase is the anhydrous salt. The monohydrate is metastable with respect to the anhydrous salt at all temperatures, and with respect to the hexahydrate below 294.0K. Linke determined the transition temperatures graphically.

The temperature dependence of the solubility product of this salt at zero ionic strength has been reported by Bousquet, Mathurin and Vermande (3). They also determined the transition temperatures graphically. At 295K the hexahydrate is changed to the monohydrate, and at 280K the hexahydrate is converted to the anhydrous salt, and the transition temperatures are in good agreement with the values reported by Linke (2). They do not state that the monohydrate is metastable.

The solubility (based on mass %) of the monohydrate salt in pure water has been reported by Linke (2) over a wide temperature range, and by Vinogradov, Azarova and Pakhomov (6) at 323K only. The result reported in (6) obtained by Vinogradov, Azarova and Pakhomov is considerably lower than the interpolated value of Linke (2) who studied the solubility of the monohydrate salt in water systematically. Therefore in curve fitting, the value reported in (6) was not used.

Solubilities of the anhydrous salt and the hexahydrate have also been reported by Linke (2). The tentative values for the solubility of the hexahydrate, the monohydrate and the anhydrous salt are based on Linke's data. The tentative values are given in Table 2 with the values calculated by using the best fit equations.

In Table 2 m(exptl) is a tentative value and m(calcd) is a value calculated from the best fit equation.

The best fit for Linke's values gave:

$$\ln(S_0/\text{mmol kg}^{-1}) = 26.99442 - 44.11388/(T/100\text{K}) - 9.948777 \ln(T/100\text{K}) \quad : \quad \sigma = 0.092$$

$$\ln(S_1/\text{mmol kg}^{-1}) = 30.03693 - 49.35209/(T/100\text{K}) - 10.70880 \ln(T/100\text{K}) \quad : \quad \sigma = 0.033$$

$$\ln(S_6/\text{mmol kg}^{-1}) = 6.492387 - 27.61725/(T/100\text{K}) + 4.271492 \ln(T/100\text{K}) \quad : \quad \sigma = 0.032$$

where S_0, S_1 and S_6 are the solubilities of the anhydrous salt, the monohydrate and the hexahydrate, respectively.

COMPONENTS:	EVALUATOR:
(1) Strontium iodate; $Sr(IO_3)_2$; [13470-01-4] (2) Water; H_2O; [7732-18-5]	Hiroshi Miyamoto Department of Chemistry Niigata University Niigata, JAPAN March, 1982

CRITICAL EVALUATION:

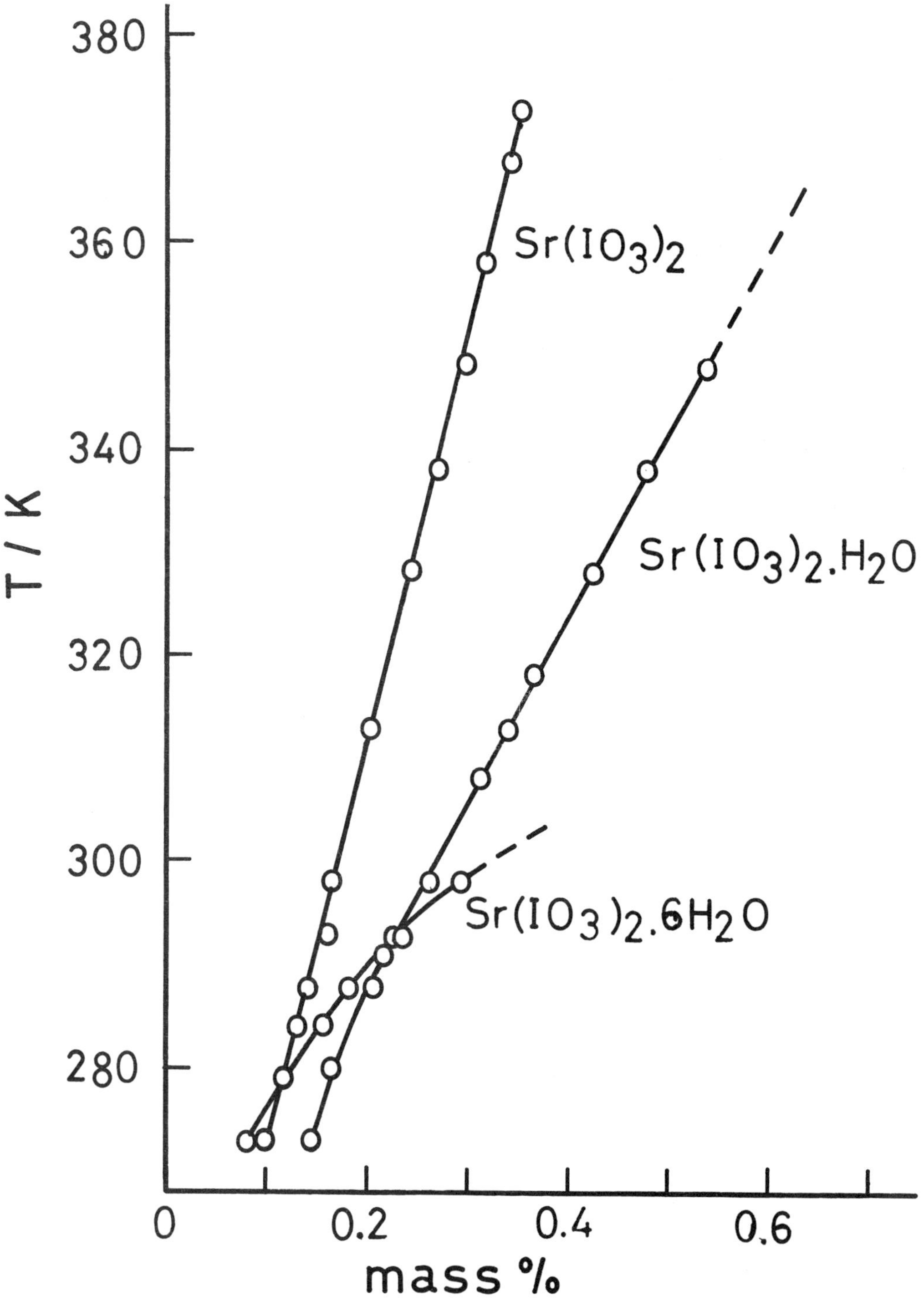

Figure 1. Strontium iodate + Water *vs.* Temperature

COMPONENTS:

(1) Strontium iodate; $Sr(IO_3)_2$; [13470-01-4]

(2) Water; H_2O; [7732-18-5]

EVALUATOR:

Hiroshi Miyamoto
Department of Chemistry
Niigata University
Niigata, Japan

March, 1982

CRITICAL EVALUATION:

Table 2 Tentative values for solubility of strontium iodate in water

T/K	10^3m_1(exptl)/mol kg^{-1}	10^3m_1(calcd)/mol kg^{-1}
	Solid phase: $Sr(IO_3)_2$	
273.2	2.243	2.336
279.2	2.678	2.663
280.2	2.747	2.719
284.2	2.976	2.947
288.2	3.159	3.181
293.2	3.687	3.480
298.2	3.778	3.785
313.2	4.650	4.718
328.2	5.545	5.638
338.2	6.120	6.224
348.2	6.787	6.774
358.2	7.224	7.280
368.2	7.799	7.734
373.2	8.029	7.941
	Solid phase: $Sr(IO_3)_2 \cdot H_2O$	
273.2	3.327	3.353
280.2	4.038	4.015
288.2	4.845	4.843
293.2	5.429	5.395
294.2	5.523	5.508
298.2	5.973	5.969
308.2	7.146	7.173
313.2	7.797	7.796
318.2	8.391	8.428
328.2	9.679	9.707
338.2	10.955	10.980
348.2	12.283	12.220
	Solid phase: $Sr(IO_3)_2 \cdot 6H_2O$	
273.2	1.970	1.967
279.2	2.678	2.682
280.2	2.815	2.821
284.2	3.446	3.443
288.2	4.173	4.183
291.2	4.891	4.826
293.2	5.274	5.301
294.2	5.523	5.554
298.2	6.681	6.673

The solubility (based on mol dm^{-3} units) of the monohydrate at 298.2K has been reported by Colman-Porter and Monk (1), Bousquet, Mathurin and Vermande (3), and Miyamoto (4). The arithmetic mean of the values reported in these publications is 5.90 mmol dm^{-3} and the standard deviation is 0.13 mmol dm^{-3}.

COMPONENTS:	EVALUATOR:
(1) Strontium iodate; $Sr(IO_3)_2$; [13470-01-4] (2) Water; H_2O; [7732-18-5]	Hiroshi Miyamoto Department of Chemistry Niigata University Niigata, Japan March, 1982

CRITICAL EVALUATION:

Solubilities (based on mol dm^{-3} units) of the anhydrous salt at 293.2, 298.2 and 303.2K have been reported by Miyamoto, Suzuki and Yanai (5). The results reported in (5) cannot be compared to these of Linke because the densities of the saturated solutions were not given in either publication.

The recommended and tentative values based on mol dm^{-3} units are given in Table 3.

Table 3 Recommended and tentative values for solubility of strontium iodate in water

T/K	10^3c/mol dm^{-3}	Solid Phase
	Recommended value	
298.2	5.90	$Sr(IO_3)_2 \cdot H_2O$
	Tentative values	
293.2	3.44	$Sr(IO_3)_2$
298.2	3.75	"
303.2	3.98	"

2. Solubility of strontium iodate in alkali solutions

Solubilities of strontium iodate in aqueous NaOH solutions have been reported by Colman-Porter and Monk (1). Solubilities increase with increasing concentration of NaOH. They calculated the dissociation constant of ion-pair from the solubility data. The dissociation constant of the ion-pair $Sr(IO_3)^+$ found to be 0.10 mol dm^{-3} taking into account the existence of $Sr(OH)^+$ and $NaIO_3$, and the value agrees with the result obtained by the conductivity method (1).

3. Solubility product of strontium iodate in aqueous solutions

The solubility product of strontium iodate in aqueous solutions reported by Bousquet, Mathurin and Vermande (3), Kolosov (7), and Fedorov, Robov, Shymyd'ko, Vorontsova and Mironov (8).

Bousquet, Mathurin and Vermande measured solubilities of strontium iodate hexahydrate, monohydrate and anhydrate in aqueous NaCl solutions over a wide temperature range, and calculated the solubility product, K°_{s0}, at zero ionic strength from the solubility data.

Fedorov, Robov, Shymyd'ko, Vorontsova and Mironov (8) measured solubilities of aqueous $LiNO_3$ solutions containing $LiClO_4$ at 298.2K. They calculated the solubility products (K°_{s0} and K_{s0}) of strontium iodate in aqueous solutions from the solubility data, and both the activity product and the concentration solubility product were reported.

Kolosov (7) measured solubilities of strontium in aqueous $HClO_4$ solutions and in aqueous $HClO_4$ solutions containing KIO_3 at 293.2K, and calculated the solubility product from the solubility data. He calculated the concentration solubility product (K_{s0}), but did not determine the thermodynamic solubility product (K°_{s0}). He also did not report the degree of hydration of the salt used.

COMPONENTS:	EVALUATOR:
(1) Strontium iodate; $Sr(IO_3)_2$; [13470-01-4] (2) Water; H_2O; [7732-18-5]	Hiroshi Miyamoto Department of Chemistry Niigata University Niigata, Japan March, 1982

CRITICAL EVALUATION:

Solubility product at 298.2K. Bousquet, Mathurin and Vermande (3) have reported the thermodynamic solubility product varying the degree of hydration at this temperature. The result obtained is 4.55×10^{-7} mol^3 dm^{-9} for the hexahydrate, 3.775×10^{-7} mol^3 dm^{-9} for the monohydrate and 1.14×10^{-7} mol^3 dm^{-9} for the anhydrous salt.
Colman-Porter and Monk (1) report that taking into account the presence of the ion-pair $Sr(IO_3)^+$, the activity solubility product is 3.289×10^{-7} mol^3 dm^{-9} at 298K. They used the monohydrate in their study, nevertheless, the result is considerably lower than that reported by Bousquet, Mathurin and Vermande.
The result reported in (8) obtained by Fedorov, Robov, Shmyd'ko, Vorontsova and Mironov is 2.94×10^{-7} mol^3 dm^{-9}. This value is also in poor agreement with the result of Bousquet, Mathurin and Vermande at 298.2K. In fitting the data to the smoothing equation, the results from (1) and (8) were not used.

The best fit for the values of Bousquet's group give:

$$\ln K^\circ_{s0}\,(1) = -162.9273 + 150.0571/(T/100\text{K}) + 89.75125 \ln (T/100\text{K}) \quad : \quad \sigma = 0.25 \times 10^{-7}$$

$$\ln K^\circ_{s0}\,(2) = 190.3871 - 314.7354/(T/100\text{K}) - 91.20536 \ln (T/100\text{K}) \quad : \quad \sigma = 0.057 \times 10^{-7}$$

$$\ln K^\circ_{s0}\,(3) = 40.60648 - 100.6973/(T/100\text{K}) - 20.93687 \ln (T/100\text{K}) \quad : \quad \sigma = 0.038 \times 10^{-7}$$

where K°_{s0} (1), K°_{s0} (2), and K°_{s0} (3) are the solubility products for the hexahydrate, the monohydrate and the anhydrous salt, respectively.

The values reported by Bousquet, Mathurin and Vermande are designated as tentative values. The tentative values are given in Table 4 along with the values calculated by using the smoothing equations. In Table 4, K°_{s0} (exptl) is the tentative value, and K°_{s0} (calcd) is the calculated value.

COMPONENTS:

(1) Strontium iodate; $Sr(IO_3)_2$; [13470-01-4]

(2) Water; H_2O; [7732-18-5]

EVALUATOR:

Hiroshi Miyamoto
Department of Chemistry
Niigata University
Niigata, Japan

March, 1982

CRITICAL EVALUATION:

Table 4 Tentative and calculated values for solubility product of strontium iodate at zero ionic strength

T/K	$10^7 K^\circ_{s0}$(exptl)/mol^3 dm^{-9}	$10^7 K^\circ_{s0}$(calcd)/mol^3 dm^{-9}
	Solid phase: $Sr(IO_3)_2 \cdot 6H_2O$	
273.2	0.1936	0.1863
281.2	0.4478	0.5206
285.2	0.976	0.875
295.2	3.452	3.246
298.2	4.55	4.82
	Solid phase: $Sr(IO_3)_2 \cdot H_2O$	
288.2	2.143	2.135
293.2	2.825	2.864
298.2	3.775	3.705
303.2	4.581	4.635
308.2	5.636	5.618
313.2	6.607	6.609
	Solid phase: $Sr(IO_3)_2$	
280.5	0.475	0.463
290.2	0.714	0.755
298.2	1.11	1.08
308.2	1.63	1.63
315.2	2.11	2.10
323.2	2.77	2.74
333.2	3.64	3.68

REFERENCES:

1. Colman-Porter, C. A.; Monk, C. B. *J. Chem. Soc.* 1952, 1312.

2. Linke, W. F. *J. Am. Chem. Soc.* 1953, *75*, 5797.

3. Bousquet, J.; Mathurin, D.; Vermande, P. *Bull. Soc. Chim. Fr.* 1969, 1111.

4. Miyamoto, H. *Nippon Kagaku Kaishi* 1972, 659.

5. Miyamoto, H.; Suzuki, K.; Yanai, K. *Nippon Kagaku Kaishi* 1978, 1050.

6. Vinogradov, E. E.; Azarova, L. A.; Pakhomov, V. I. *Zh. Neorg. Khim.* 1978, *23*, 534; *Russ. J. Inorg. Chem. (Engl. Transl.)* 1978, *23*, 297.

7. Kolosov, L. V. *Zh. Neorg. Khim.* 1965, *10*, 2200; *Russ. J. Inorg. Chem. (Engl. Transl.)* 1965, *10*, 1197.

8. Fedorov, V. A.; Robov, A. M.; Shmyd'ko, I. I.; Vorontsova, N. A.; Mironov, V. E. *Zh. Neorg. Khim.* 1974, *19*, 1946; *Russ. J. Inorg. Chem. (Engl. Transl.)* 1974, *19*, 950.

COMPONENTS:

(1) Strontium iodate; $Sr(IO_3)_2$; [13470-01-4]

(2) Water; H_2O; [7732-18-5]

ORIGINAL MEASUREMENTS:

Linke, W. F.

J. Am. Chem. Soc. 1953, *75*, 5797-800.

VARIABLES:

T/K = 273 - 368

PREPARED BY:

Hiroshi Miyamoto

EXPERIMENTAL VALUES:

t/°C	Strontium Iodate					
	Anhydrate[a]		Monohydrate[a]		Hexahydrate[a]	
	mass %	$10^3 m_1$/mol kg^{-1} (compiler)	mass %	$10^3 m_1$/mol kg^{-1} (compiler)	mass %	$10^3 m_1$/mol kg^{-1} (compiler)
0	0.098m	2.243	0.1453m	3.327	0.0861	1.970
t 6±1g	0.117g	2.678	--	--	0.117g	2.678
7	0.120	2.747	0.1763m	4.038	0.1230m	2.815
11	0.130	2.976	--	--	0.1505m	3.446
15	0.138	3.159	0.2115m	4.845	0.1822m	4.173
18	--	--	--	--	0.2135m	4.891
20	0.161	3.687	0.2369m	5.429	0.2302m	5.274
t20.8 ±0.5g	--	--	0.241m,g	5.523	0.241m	5.523
25	0.165	3.778	0.2606m	5.973	0.2914m	6.681
35	--	--	0.3116m	7.146	--	--
40	0.203	4.650	0.3399m	7.797	--	--
45	--	--	0.3657m	8.391	--	--
55	0.242	5.545	0.4216m	9.679	--	--
65	0.267	6.120	0.4769m	10.955	--	--
75	0.296	6.787	0.5344m	12.283	--	--
85	0.315	7.224	--	--	--	--
95	0.340	7.799	--	--	--	--
100	0.350g	8.029	0.68m,g	15.651	--	--

[a] Nature of the solid phase.

m = metastable; g = determined graphically; t = transition temperature.

AUXILIARY INFORMATION

METHOD/APPARATUS/PROCEDURE:

Equilibrium was attained after 1-2 hours of stirring when either hexahydrate or monohydrate was the saturating phase. The anhydrous salt reached equilibrium much more slowly, and in one case at a low temperature had not reached equilibrium after 6 hours of stirring. Representative points were checked by approach from supersaturation. Each reported value is the average of at least two closely agreeing determinations. Filtered samples of the solution were withdrawn with preheated calibrated pipets, and approximate densities were calculated. The iodate content was determined iodometrically.

SOURCE AND PURITY OF MATERIALS:

$Sr(IO_3)_2 \cdot H_2O$ was prepared by adding solns containing equivalent quantities of c.p. grade $SrCl_2 \cdot 6H_2O$ and HIO_3 in a large volume of water at 24°C. White finely crystalline $Sr(IO_3)_2 \cdot H_2O$ settled rapidly and was washed by decantation until no test for Cl^- ions was obtained. The salt air-dired to a fluffy white powder, and analysis by iodometry and loss in weight at 110°C showed 96.0% $Sr(IO_3)_2$ in $Sr(IO_3)_2 \cdot H_2O$.
$Sr(IO_3)_2 \cdot 6H_2O$ was prepared in the same manner as the monohydrate, except that all solutions and washed water were cooled with ice. Drying with acetone produced a fluffy powder which contained 79.85% $Sr(IO_3)_2$ (calcd for $Sr(IO_3)_2 \cdot 6H_2O$ 80.18%). Anhydrous $Sr(IO_3)_2$ was obtained by heating either the mono- or hexahydrate to 110°C for a few hours, or by boiling them with water.

ESTIMATED ERROR:

Soly: the estimated accuracy is ± 0.005 mass % for solns saturated with the anhydrous salt, and ± 0.0005 mass% when the hydrates were present.

Temp: the maximum variation never exceed ± 0.1°C, and was usually much less.

COMPONENTS:	ORIGINAL MEASUREMENTS:
(1) Strontium iodate; $Sr(IO_3)_2$; [13470-01-4]	Bousqet, J.; Mathurin, D.; Vermande, P.
(2) Sodium chloride; NaCl; [7647-14-5]	*Bull. Soc. Chim. Fr.* 1969, 1111-5.
(3) Water; H_2O; [7732-18-5]	

EXPERIMENTAL VALUES:

t/°C	Sodium Chloride	Strontium Iodate				
	c_2/mol dm^{-3}	10^3c_1/mol dm^{-3}	(1/3)log4c_1	I	$Z_+Z_-AI^{1/2}$	y
25	0	6.04	-2.0182	0.0181	0.1375	-2.1557
	0.0125	6.58	-1.9811	0.0322	0.184	-2.1651
	0.0250	6.92	-1.9592	0.0458	0.219	-2.1782
	0.0500	7.47	-1.9260	0.0724	0.275	-2.201
	0.100	8.25	-1.8828	0.1248	0.3615	-2.2443

y = -BI + 1/3 log K°_{s0}, where I = ionic strength, K°_{s0} = activity solubility product.

t/°C	Strontium Iodate Solubility Product		
	Hexahydrate $10^8K^\circ_{s0}$/mol^{3}dm^{-9}	Monohydrate $10^7K^\circ_{s0}$/mol^{3}dm^{-9}	Anhydrate $10^8K^\circ_{s0}$/mol^{3}dm^{-9}
0	1.936	--	--
7.3	--	--	4.75
8	4.478	--	--
12	9.76	--	--
15	--	2.143	--
17	--	--	7.14
20	--	2.825	--
22	34.52	--	--
25	45.5	3.775	11.1
30		4.581	--
35		5.636	16.3
40		6.607	--
42			21.1
50			27.7
60			36.4

The solubility product, K°_{s0}, of $Sr(IO_3)_2 \cdot xH_2O$ was given in the following:

$$K^\circ_{s0} = (C_{Sr^{2+}} \times C^2_{IO_3^-})(y_{Sr^{2+}} \times y^2_{IO_3^-}) = 4S^3y^3_\pm \quad (1)$$

where S represents solubility of iodate, $y_\pm$ is an activity coefficient, and is given by modified Debye-Hückel equation

$$-\log y_\pm = Z_+Z_- A \sqrt{I} - BI \quad (2)$$

From (1) and (2)

$$Y = -BI + 1/3 \log K^\circ_{s0} \quad (3)$$

where $Y = 1/3 \log (4S^3) - Z_+Z_-A\sqrt{I}$, and A = 0.5115 at 25°C

Solubility product (K°_{s0}) and unknown constant (B) are evaluated from the intercept and the slope of Y vs I plots, respectively.

COMPONENTS:

(1) Strontium iodate; $Sr(IO_3)_2$; [13470-01-4]

(2) Sodium chloride; NaCl; [7647-14-5]

(3) Water; H_2O; [7732-18-5]

ORIGINAL MEASUREMENTS:

Bousqet, J.; Mathurin, D.; Vermande, P.

Bull. Soc. Chim. Fr. 1969, 1111-5.

VARIABLES:

c_2/mol dm^{-3} = 0 to 0.100

T/K = 273 to 333

PREPARED BY:

Hiroshi Miyamoto

EXPERIMENTAL VALUES:

AUXILIARY INFORMATION

METHOD/APPARATUS/PROCEDURE:

The aqueous NaCl solutions and desired hydrate crystals were placed in sintered glass-stoppered Erlenmeyer flasks. The flasks were stirred in a thermostat for 1-15 hours. The iodate content was determined iodometrically.

SOURCE AND PURITY OF MATERIALS:

$Sr(IO_3)_2 \cdot 6H_2O$ was prepared by mixing dilute solutions of strontium chloride and KIO_3 at 6°C or lower. The precipitates were washed with water. $Sr(IO_3)_2 \cdot H_2O$ was prepared similarly at 25°C, and the anhydrate was prepared from the hydrate salt by dehydration at 200°C.

ESTIMATED ERROR:

Soly: nothing specified

Temp: ± 0.05°C (authors)

REFERENCES:

COMPONENTS:

(1) Strontium iodate; $Sr(IO_3)_2$; [13470-01-4]

(2) Sodium hydroxide; NaOH; [1310-73-2]

(3) Water; H_2O; [7732-18-5]

ORIGINAL MEASUREMENTS:

Colman-Porter, C. A.; Monk, C. B.

J. Chem. Soc. 1952, 1312-4.

VARIABLES:

$T/K = 298$

$10^3 c_2/\text{mol dm}^{-3} = 0$ to 45.99

PREPARED BY:

Hiroshi Miyamoto

EXPERIMENTAL VALUES:

1. The solubilities of $Sr(IO_3)_2$ in aqueous NaOH solutions at 25°C are:

Concn of NaOH $10^3 c_2/\text{mol dm}^{-3}$	Soly of $Sr(IO_3)_2$ $10^3 c_1/\text{mol dm}^{-3}$
0	5.87
16.70	6.55
22.31	6.73
30.21	6.98
45.99	7.40

2. The activity solubility product of $Sr(IO_3)_2$ is 3.289×10^{-7} $\text{mol}^3 \text{ dm}^{-9}$. In this calculation, the following activity coefficient expression used by Davies (1) was applied

$$-\log y_{\pm} = 0.5Z^2\{ I^{1/2}/(1 + I^{1/2}) - 0.2I \}$$

where I is the ionic strength.

AUXILIARY INFORMATION

METHOD/APPARATUS/PROCEDURE:

A saturating column method was used as described by Money and Davies (2). The concentration of strontium iodate in the saturated solution was determined iodometrically.

SOURCE AND PURITY OF MATERIALS:

Strontium iodate was formed by allowing about 0.1 mol dm^{-3} solutions of strontium chloride (AnalaR) and of KIO_3 to drip very slowly into a distilled water at room temperature. The product was dried by washing with acetone. Heating a sample of the crystals to 150°C showed these to be the monohydrate $Sr(IO_3)_2 \cdot H_2O$. A stock solution of carbonate-free NaOH (AnalaR) was used.

ESTIMATED ERROR:

Nothing specified

REFERENCES:

1. Davies, C. W. *J. Chem. Soc.* 1938, 2093.

2. Money, R. W.; Davies, C. W. *J. Chem. Soc.* 1934, 400.

COMPONENTS:

(1) Strontium iodate; $Sr(IO_3)_2$; [13470-01-4]
(2) Lithium perchlorate; $LiClO_4$; [7791-03-9]
(3) Lithium nitrate; $LiNO_3$; [7790-69-4]
(4) Water; H_2O; [7732-18-5]

ORIGINAL MEASUREMENTS:

Fedorov, V. A.; Robov, A. M.; Shmyd'ko, I. I.; Vorontsova, N. A.; Mironov, V. E.

Zh. Neorg. Khim. 1974, *19*, 1746-50; *Russ. J. Inorg. Chem. (Engl. Transl.)* 1974, *19*, 950-3.

VARIABLES:

T/K = 298
Concentration of $LiClO_4$ and $LiNO_3$

PREPARED BY:

Hiroshi Miyamoto

EXPERIMENTAL VALUES:

t/°C	Lithium Nitrate c_2/mol dm^{-3}	Strontium Iodate, 10^2c_1/mol dm^{-3}				
		Ionic Strength[a]				
		0.5	1.0	2.0	3.0	4.0
25	0	0.995	1.09	1.08	1.02	0.890
	0.1	1.03	--	--	--	--
	0.2	1.06	1.18	1.17	--	--
	0.3	1.09	--	--	--	--
	0.4	1.12	1.24	--	1.17	1.05
	0.5	1.15	--	1.30	--	--
	0.6		1.32	--	--	--
	0.8		1.36	1.43	1.32	1.22
	1.0		1.51	1.53	--	--
	1.2			--	1.48	1.42
	1.3			1.67	--	--
	1.5			1.75	--	--
	1.6			--	1.66	1.60
	1.8			1.89	--	--
	2.0			1.97	1.86	1.80
	2.4				2.08	2.05
	2.8				2.35	2.25
	3.0				2.50	--
	3.2					2.46
	3.6					2.68
	4.0					2.94

[a] The ionic strength adjusted by addition of lithium perchlorate to the lithium nitrate concentration given above.

AUXILIARY INFORMATION

METHOD/APPARATUS/PROCEDURE:

Equilibrium between $Sr(IO_3)_2$ crystals and the solution was reached by vigorous agitation with a magnetic stirrer in stoppered vessels in a thermostat. Equilibrium was established after stirring for 4-6 hours and was checked by removing specimens after equal intervals of time. The total concentration of $Sr(IO_3)_2$ in the saturated solutions was determined iodometrically.

SOURCE AND PURITY OF MATERIALS:

$Sr(IO_3)_2 \cdot H_2O$ was prepared by mixing solutions of $SrCl_2$ and KIO_3 at the temperature below 6°C. The product was washed with water.
$LiClO_4$ and $LiNO_3$ used were prepared from the chemically pure grade materials by recrystallization from twice-distilled water. Before recrystallization the solutions were boiled with active carbon.

ESTIMATED ERROR:

Soly: the reproducibility of the results averages 1.5 - 2%
Temp: not given

REFERENCES:

COMPONENTS:

(1) Strontium iodate; $Sr(IO_3)_2$; [13470-01-4]
(2) Potassium iodate; KIO_3; [7758-05-6]
(3) Perchloric acid; $HClO_4$; [7601-90-3]
(4) Water; H_2O; [7732-18-5]

ORIGINAL MEASUREMENTS:

Kolosov, I. V.

Zh. Neorg. Khim. 1965, *10*, 2200-2; *Russ. J. Inorg. Chem.* (*Engl. Transl.*) 1965, *10*, 1197-9.

VARIABLES:

T/K = 293
c_2/mol dm^{-3} = 0 - 3.74

PREPARED BY:

Hiroshi Miyamoto

EXPERIMENTAL VALUES:

t/°C	Perchloric Acid c_3/mol dm^{-3}	Potassium Iodate c_2/mol dm^{-3}		Strontium Iodate mg dm^{-3}	$10^3 c_1$/mol dm^{-3} [c]	$10^6 K_{s0}$/mol^{3}dm^{-9}
20	1.00	0.00		4.40	10.06	
				4.24	9.69	
				4.18	9.56	
			(Av)	4.27[a]	9.76	3.7
	1.00	1.88		2.14	4.89	
				2.03	4.64	
				2.20	5.03	
			(Av)	2.12[b]	4.85	1.7
	1.00	3.74		0.98	2.24	3.1

The solubility product (K_{s0} = $[Sr^{2+}][IO_3^-]^2$) of $Sr(IO_3)_2$ given by the author was 2.8 x 10^{-6} mol^3 dm^{-9}, which was the mean of values shown in Table.

[a] σ_n = 0.11, n = 3.

[b] σ_n = 0.09, n = 2.

[c] Molarities calculated by the compiler.

AUXILIARY INFORMATION

METHOD/APPARATUS/PROCEDURE:

The solubility of $Sr(IO_3)_2$ in 1.00 mol dm^{-3} $HClO_4$ solution containing KIO_3 at 20°C was studied. $HClO_4$ was used to adjust the ionic strength = 1.00. The solutions and $Sr(IO_3)_2$ crystals were mixed by a screw stirrer for 1-2 hours.
^{89}Sr like ^{90}Sr is practically a pure β emitter and the compiler presumes the method of analysis involves radio-assay techniques.

SOURCE AND PURITY OF MATERIALS:

$Sr(IO_3)_2$ was made from c.p. grade $SrCl_2 \cdot 6H_2O$ and KIO_3. The product was recrystallized twice from distilled water, and the precipitates were washed with water until the electrical conductivity of the wash water remained constant for two months. The radioactive isotope ^{89}Sr was introduced during the precipitation. The method of introduction was not given in the original paper.

ESTIMATED ERROR:

Soly: the relative square error of the observed value was within 2.3%.
Temp: not given

REFERENCES:

COMPONENTS:

(1) Strontium iodate; $Sr(IO_3)_2$; [13470-01-4]

(2) Iodic acid; HIO_3; [7782-68-5]

(3) Water; H_2O; [7732-18-5]

ORIGINAL MEASUREMENTS:

Vinogradov, E. E.; Azarova, L. A.; Pakhomov, V. I.

Zh. Neorg. Khim. 1978, *23*, 534-7; *Russ. J. Inorg. Chem. (Engl. Transl.)* 1978, *23*, 297-9.

VARIABLES:

T/K = 323

HIO_3/mass % = 0 - 78

PREPARED BY:

Hiroshi Miyamoto

EXPERIMENTAL VALUES:

t/°C	Composition of Saturated Solutions				Nature of the Solid Phase[a]
	Iodic Acid		Strontium Iodate		
	mass %	mol % (compiler)	mass %	mol % (compiler)	
50	--	--	0.263[b]	0.0109	A
	2.60	0.273	0.247	0.0104	A
	5.20	0.559	0.019	0.00082	A + B
	6.01	0.0651	0.020	0.00087	A + B
	8.44	0.936	0.036	0.0016	B
	20.31	2.545	0.060	0.0030	B
	47.46	8.477	0.064	0.0046	B
	65.39	16.245	0.086	0.0086	B
	71.97	20.866	0.080	0.0C93	B + C
	73.62	22.275	0.075	0.0091	C
	78.62	27.357	--	--	C

[a] A = $Sr(IO_3)_2 \cdot H_2O$; B = $Sr(IO_3)_2 \cdot HIO_3 \cdot H_2O$; C = HIO_3.

[b] For binary system the compiler computes the following

Soly $Sr(IO_3)_2$ = 6.03×10^{-3} mol kg^{-1} at 50°C

AUXILIARY INFORMATION

METHOD/APPARATUS/PROCEDURE:

The solubility study was carried out isothermally. The equilibrium of $Sr(IO_3)_2$-HIO_3-H_2O system was established in 14 days.
The concentration of strontium in the liquid phase was determined complexometrically, and the total concentration of the iodate was determined by iodometric titration. The composition and nature of the solid phases were found by Schreinemakers' method of "residue," X-ray diffraction, and the thermal analysis. The X-ray diffraction proved to form of the double compound of the type $Sr(IO_3)_2 \cdot HIO_3 \cdot H_2O$.

SOURCE AND PURITY OF MATERIALS:

Chemically pure grade HIO_3 was used. Strontium iodate was made from HIO_3 and strontium carbonate. The purity of the product was checked by chemical, X-ray diffraction and thermal analyses. The initial strontium iodate had the formula $Sr(IO_3)_2 \cdot H_2O$.

ESTIMATED ERROR:

Nothing specified

REFERENCES:

COMPONENTS:

(1) Strontium iodate; $Sr(IO_3)_2$; [13470-01-4]

(2) Iodic acid; HIO_3; [7782-68-5]

(3) Water; H_2O; [7732-18-5]

ORIGINAL MEASUREMENTS:

Vinogradov, E. E.; Azarova, L. A.; Pakhomov, V. I.

Zh. Neorg. Khim. 1978, *23*, 534-7; *Russ. J. Inorg. Chem. (Engl. Transl.)* 1978, *23*, 297-9.

COMMENTS AND/OR ADDITIONAL DATA:

The phase diagram is given below (based on mass%).

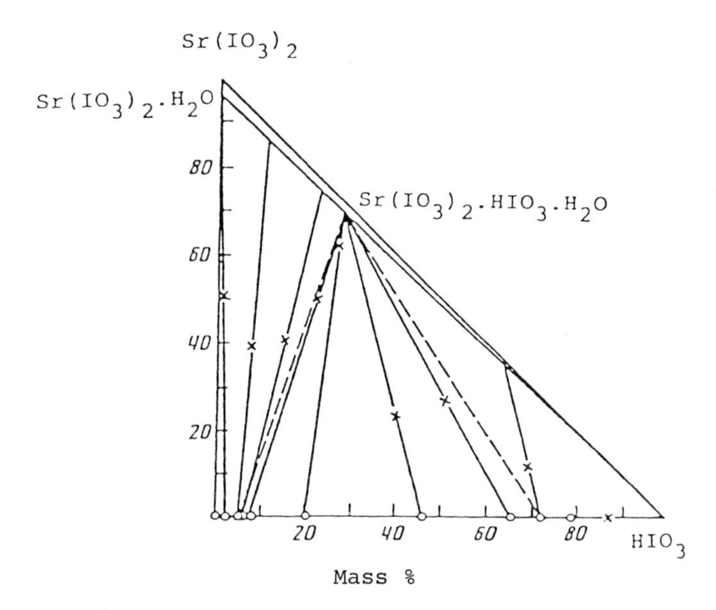

50 °C solubility isotherm in the $Sr(IO_3)_2$-HIO_3-H_2O system.

AUXILIARY INFORMATION

METHOD/APPARATUS/PROCEDURE:

SOURCE AND PURITY OF MATERIALS:

ESTIMATED ERROR:

ACKNOWLEDGEMENT:

The figure reprinted from *Zh. Neorg. Khim.* by permission of the copyright owners, VAAP, The copyright agency of the USSR.

COMPONENTS:

(1) Strontium iodate; $Sr(IO_3)_2$; [13470-01-4]

(2) Tetrahydrofuran; C_4H_8O; [109-99-9]

(3) Water; H_2O; [7732-18-5]

ORIGINAL MEASUREMENTS:

Miyamoto, H.

Nippon Kagaku Kaishi 1972, 659-61.

VARIABLES:

T/K = 298
Tetrahydrofuran/mass % = 0 - 40

PREPARED BY:

Hiroshi Miyamoto

EXPERIMENTAL VALUES:

t/°C	Tetrahydrofuran mass %	Tetrahydrofuran mol % (compiler)	Strontium Iodate 10^3c_1/mol dm^{-3}
25	0	0	5.78
	5	1.3	3.98
	10	2.7	2.78
	15	4.2	1.96
	20	5.9	1.34
	25	7.7	0.96
	30	9.7	0.68
	40	14.3	0.36

AUXILIARY INFORMATION

METHOD/APPARATUS/PROCEDURE:

Excess of $Sr(IO_3)_2 \cdot H_2O$ and solvent mixtures were placed in glass-stoppered bottles. The bottles were rotated in a thermostat at 25°C for 48 hours.
After the saturated solution settled, the solution was withdrawn through a siphon tube equipped with a glass-sintered filter.
The iodate content was determined iodometrically.

SOURCE AND PURITY OF MATERIALS:

$Sr(IO_3)_2 \cdot H_2O$ was prepared by adding solutions of $SrCl_2 \cdot 6H_2O$ (Wako Co guarantee reagent) and HIO_3 (Wako Co guarantee reagent) to a large volume of water containing KNO_3.
The precipitate was filtered off, washed and dried under reduced pressure.
Tetrahydrofuran was distilled from NaOH and then redistilled from sodium metal.

ESTIMATED ERROR:

Soly: nothing specified
Temp: ± 0.02°C (author)

REFERENCES:

COMPONENTS:

(1) Strontium iodate; $Sr(IO_3)_2$; [13470-01-4]

(2) N,N-Dimethylformamide; C_3H_7NO; [68-12-2]

(3) Water; H_2O; [7732-18-5]

ORIGINAL MEASUREMENTS:

Miyamoto, H.; Suzuki, K.; Yanai, K.

Nippon Kagaku Kaishi 1978, 1050-2.

VARIABLES:

T/K = 293, 298 and 303
Dimethylformamide/mass % = 0 to 41

PREPARED BY:

Hiroshi Miyamoto

EXPERIMENTAL VALUES:

t/°C	Dimethylformamide		Strontium Iodate
	mass %	mol % (compiler)	$10^3 c_1$/mol dm^{-3}
20	0	0	3.44
	4.92	1.26	2.62
	10.19	2.72	1.93
	15.12	4.21	1.39
	20.01	5.81	1.04
	25.17	7.66	0.78
	29.94	9.53	0.58
	40.71	14.47	0.29_2
25	0	0	3.75
	5.18	1.33	2.78
	9.97	2.66	2.14
	15.30	4.26	1.57
	19.96	5.79	1.21
	25.20	7.67	0.85
	29.90	9.51	0.64
	40.37	14.30	0.29_6
30	0	0	3.98
	4.91	1.26	3.03
	10.05	2.68	2.26
	14.74	4.09	1.74
	20.11	5.84	1.28
	25.68	7.85	0.88
	30.08	9.59	0.67
	40.05	14.14	0.31_6

AUXILIARY INFORMATION

METHOD/APPARATUS/PROCEDURE:

$Sr(IO_3)_2$ crystals and solvent mixtures were placed in glass-stoppered bottles. The bottles were placed in a thermostat at a given temperature, and rotated for 72 hours. The saturated solutions were separated from the solid phase using a sintered glass filter. After the saturated solutions were diluted with water, the concentrations of iodate were determined iodometrically. The solubility of $Sr(IO_3)_2$ was calculated from the observed values.

SOURCE AND PURITY OF MATERIALS:

Strontium iodate was prepared by adding dilute solutions of $SrCl_2$ and HIO_3 to a boiled water. The product was washed and dried at room temperature. $Sr(IO_3)_2$ was obtained.
DMF (from Mitsubishi Gas Co) was distilled under reduced pressure. After the product was dried over Na_2CO_3, the distillation of the solvent was repeated 3 times.

ESTIMATED ERROR:

Soly: the probable errors of the observed mean value were within ± 0.03 x 10^{-5} mol dm^{-3}.
Temp: ± 0.02°C (authors)

REFERENCES:

COMPONENTS:

(1) Barium chlorate; $Ba(ClO_3)_2$; [13477-00-4]

(2) Water; H_2O; [7732-18-5]

EVALUATOR:

Hiroshi Miyamoto
Department of Chemistry
Niigata University
Niigata, Japan

July, 1982

CRITICAL EVALUATION:

1. Solubility in the binary $Ba(ClO_3)_2$-H_2O system

Solubilities in the binary $Ba(ClO_3)_2$-H_2O system have been reported in 7 publications (1-7).

Trautz and Anschütz (1) determined solubilities of barium chlorate in pure water by a gravimetric method over the temperature range of 270 to 373K.

Di Capua and Bertoni (2), Foote and Hickey (3) and Ricci and Freedman (4,5) studied ternary systems, and the solubility in the binary $Ba(ClO_3)_2$-H_2O system was given as one point on a phase diagram.

Remy-Genneté and Durand (6) studied solubilities of barium chlorate in mixtures of ethanol and water, and also measured the solubility of barium chlorate in pure water.

None of these investigations (1-7) report experimental errors.

Depending upon temperature and composition, equilibrated solid phases of varying the degrees of hydration have been reported. The following solid phases have been identified.

$Ba(ClO_3)_2 \cdot H_2O$ [10294-38-9]

$Ba(ClO_3)_2$ [13477-00-4]

The relation between the solubility of barium chlorate in water and temperature is given in Fig. 1.

The data to be considered in the critical evaluation are summarized in Table 1.

Table 1 Summary of solubilities in the binary $Ba(ClO_3)_2$-H_2O

T/K	m_1/mol kg^{-1}	ref
270.401	0.5928	(1)
273.2	0.6685	(1)
283.2	0.8854	(4)
"	0.8859	(1)
293.2	0.904	(6)
"	1.024	(2)
"	1.129	(1)
298.1	1.249	(7)
298.2	1.242	(4)
"	1.242	(5)
"	1.249	(1)
"	1.252	(3)
303.2	1.371	(1)
313.2	1.631	(1)

T/K	m_1/mol kg^{-1}	ref
318.2	1.762	(4)
323.2	1.905	(1)
333.2	2.196	(1)
343.2	2.484	(1)
353.2	2.789	(1)
363.2	3.120	(1)
372.3	3.444	(1)
378.3	3.658	(1)

COMPONENTS:

(1) Barium chlorate; $Ba(ClO_3)_2$; [13477-00-4]

(2) Water; H_2O; [7732-18-5]

EVALUATOR:

Hiroshi Miyamoto
Department of Chemistry
Niigata University
Niigata, Japan

July, 1982

CRITICAL EVALUATION:

Solubility at 283.2K. This value has been reported in 2 publications (1,4). The result of Ricci and Freedman (4) is in good agreement with that of Trautz and Anschütz. Converting to molality units, the arithmetic mean of the two results is 0.8857 mol kg^{-1}, and the standard deviation is 0.0003 mol kg^{-1}. The mean is designated as a recommended value.

Solubility at 293.2K. This value has been reported in 3 publications (1,2,6). The result of Di Capua and Bertoni (2) is considerably larger than that of Remy-Genneté and Durand (6), and lower than that of Trautz and Anschütz (1).
The arithmetic mean of three results is 1.019 mol kg^{-1}, and the standard deviation is 0.11 mol kg^{-1}. The mean is designated as a tentative value.

Solubility of 298.2K. The solubility at this temperature has been reported by Trautz and Anschütz (1), Foote and Hickey (3), and Ricci and Freedman (4,5). All investigators report that the solid phase is the monohydrate. The result of Ricci and Smiley (7) at 298.1K can be used.The arithmetic mean of five results (1,3,4,5,7) is 1.247 mol kg^{-1}, and the standard deviation is 0.005 mol kg^{-1}. The mean is designated as a recommended value.

Solubility at 318.2K. Only one result has been reported by Ricci and Freedman (4). The value of 1.762 mol kg^{-1} is taken as a tentative value.

Solubility at other temperatures. Only one publication (1) is available for the solubility of barium chlorate at other temperatures. These results of Trautz and Anschütz (1) are designated as tentative values.

The recommended and tentative values are given in Table 2. The data in Table 2 were fitted to the following equation:

$$\ln(S/\text{mol kg}^{-1}) = 24.93084 - 44.97886/(T/100\text{K}) - 8.839210 \ln(T/100\text{K}) \quad : \quad \sigma = 0.038$$

where S is the solubility of barium chlorate in pure water. The values calculated from the smoothing equation are also given in Table 2.

Table 2 Recommended and tentative values for the solubility of barium chlorate in water

T/K	m_1/mol kg^{-1}	m_1'/mol kg^{-1}	Solid Phase
270.4	0.5928	0.6090	--
273.2	0.6685	0.6593	--
283.2	0.8857[a]	0.8581	$Ba(ClO_3)_2\cdot H_2O$
293.2	1.019	1.085	--
298.2	1.247[a]	1.209	$Ba(ClO_3)_2\cdot H_2O$
303.2	1.371	1.338	--
313.2	1.631	1.613	--
318.2	1.762	1.758	$Ba(ClO_3)_2\cdot H_2O$
323.2	1.905	1.905	$Ba(ClO_3)_2\cdot H_2O$
333.2	2.196	2.210	--
343.2	2.484	2.522	--
353.2	2.789	2.836	--
363.2	3.120	3.146	--
372.3	3.444	3.422	--
378.3	3.658	3.598	--

m_1: experimental value

m_1': calculated value

a: recommended value

COMPONENTS:

(1) Barium chlorate; $Ba(ClO_3)_2$; [13477-00-4]

(2) Water; H_2O; [7732-18-5]

EVALUATOR:

Hiroshi Miyamoto
Department of Chemistry
Niigata University
Niigata, Japan

July, 1982

CRITICAL EVALUATION:

2. Ternary and quarternary systems

A summary of the ternary systems is given in Table 3.

Table 3 Summary of the ternary systems

System	t/°C	Reference
$Ba(ClO_3)_2$-NaCl-H_2O	293.2	Di Capua; Bertoni (2)
$Ba(ClO_3)_2$-$NaClO_3$-H_2O	293.2	Di Capua; Bertoni (2)
$Ba(ClO_3)_2$-$BaCl_2$-H_2O	293.2	Di Capua; Bertoni (2)
	298.2	Ricci; Freedman (4)
$Ba(ClO_3)_2$-$BaBr_2$-H_2O	288.2	Ricci; Freedman (4)
	298.2	"
$Ba(ClO_3)_2$-$Ba(NO_3)_2$-H_2O	283.2	Ricci; Freedman (4)
	298.2	"
	318.2	"
$Ba(ClO_3)_2$-$Ba(OH)_2$-H_2O	298.2	Foote; Hickey (3)
$Ba(ClO_3)_2$-$Ba(BrO_3)_2$-H_2O	288.1	Ricci; Smiley (7)

In the ternary $Ba(ClO_3)_2$-$Ba(NO_3)_2$-H_2O system, the double salt, $Ba(ClO_3)_2 \cdot 6Ba(NO_3)_2 \cdot 12H_2O$ is formed. The field of crystallization of the double is larger at 283.2K than 298.2K, and the compound does not appear at 318.2K. In other ternary systems, the existence of double salts was not reported.
Ricci and Freedman (4) reported solubilities in the quarternary $Ba(ClO_3)_2$-$BaBr_2$-$Ba(NO_3)_2$-H_2O system at 283.2K; the only congruent drying-up point of the isotherm is a solution saturated with the solid $Ba(ClO_3)_2 \cdot H_2O$, $BaBr_2 \cdot 2H_2O$ and $Ba(ClO_3)_2 \cdot 6Ba(NO_3)_2 \cdot 12H_2O$.

3. Solubility of barium chlorate in ethanol-water mixed solvent

Remy-Gennetė and Durand (6) have reported solubilities of barium chlorate in the mixtures of ethanol and water. They used the monohydrate as the initial starting salt.

The solubility of barium chlorate in ethanol-water mixed solvent decreases with increasing concentration of ethanol.

The results of Remy-Gennetė and Durand are designated as tentative values.

COMPONENTS:	EVALUATOR:
(1) Barium chlorate; $Ba(ClO_3)_2$; [13477-00-4] (2) Water; H_2O; [7732-18-5]	Hiroshi Miyamoto Department of Chemistry Niigata University Niigata, JAPAN July 1982

CRITICAL EVALUATION:

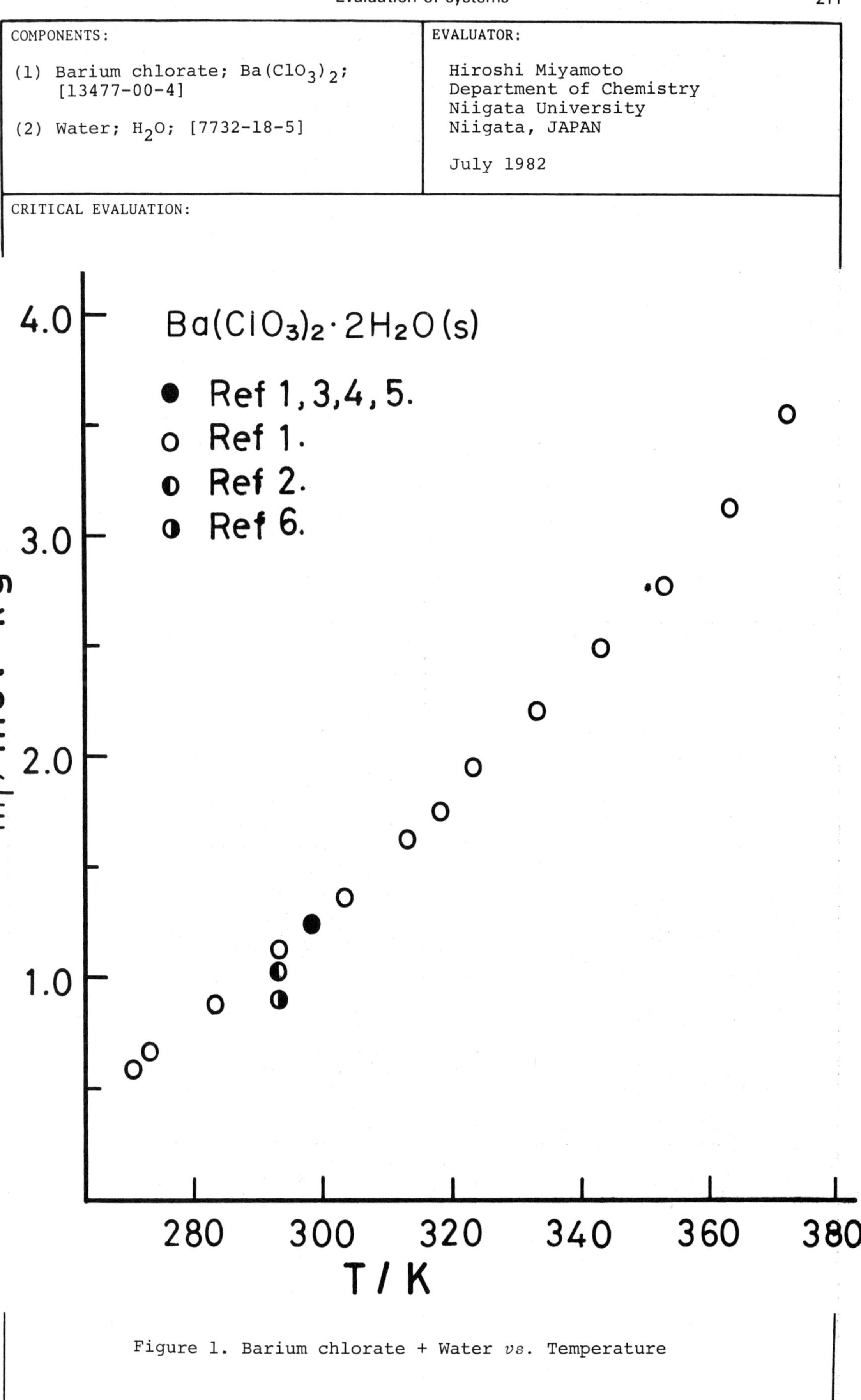

Figure 1. Barium chlorate + Water *vs.* Temperature

COMPONENTS:

(1) Barium chlorate; $Ba(ClO_3)_2$; [13477-00-4]

(2) Water; H_2O; [7732-18-5]

EVALUATOR:

Hiroshi Miyamoto
Department of Chemistry
Niigata University
Niigata, Japan

CRITICAL EVALUATION:

REFERENCES:

1. Trautz, M.; Anschütz, A. *Z. Phys. Chem.* 1906, *56*, 236.

2. Di Capua, C.; Bertoni, A. *Gazz. Chim. Ital.* 1928, *58*, 249.

3. Foote, H. W.; Hickey, F. C. *J. Am. Chem. Soc.* 1937, *59*, 648.

4. Ricci, J. E.; Freedman, A. J. *J. Am. Chem. Soc.* 1952, *74*, 1765.

5. Ricci, J. E.; Freedman, A. J. *J. Am. Chem. Soc.* 1952, *74*, 1769.

6. Remy-Gennetté, P.; Durand, G. *Bull. Soc. Chim. Fr.* 1955, 1059.

7. Ricci, J. E.; Smiley, S. H. *J. Am. Chem. Soc.* 1944, *66*, 1011.

COMPONENTS:	ORIGINAL MEASUREMENTS:
(1) Barium chlorate; $Ba(ClO_3)_2$; [13477-00-4]	Trautz, M.; Anschütz, A.
(2) Water; H_2O; [7732-18-5]	*Z. Phys. Chem.* 1906, *56*, 236-42.

EXPERIMENTAL VALUES:

t/°C	Shaking time t/h	Barium Chlorate mass %	mol % (compiler)	m_1/mol kg^{-1} (compiler)	Nature of the Solid Phase
-2.749 ± 0.004	4	15.32			
	4	15.24			
		(Av) 15.28	1.057	0.5928	--
0	7	16.88			
	7	16.91			
		(Av) 16.90	1.190	0.6685	--
10	14	21.24			
	14	21.22			
		(Av) 21.23	1.571	0.8859	--
20	40	25.54			
	14	25.57			
		(Av) 25.56	1.993	1.129	--
25	22	27.53			
	12	27.52			
		(Av) 27.53	2.200	1.249	$Ba(ClO_3)_2 \cdot H_2O$
30	24	29.45			
	24	29.40			
		(Av) 29.43	2.410	1.371	--
40	36	33.16			
	36	33.15			
		(Av) 33.16	2.854	1.631	--
50	14	36.68			
	14	36.70			
		(Av) 36.69	3.318	1.905	--
60	14	40.07			
	14	40.02			
		(Av) 40.05	3.805	2.196	$Ba(ClO_3)_2 \cdot H_2O$
70	14	43.04			
	14	43.05			
		(Av) 43.04	4.283	2.484	--
80	7	45.93			
	7	45.88			
		(Av) 45.90	4.784	2.789	--
90	8	48.67			
	8	48.73			
		(Av) 48.70	5.322	3.120	--
99.1	6	51.15			
	6	51.18			
		(Av) 51.17	5.843	3.444	--
104.6/740 mmHg	4	52.63			
(ca. 105.0	4	52.70			
/760 mmHg)		(Av) 52.67	6.182	3.658	--

COMPONENTS:

(1) Barium chlorate; $Ba(ClO_3)_2$; [13477-00-4]

(2) Water; H_2O; [7732-18-5]

ORIGINAL MEASUREMENTS:

Trautz, M.; Anschütz, A.

Z. Phys. Chem. 1906, *56*, 236-42.

VARIABLES:

T/K = 270.401 to 377.8

PREPARED BY:

Hiroshi Miyamoto

EXPERIMENTAL VALUES:

AUXILIARY INFORMATION

METHOD/APPARATUS/PROCEDURE:

$Ba(ClO_3)_2$ crystals and water were shaken in a thermostat at 10-90°C for 14 hours.
Equilibrium at 100°C was established in a vapor of boiling water for 6-7 hours, (the temperature was checked the boiling point of pure water). Aliquots of saturated solution were removed by means of a pipet fitted with cotton wool. The solution was placed in a stoppered tube, and the sample was weighed. $Ba(ClO_3)_2$ was determined gravimetrically by evaporation of the solvent. After the solution saturated with the barium chlorate was frozen at near 0°C, the melted part of the solution was analyzed for the chlorate content, and the melting point of the frozen part was measured by using a Beckman thermometer. The chlorate content of solid phases at both 25 and 60°C was also determined.

SOURCE AND PURITY OF MATERIALS:

Barium chlorate was purchased, and recrystallized several times.

ESTIMATED ERROR:

Nothing specified

REFERENCES:

COMPONENTS:

(1) Barium chlorate; $Ba(ClO_3)_2$; [13477-00-4]

(2) Sodium chloride; NaCl; [7647-14-5]

(3) Water; H_2O; [7732-18-5]

ORIGINAL MEASUREMENTS:

Di Capua, C.; Bertoni, A.

Gazz. Chim. Ital. 1928, *58*, 249-53.

VARIABLES:

T/K = 293

PREPARED BY:

Bruno Scrosati
Hiroshi Miyamoto

EXPERIMENTAL VALUES:

The solubility at 20°C is given as follows:

	gram mol salt in 1000[a] H_2O
$Ba(ClO_3)_2$	1.068
NaCl	6.127

[a] Neither g nor cm^{-3} were given in the original paper. Therefore, the compiler can not judge that the units of the solubility are either mol kg^{-1} or mol dm^{-3}.

AUXILIARY INFORMATION

METHOD/APPARATUS/PROCEDURE:

The method and the procedure for preparing the saturated solutions are not reported in the original paper.
The residue Cl^- was determined by the Mohr method. For the residue ClO_3^- the Volhard method was used, after reduction with Zn and acetic acid. The barium contents were determined as $BaSO_4$.
The sodium contents were evaluated by difference after the determination of the water weight.

SOURCE AND PURITY OF MATERIALS:

Not reported

ESTIMATED ERROR:

Not possible to estimate due to insufficient details.

REFERENCES:

COMPONENTS:

(1) Sodium chlorate; $NaClO_3$; [7775-09-9]

(2) Barium chlorate; $Ba(ClO_3)_2$; [13477-00-4]

(3) Water; H_2O; [7732-18-5]

ORIGINAL MEASUREMENTS:

Di Capua, C.; Bertoni, A.

Gazz. Chim. Ital. 1928, *58*, 249-53.

VARIABLES:

T/K = 293
composition

PREPARED BY:

Bruno Scrosati
Hiroshi Miyamoto

EXPERIMENTAL VALUES:

t/°C	Composition of Saturated Solutions[a]			
	Barium Chlorate		Sodium Chlorate	
	mass %	mol % (compiler)	mass %	mol % (compiler)
20	0	0	49.7[c]	14.3
	1.05	0.101	45	12
	2.73	0.263	43.2	11.9
	3.30	0.294	36.5	9.28
	4.73	0.394	29.52	7.034
	6.13	0.496	25.32	5.855
	8.05	0.599	15.52	3.303
	10.29	0.7319	8.5	1.7
	16.91	1.246	4.52	0.952
	23.75[b]	1.811	0	0

[a] For the binary system the compiler computes the following:

[b] Soly of $Ba(ClO_3)_2$ = 1.024 mol kg^{-1} (The solid phase is probably the monohydrate, compiler).

[c] Soly of $NaClO_3$ = 9.28 mol kg^{-1} (The solid phase is probably the anhydrous salt, compiler).

AUXILIARY INFORMATION

METHOD/APPARATUS/PROCEDURE:

The method and the procedure for preparing the saturated solutions are not reported in the original paper.
The residue Cl^- was determined by the Mohr method. For the residue ClO_3^- the Volhard method was used after reduction with Zn and acetic acid. The barium contents were determined as $BaSO_4$.
The sodium contents were evaluated by difference after the determination of the water content.

SOURCE AND PURITY OF MATERIALS:

Not reported

ESTIMATED ERROR:

Not possible to estimate due to insufficient details.

COMPONENTS:

(1) Barium chlorate; $Ba(ClO_3)_2$; [13477-00-4]

(2) Barium chloride; $BaCl_2$; [10361-37-2]

(3) Water; H_2O; [7732-18-5]

ORIGINAL MEASUREMENTS:

Di Capua, C.; Bertoni, A.

Gazz. Chim. Ital. 1928, *58*, 249-53.

VARIABLES:

T/K = 293
composition

PREPARED BY:

Bruno Scrosati
Hiroshi Miyamoto

EXPERIMENTAL VALUES:

t/°C	Composition of Saturated Solutions			
	Barium Chlorate		Barium Chloride	
	mass %	mol % (compiler)	mass %	mol % (compiler)
20	0	0	22.34	2.428
	5.20	0.403	20.55	2.329
	8.80	0.714	20.46	2.424
	12.30	1.022	18.77	2.278
	16.50	1.410	16.63	2.076
	18.13	1.585	16.67	2.130
	21.08	1.865	14.50	1.874
	23.02	1.884	6.56	0.784
	23.75[a]	1.811	0	0

[a] For the binary system the compiler computes the following

Soly of $Ba(ClO_3)_2$ = 1.024 mol kg^{-1}

Solid phase is probably the monohydrate, (compiler).

AUXILIARY INFORMATION

METHOD/APPARATUS/PROCEDURE:

The method and the procedure for preparing the saturated solutions are not reported in the original paper.
The residue Cl^- was determined by the Mohr method. For the residue ClO_3^- the Volhard method was used after reduction with Zn and acetic acid. The barium contents were determined as $BaSO_4$.

SOURCE AND PURITY OF MATERIALS:

Not reported

ESTIMATED ERROR:

Not possible to estimate due to insufficient details.

COMPONENTS:

(1) Barium chloride; $BaCl_2$; [10361-37-2]

(2) Barium chlorate; $Ba(ClO_3)_2$; [13477-00-4]

(3) Water; H_2O; [7732-18-5]

ORIGINAL MEASUREMENTS:

Ricci, J. E.; Freedman, A. J.

J. Am. Chem. Soc. 1952, *74*, 1769-73.

VARIABLES:

$T/K = 298$
Composition

PREPARED BY:

Hiroshi Miyamoto

EXPERIMENTAL VALUES:

t/°C	Composition of Saturated Solutions				Density	Nature of the Solid Phase[a]
	Barium chlorate		Barium chloride		ρ/g cm^{-3}	
	mass %	mol % (compiler)	mass %	mol % (compiler)		
25	27.42[b]	2.188	0.00	0.00	1.263	A
	23.06	1.883	6.35	0.758	1.294	"
	18.94	1.599	13.16	1.623	1.338	"
	15.74	1.361	18.29	2.310	1.373	"
	(Av) 14.73	1.304	(Av) 21.06	2.723	1.398	A + B
	11.22	0.9642	22.48	2.822	1.373	B
	11.21	0.9628	22.45	2.817	1.371	"
	5.22	0.427	24.90	2.978	1.327	"

[a] A = $Ba(ClO_3)_2 \cdot H_2O$; B = $BaCl_2 \cdot 2H_2O$

[b] For binary system the compiler computes the following

Soly of $Ba(ClO_3)_2$ = 1.242 mol kg^{-1} at 25°C

AUXILIARY INFORMATION

METHOD/APPARATUS/PROCEDURE:

Complexes were made up from water, $Ba(ClO_3)_2 \cdot H_2O$ and $BaCl_2 \cdot 2H_2O$. The details of equilibrium procedure were not given in the paper, but probably the isothermal method was used as in previous researches by the senior author. The analysis for the system involved the determination of total solid and the Volhard method for chloride.

SOURCE AND PURITY OF MATERIALS:

C.p. grade $Ba(ClO_3)_2 \cdot H_2O$ was used as received. The salt was found to contain 5.69 mass % water (by drying at 110°C) as compared with the theoretical 5.60 mass %. Barium chloride dihydrate was a commercial c.p. product used without further purification. The purity was checked, however the results were not given.

ESTIMATED ERROR:

Nothing specified.

REFERENCES:

COMPONENTS:

(1) Barium nitrate; $Ba(NO_3)_2$; [10022-31-8]

(2) Barium chlorate; $Ba(ClO_3)_2$; [13477-00-4]

(3) Water; H_2O; [7732-18-5]

ORIGINAL MEASUREMENTS:

Ricci, J. E.; Freedman, A. J.

J. Am. Chem. Soc. 1952, *74*, 1765-9.

EXPERIMENTAL VALUES:

t/°C	Composition of Saturated Solutions				Density ρ/g cm^{-3}	Nature of the Solid Phase[a]
	Barium Chlorate		Barium Nitrate			
	mass %	mol % (compiler)	mass %	mol % (compiler)		
10	21.22[b]	1.570	0.00	0.00	1.198	A
	21.38	1.673	4.52	0.412	1.249	A + E
	20.97	1.633	4.56	0.413	1.245	E
	18.90	1.436	4.60	0.407	1.224	"
	16.55	1.224	4.67	0.402	1.197	"
	14.15	1.019	4.83	0.405	1.174	"
	9.51	0.653	5.18	0.414	1.128	"
	4.24	0.278	6.00	0.457	1.088	"
	3.95	0.258	6.03	0.458	1.084	"
	3.94	0.257	6.06	0.461	1.087	"
	3.47	0.226	6.16	0.467	1.083	E + C
	0.00	0.00	6.361	0.4661	1.051	C
25	27.42[b]	2.188	0.00	0.00	1.263	A
	26.82	2.357	7.94	0.812	1.362	A + C(m)[c]
	26.67	2.318	7.28	0.737	1.347	A + E
	26.12	2.252	7.25	0.728	1.344	E
	25.91	2.227	7.23	0.723	1.337	"
	25.39	2.168	7.28	0.724	1.333	"
	25.32	2.160	7.28	0.723	1.332	"
	24.27	2.044	7.36	0.722	1.318	"
	22.48	1.853	7.53	0.723	1.301	"
	20.24	1.626	7.81	0.731	1.277	"
	20.20	1.622	7.78	0.727	1.275	"
	18.84	1.488	7.84	0.721	1.258	"
	18.29	1.437	7.96	0.728	1.254	"
	17.32	1.346	8.09	0.732	1.243	"
	15.08	1.145	8.39	0.741	1.222	E + C
	0.00	0.00	9.246	0.697	1.079	C
45	34.90[b]	3.077	0.00	0.00	1.347	A
	32.48	3.206	10.16	1.168	1.465	A + C
	0.00	0.00	13.60	1.073	1.110	C

[a] A = $Ba(ClO_3)_2 \cdot H_2O$; C = $Ba(NO_3)_2$; E = $Ba(ClO_3)_2 \cdot 6Ba(NO_3)_2 \cdot 12H_2O$

[b] For binary systems the compiler computes the following

Soly of $Ba(ClO_3)_2$ = 0.8854 mol kg^{-1} at 10°C
= 1.242 mol kg^{-1} at 25°C
= 1.762 mol kg^{-1} at 45°C

[c] (m) = metastable

COMPONENTS:

(1) Barium nitrate; $Ba(NO_3)_2$; [10022-31-8]

(2) Barium chlorate; $Ba(ClO_3)_2$; [13477-00-4]

(3) Water; H_2O; [7732-18-5]

ORIGINAL MEASUREMENTS:

Ricci, J. E.; Freedman, A. J.

J. Am. Chem. Soc. 1952, *74*, 1765-9.

VARIABLES:

T/K = 283, 298, and 318

$Ba(NO_3)_2$/mass % = 0 to 34.90

PREPARED BY:

Hiroshi Miyamoto

COMMENTS AND/OR ADDITIONAL DATA:

The phase diagram is given below (based on mass %):

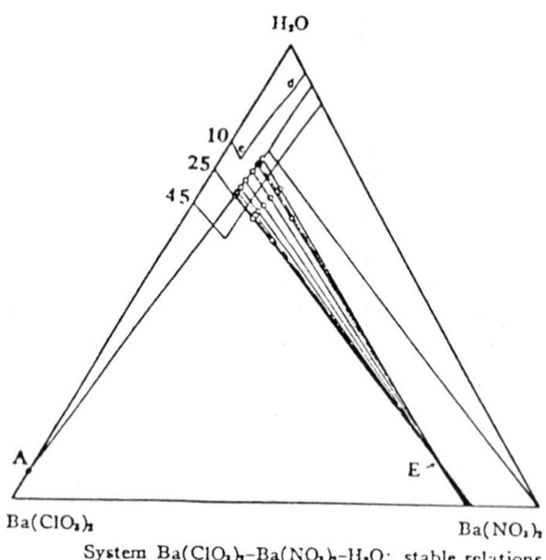

System $Ba(ClO_3)_2$-$Ba(NO_3)_2$-H_2O; stable relations only.

AUXILIARY INFORMATION

METHOD/APPARATUS/PROCEDURE:

Complexes of known composition were sealed in Pyrex tubes and rotated in a constant temperature water bath. Calibrated pipets with filter paper tips were used in sampling the saturated solutions for analysis, and approximate densities were reported. The total salt content was determined by evaporation to constant weight at 300°C. The chlorate was determined as chloride after reduction with sodium nitrite. To the sample, diluted to 100 cm^3, were added a measured excess of standard $AgNO_3$ and a solution of 2g sodium nitrite in 10 cm^3 water. The solution was warmed until the silver nitrate first precipitated had dissolved; 10 cm^3 of 6N nitric acid was then added, with vigorous shaking. After boiling to remove oxides of nitrogen, the silver chloride was filtered off and the excess of silver nitrate was titrated by the Volhard method.

SOURCE AND PURITY OF MATERIALS:

C.p. grade barium chlorate monohydrate was used. C.p. grade barium nitrate was found to be 99.9% pure by determination of barium both as barium sulfate and by precipitation with excess of standard potassium iodate.

ESTIMATED ERROR:

Nothing specified.

ACKNOWLEDGEMENT:

The figure reprinted from the *J. Am. Chem. Soc.* by permission of the copyright owners, The American Chemical Society.

COMPONENTS:	ORIGINAL MEASUREMENTS:
(1) Barium nitrate; $Ba(NO_3)_2$; [10022-31-8] (2) Barium chlorate; $Ba(ClO_3)_2$; [13477-00-4] (3) Barium bromide; $BaBr_2$; [10553-31-8] (4) Water; H_2O; [7732-18-5]	Ricci, J. E.; Freedman, A. J. *J. Am. Chem. Soc.* 1952, *74*, 1765-9.

EXPERIMENTAL VALUES:

Curve or point	Composition of Saturated Solutions						Density ρ/g cm⁻³	Nature of the Solid Phase[a]
	Barium Chlorate		Barium Bromide		Barium Nitrate			
	mass %	mol % (compiler)	mass %	mol % (compiler)	mass %	mol % (compiler)		
e 3	20.65	1.627	1.33	0.107	4.46	0.409	---	A + E
"	19.93	1.584	2.81	0.229	4.39	0.406	1.267	"
"	19.13	1.533	4.29	0.352	4.36	0.407	1.277	"
"	17.18	1.411	8.26	0.694	4.25	0.406	1.307	"
"	15.11	1.279	12.76	1.106	4.11	0.405	1.342	"
"	15.26	1.305	13.42	1.175	4.08	0.406	1.350	"
"	12.86	1.130	17.81	1.603	4.07	0.416	1.390	"
"	10.61	0.9841	24.08	2.287	3.82	0.412	1.456	"
"	10.30	0.9617	24.87	2.378	3.79	0.412	1.467	"
"	7.46	0.757	33.28	3.460	3.65	0.431	1.576	"
"	4.58	0.535	44.10	5.278	3.87	0.527	1.761	"
3	4.50	0.531	44.68	5.400	3.90	0.536	1.779	A + E + B
d 1	3.34	0.218	0.61	0.041	6.14	0.467	1.085	E + C
"	3.19	0.210	1.58	0.107	5.98	0.458	1.094	"
"	3.29	0.219	2.45	0.167	6.04	0.468	1.113	"
"	3.27	0.219	3.23	0.221	5.77	0.449	1.108	"
"	3.25	0.221	5.00	0.348	5.55	0.439	1.127	"
"	3.17	0.221	7.45	0.531	5.32	0.431	1.152	"
"	3.07	0.221	10.62	0.7810	5.04	0.421	1.183	"
"	2.60	0.202	18.06	1.434	4.55	0.411	1.266	"
"	2.17	0.179	23.66	1.999	4.27	0.410	1.335	"
1 a	2.0	0.17	24.3	2.06	4.2	0.41	---	E + C + D
c 1	0.40	0.033	24.10	2.015	4.80	0.456	1.326	C + D
"	0.73	0.060	24.05	2.014	4.65	0.443	1.327	"
"	0.87	0.071	23.93	2.002	4.53	0.431	---	"
"	1.10	0.0901	23.98	2.011	4.45	0.424	1.327	"
"	1.45	0.119	24.05	2.026	4.36	0.418	1.333	"
"	1.45	0.120	24.04	2.030	4.57	0.439	1.335	"
1 2	2.07	0.177	26.58	2.328	4.15	0.413	1.377	D + E
"	2.07	0.178	26.79	2.353	4.15	0.414	1.380	"
"	2.04	0.184	30.58	2.825	4.00	0.420	1.438	"
"	1.97	0.186	33.72	3.252	3.88	0.426	1.491	"
2	1.59	0.184	46.57	5.529	3.96	0.535	1.761	B + D + E
"	1.61	0.187	46.59	5.530	3.91	0.528	1.763	"
b 2	1.02	0.118	46.85	5.534	3.98	0.535	1.759	B + D
2 3	2.17	0.252	46.13	5.487	3.92	0.530	1.763	B + E
"	3.52	0.413	45.27	5.438	3.96	0.541	1.775	"
3	4.33	0.511	44.74	5.407	4.01	0.551	1.780	A + B + E
a 3	4.49	0.511	45.66	5.317	0.88	0.117	1.738	A + B
"	4.40	0.507	45.41	5.353	1.90	0.255	1.753	"
"	4.30	0.502	45.07	5.390	3.14	0.427	1.766	"
3	4.39	0.518	44.69	5.399	3.98	0.547	1.779	A + B + E
3 (Av)	4.42	0.522	44.71	5.405	3.96	0.544	1.780	"

[a] A = $Ba(ClO_3)_2 \cdot H_2O$; B = $BaBr_2 \cdot 2H_2O$; C = $Ba(NO_3)_2$;

D = $BaBr_2 \cdot 8Ba(NO_3)_2 \cdot 12H_2O$; E = $Ba(ClO_3)_2 \cdot 6Ba(NO_3)_2 \cdot 12H_2O$

COMPONENTS:

(1) Barium nitrate; $Ba(NO_3)_2$; [10022-31-8]

(2) Barium chlorate; $Ba(ClO_3)_2$; [13477-00-4]

(3) Barium bromide; $BaBr_2$; [10553-31-8]

(4) Water; H_2O; [7732-18-5]

ORIGINAL MEASUREMENTS:

Ricci, J. E.; Freedman, A. J.

J. Am. Chem. Soc. 1952, *74*, 1765-9.

VARIABLES:

T/K = 283
composition

PREPARED BY:

Hiroshi Miyamoto

METHOD/APPARATUS/PROCEDURE:

Complexes of known composition sealed in Pyrex tubes were rotated in a constant temperature water bath. Calibrated pipets with filter paper tips were used in sampling the saturated solutions for analysis so that approximate densities are also reported.
The total salt content was determined by evaporation to constant weight at 300°C. Bromide was determined by the Volhard method with filtration of the silver bromide. The chlorate was determined as chloride after reduction with sodium nitrite. Excess standard silver nitrate and a solution of 2g sodium nitrite in 10 cm^3 water was added to a diluted aliquot of saturated solution. The solution was warmed until the silver nitrite first precipitated had dissolved; 10 cm^3 of 6N nitric acid was then added and the mixture shaken vigorously. After boiling to remove oxides of nitrogen, the silver chloride was filtered off and the excess of silver nitrate was titrated by the Volhard method. In the quaternary system, this procedure was used to determine total halogen or the total number of equivalents of chlorate and bromide. The identity of solid phases was established indirectly by means of the tie-lines fixed by the compositions of saturated solution as determined by analysis and of total complex or mixture as prepared synthetically.

ACKNOWLEDGEMENT: The figure reprinted from the *J. Am. Chem. Soc.* by permission of the copyright owners, The American Chemical Society.

AUXILIARY INFORMATION

SOURCE AND PURITY OF MATERIALS:

C.p. grade barium chlorate monohydrate was used. The c.p. grade barium nitrate was found to be 99.9 % pure by determination of barium both as barium sulfate and by precipitation with excess of standard KIO_3. The barium bromide was used as the dihydrate. The material was made from c.p. grade $Ba(OH)_2 \cdot 8H_2O$, which was first recrystallized from water and then neutralized with 40% aqueous HBr. The resulting solution was evaporated to near dryness and cooled in ice; then the residue was filtered and the last portion of mother liquor discarded. The crystals were redissolved and the evaporation and filtration were repeated. The resulting pure crystals were dried in air.
The water content of the product was somewhat higher (11.41 %) than the theoretical value of 10.81 % for the dihydrate.

COMMENT AND/OR ADDITIONAL DATA:

The phase diagram is given below (based on mass %).

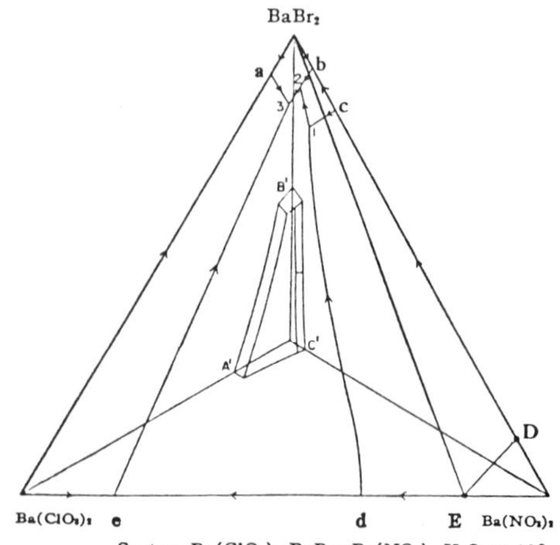

System $Ba(ClO_3)_2$–$BaBr_2$–$Ba(NO_3)_2$–H_2O at 10°. Jänecke diagram. The orthogonal diagram is enclosed, with primed letters.

ESTIMATED ERROR:

Nothing specified

COMPONENTS:	ORIGINAL MEASUREMENTS:
(1) Barium chlorate; $Ba(ClO_3)_2$; [13477-00-4] (2) Barium hydroxide; $Ba(OH)_2$; [17194-00-2] (3) Water; H_2O; [7732-18-5]	Foote, H. W.; Hickey, F. C. *J. Am. Chem. Soc.* 1937, *59*, 648-50.
VARIABLES:	PREPARED BY:
T/K = 298 $Ba(OH)_2$/mass % = 0 to 4.489	Hiroshi Miyamoto

EXPERIMENTAL VALUES:

t/°C	Composition of the Saturated Solutions				Nature of the Solid Phase[a]
	Barium Hydroxide		Barium Chlorate		
	mass %	mol % (compiler)	mass %	mol % (compiler)	
25	4.489	0.4917	0	0	A
	4.02	0.480	8.79	0.591	"
	3.85	0.497	15.98	1.161	"
	3.77	0.521	21.85	1.701	"
	3.72	0.546	26.55	2.193	A + B
	3.71	0.545	26.62	2.200	"
	1.87	0.270	27.17	2.211	B
	0	0	27.58[b]	2.205	"

[a] A = $Ba(OH)_2 \cdot 8H_2O$; B = $Ba(ClO_3)_2$

[b] For binary system the compiler computes the following

Soly of $Ba(ClO_3)_2$ = 1.252 mol kg^{-1} at 25°C

ACKNOWLEDGEMENT: The figure reprinted from the *J. Am. Chem. Soc.* by permission of the copyright owners, The American Chemical Society.

AUXILIARY INFORMATION

METHOD/APPARATUS/PROCEDURE:

Mixtures of the three components in suitable proportions were rotated in a thermostat for several days. Samples of solution for analysis were drawn off through asbestos or glass wool filters, or when the solid settled properly without filtering, and weighed.
Barium hydroxide was determined by titration with standard hydrochloric acid using nitrazine yellow indicator. The acid was standardized by titrating barium hydroxide whose concentration was determined by converting the hydroxide to the chloride, evaporating and drying to constant weight. Total barium was determined by evaporating with HCl and weighing as $BaCl_2$.

ESTIMATED ERROR:

Soly: nothing specified
Temp: ± 0.03°C (authors)

SOURCE AND PURITY OF MATERIALS:

Barium hydroxide and chlorate were recrystallized before use.

COMMENTS AND/OR ADDITIONAL DATA:

The phase diagram is given below (based on mass %)

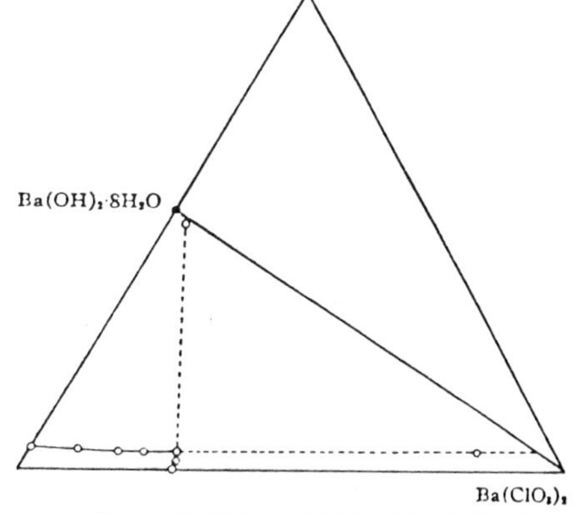

System $Ba(OH)_2$–$Ba(ClO_3)_2$–H_2O. Solid phases are $Ba(OH)_2 \cdot 8H_2O$ and $Ba(ClO_3)_2$.

COMPONENTS:

(1) Barium chlorate; $Ba(ClO_3)_2$; [13477-00-4]

(2) Ethanol; C_2H_6O; [64-17-5]

(3) Water; H_2O; [7732-18-5]

ORIGINAL MEASUREMENTS:

Remy-Gennete, P.; Durand, G.

Bull. Soc. Chem. Fr. 1955, 1059-60.

VARIABLES:

$T/K = 293$

$c_2/\text{vol } \% = 0 - 100$

PREPARED BY:

Hiroshi Miyamoto

EXPERIMENTAL VALUES:

t/°C	Ethanol vol %	Barium Chlorate		Specific Gravity	
		S/(g/100 ml)	c_1/mol dm^{-3} (compiler)	solvent	solution
20	0	26.7	0.878	0.996	1.238
	10.07	20.7	0.680	0.986	1.154
	30.9	11.5	0.378	0.964	1.050
	52.2	6.25	0.205	0.930	0.982
	73	2.52	0.0828	0.882	0.895
	83	1.0	0.033	0.856	0.870
	92.2	0.2	0.007	0.826	0.824
	100	0.0	0.000	0.794	0.788

At 20°C the solubility, S, of $Ba(ClO_3)_2$ in mixtures of ethanol and water is given by

$$S/(g/100\ ml) = 26.7 - 54.1\phi - 27.4\phi^2$$

where ϕ is the % EtOH by volume.

AUXILIARY INFORMATION

METHOD/APPARATUS/PROCEDURE:

The solubility determination of $Ba(ClO_3)_2$ in the mixtures of water and ethanol was studied by the method as described in ref (1). Analyses and solubility measurements were performed in duplicate. The densities of the mixtures of ethanol and water, and the saturated solutions were also determined.

SOURCE AND PURITY OF MATERIALS:

Nothing specified.

ESTIMATED ERROR:

Soly: nothing specified

Temp: ± 0.1°C (authors)

REFERENCES:

1. Hering, H. *Bull. Soc. Chem. Fr.* 1947, 333.

COMPONENTS:	EVALUATOR:
(1) Barium bromate; $Ba(BrO_3)_2$; [13967-90-3] (2) Water; H_2O; [7732-18-5]	Hiroshi Miyamoto Department of Chemistry Niigata University Niigata, Japan May, 1982

CRITICAL EVALUATION:

1. The binary $Ba(BrO_3)_2$-H_2O system

Solubilities in the binary $Ba(BrO_3)_2$-H_2O system have been reported in 5 publications (1-5).

Trautz and Anschütz (1) determined the solubility of barium bromate in water over the temperature range 273 to 373K using iodometric titration for analysing saturated solutions.

Harkins and Winninghoff (2), and Keefer, Reiber and Bisson (4) measured solubilities of barium bromate in various salt solutions at 298K. The solubility in the binary $Ba(BrO_3)_2$-H_2O is given as one point in a series of solubility values of varying salt concentration.

Ricci and Smiley (3), and Ricci and Freedman (5) studied the ternary system, and give the solubility in the binary $Ba(BrO_3)_2$-H_2O system as one point on a phase diagram.

Popiel and Rustom (6) state that the solubilities of $Ba(BrO_3)_2 \cdot H_2O$ in water and in solutions on sodium chloride were measured in order to determine the thermodynamic solubility product, but data for the solubilities are not reported in (6).

Depending upon temperature and composition equilibrated solid phases of varying degrees of hydration have been reported. The following solid phases have been identified:

$Ba(BrO_3)_2 \cdot H_2O$ [10326-26-8]

$Ba(BrO_3)_2$ [13967-90-3]

The temperature of transition from the monohydrate to the anhydrous salt was not reported from the determination of the solubility.

The data to be considered in this critical evaluation are summarized in Table 1.

Table 1 Summary of solubilities in the binary $Ba(BrO_3)_2$-H_2O system (the solid phase is the monohydrate, $Ba(BrO_3)_2 \cdot H_2O$).

T/K	m/mmol kg^{-1}	ref
273.116	7.14	(1)
273.2	7.30	(1)
283.2	11.2	(1)
"	11.65	(5)
293.2	16.7	(1)
298.1	20.28	(3)
298.2	20.07	(5)
"	20.08	(4)
"	20.2	(1)
"	20.33	(2)
303.2	24.4	(1)
313.2	33.8	(1)

T/K	m/mmol kg^{-1}	ref
318.2	39.50	(5)
323.2	44.5	(1)
333.2	59.11	(1)
343.2	76.56	(1)
353.2	92.83	(1)
363.2	113	(1)
371.9	141.1	(1)
372.80	145	(1)

COMPONENTS:	EVALUATOR:
(1) Barium bromate; $Ba(BrO_3)_2$; [13967-90-3] (2) Water; H_2O; [7732-18-5]	Hiroshi Miyamoto Department of Chemistry Niigata University Niigata, Japan May, 1982

CRITICAL EVALUATION:

EVALUATION OF THE DATA

Solubility at 283.2K. This result has been reported in 2 publications (1,5).

The solubility of barium bromate has been determined by Trautz and Anschütz (1) gravimetrically, and by Ricci and Freedman (5) iodometrically. Within the estimated errors the difference in solubility between both results is significant. The result of Ricci and Freedman is slightly higher than that of Trautz and Anschütz. The arithmetic mean of the two results is 11.4 mmol kg^{-1}, and the standard deviation is 0.3 mmol kg^{-1}. The mean is designated as a tentative value.

Solubility at 298.2K. The results reported in 5 publications (1-5) were used for the critical evaluation. The result of Harkins (2) was given in units of mol dm^{-3}, and the evaluator converted to mol kg^{-1} units using the density of the saturated solution given as 1.0038 kg m^{-3} by the author. The result of Ricci and Smiley (3) at 298.10K can be used in the evaluation since the error in temperature measurement by these authors was reported as ± 0.02K compared to ± 0.04K reported by Trautz and Anschütz (1): i.e. the recommended value is taken as the average from ref (1-5), and at 298.2 ± 0.05K, the value is 0.0202 mol kg^{-1}. The standard deviation of the average value is 0.0001 mol kg^{-1}

Solubility at other temperatures. The results of Trautz and Anschütz (1), and Ricci and Freedman (5) are designated as tentative values.

Solubility at 298.2K. Only one result has been reported by Harkins (2). The value is 20.25 mmol dm^{-3} which is designated as a tentative value.

The recommended and tentative values are given in Table 2. The data in Table 2 were fitted to the following smoothing equation:

$$\ln(S/\text{mmol kg}^{-1}) = 32.17468 - 58.29317/(T/100\text{K}) - 8.807149 \ln(T/100\text{K}) \quad : \quad \sigma = 0.021$$

where S is the solubility of barium bromate in water. The values calculated from the smoothing equation are also given in Table 2.

Table 2 Recommended and tentative values for the solubility of barium bromate

T/K	$10^3 m_1$/mol kg^{-1}	$10^3 m_1'$/mol kg^{-1}	Solid Phase
273.116	7.14	7.26	$Ba(BrO_3)_2 \cdot H_2O$
273.2	7.30	7.29	"
283.2	11.4	11.3	"
293.2	16.7	16.8	"
298.2	20.2[a]	20.2	"
303.2	24.4	24.0	"
313.2	33.8	33.4	"
323.2	44.5	45.0	"
333.2	59.11	59.1	"
343.2	76.56	75.9	"
353.2	92.83	95.3	"
363.2	113	117	"
371.9	141.1	139	"
372.80	145	141	"

m_1: experimental value

m_1': calculated value

a: recommended value

COMPONENTS:

(1) Barium bromate; $Ba(BrO_3)_2$; [13967-90-3]

(2) Water; H_2O; [7732-18-5]

EVALUATOR:

Hiroshi Miyamoto
Department of Chemistry
Niigata University
Niigata, Japan

May, 1982

CRITICAL EVALUATION:

2. Solubility product of barium bromate

Only one study (6) reports the thermodynamic solubility product of $Ba(BrO_3)_2 \cdot H_2O$ in water, and in solutions of sodium chloride.

The tentative values for the solubility product based on the results reported in (6) are given in the compilation along with a smoothing equation.

3. Ternary systems

Solubilities in the ternary $Ba(BrO_3)_2$-$Ba(ClO_3)_2$-H_2O system at 298K and the $Ba(BrO_3)_2$-$BaCl_2$-H_2O system at 283, 298 and 318K have been studied by Ricci and Smiley (3), and Ricci and Freedman (5), respectively.

In the system $Ba(BrO_3)_2$-$BaCl_2$-H_2O the double salt $Ba(BrO_3)_2 \cdot BaCl_2 \cdot 2H_2O$ forms at each of the temperatures studied, but in the system $Ba(BrO_3)_2$-$Ba(ClO_3)_2$-H_2O no double salts were reported.

REFERENCES:

1. Trautz, M; Anschütz, A. *Z. Phys. Chem.* 1906, *56*, 236.

2. Harkins, W. D. *J. Am. Chem. Soc.* 1911, *33*, 1807.

3. Ricci, J. E.; Smiley, S. H. *J. Am. Chem. Soc.* 1944, *66*, 1011.

4. Keefer, R. M.; Reiber, H. G.; Bisson, C. S. *J. Am. Chem. Soc.* 1940, *62*, 2951.

5. Ricci, J. E.; Freedman, A. J. *J. Am. Chem. Soc.* 1952, *74*, 1769.

6. Popiel, W. J.; Rustom, M. S. *Chem. Ind. (London)* 1971, 543.

COMPONENTS:

(1) Barium bromate; $Ba(BrO_3)_2$; [13967-90-3]

(2) Water; H_2O; [7732-18-5]

ORIGINAL MEASUREMENTS:

Trautz, M.; Anschütz, A.

Z. Phys. Chem. 1906, *56*, 236-42.

VARIABLES:

T/K = 273.116 to 372.80

PREPARED BY:

Hiroshi Miyamoto

EXPERIMENTAL VALUES:

t/°C	Barium Bromate	
	mass %	m_1/mol kg^{-1} [b]
-0.034 ± 0.002[a]	0.280	0.00714
0	0.286	0.00730
+10	0.439	0.0112
20	0.652	0.0167
25	0.788	0.0202
30	0.95	0.0244
40	1.31	0.0338
50	1.72	0.0445
60	2.271	0.05911
70	2.922	0.07656
80	3.521	0.09283
90	4.26	0.113
98.7	5.256	0.1411
99.65	5.39	0.145

[a] Eutectic point

[b] Molalities calculated by the compiler

AUXILIARY INFORMATION

METHOD/APPARATUS/PROCEDURE:

$Ba(BrO_3)_2$ crystals and water were shaken in a thermostat at 10-90°C for 14 hours.
Equilibrium at 100°C was established in a vapor of boiling water for 6-7 hours (the temperature was checked against the boiling point of pure water). The saturated solution was permitted to settle, and the solution was removed by means of a pipet fitted with cotton wool. The solution was placed in a stoppered tube, dried and weighed for determination of the barium bromate content. After the solution saturated with the barium bromate was frozen at near 0°C, the melted part of the solution was analyzed for the bromate content, and the melting point of the frozen part was measured by using a Beckmann thermometer.

SOURCE AND PURITY OF MATERIALS:

Barium bromate was recrystallized from water. The number of hydrated waters was not given.

ESTIMATED ERROR:

Soly: the deviations from the mean were about ± 5%.

Temp: ± 0.04°C except eutectic point (authors)

REFERENCES:

COMPONENTS:

(1) Barium bromate; $Ba(BrO_3)_2$; [13967-90-3]

(2) Water; H_2O; [7732-18-5]

ORIGINAL MEASUREMENTS:

Popiel, W. J.; Rustom, M. S.

Chem. Ind. (London) 1971, 543.

VARIABLES:

$T/K = 288$ to 318

PREPARED BY:

Hiroshi Miyamoto

EXPERIMENTAL VALUES:

t/°C	$10^5\ K^\circ_{s0}$ [a]
15	0.37
20	0.54
25	0.78
30	1.11
35	1.58
40	2.20
45	3.05

[a] From $K^\circ_{s0} = 4S^3y_\pm^3$, using $-\log y_\pm = z_1z_2AI^{1/2} - BI$ where A is from ref (1), and B 0.97 at 15°C, 0.85 at 30°C and 0.73 at 45°C.

The K°_{s0} data were fitted to the following equation

$$\log(K^\circ_{s0}/\text{mol}^3\ \text{dm}^{-9}) = -2805/(T/\text{K}) + 4.304$$

AUXILIARY INFORMATION

METHOD/APPARATUS/PROCEDURE:

The solubilities of $Ba(BrO_3)_2 \cdot H_2O$ in water and in solutions of sodium chloride (0.015-0.50 mol dm^{-3}) were measured at twelve temperatures in the range 15-45°C in order to obtain the thermodynamic solubility products.

Saturated solutions of barium bromate were prepared by "static" method as described in ref (2), and were analyzed by titrating the barium ion with EDTA.

SOURCE AND PURITY OF MATERIALS:

No information.

ESTIMATED ERROR:

Nothing specified

REFERENCES:

1. Robinson, R. A.; Stokes, R. H. *Electrolyte Solutions* 1965, 468, Butterworths, London.

2. Nezzal, G.; Popiel, W. J.; Vermande, P. *Chem. Ind. (London)* 1971, *15*, 543.

COMPONENTS:

(1) Barium bromate; $Ba(BrO_3)_2$; [13967-90-3]

(2) Potassium chloride; KCl; [7447-40-7]

(3) Water; H_2O; [7732-18-5]

ORIGINAL MEASUREMENTS:

Keefer, R. M.; Reiber, H. G.; Bisson, C. S.

J. Am. Chem. Soc. 1940, *62*, 2951-5.

VARIABLES:

$T/K = 298$

$m_2/mol\ kg^{-1} = 0$ to 0.1007

PREPARED BY:

Hiroshi Miyamoto

EXPERIMENTAL VALUES:

$t/^0C$	Potassium Chloride $m_2/mol\ kg^{-1}$	Barium Bromate $10^2 m_1/mol\ kg^{-1}$
25	0	2.008
	0.02008	2.135
	0.04019	2.233
	0.06034	2.330
	0.08050	2.416
	0.1007	2.483

AUXILIARY INFORMATION

METHOD/APPARATUS/PROCEDURE:

An excess of $Ba(BrO_3)_2$ crystals and aqueous KCl solution was placed in glass-stoppered Pyrex flasks. The flasks were rotated in a thermostat at 25°C for at least 12 hours. The solutions saturated with barium bromate were analyzed for bromate by iodometric analysis, and by precipitating the barium as $BaSO_4$ after the bromate ion was removed by evaporating the saturated barium bromate solution to dryness in the presence of HCl and KBr.

SOURCE AND PURITY OF MATERIALS:

Barium bromate was prepared by adding with stirring an equivalent amount of 0.20 M barium chloride solution to 1000 ml of 0.16 M potassium bromate. The precipitate was filtered, washed, and then air-dried at room temperature. The number of hydrated waters for the barium bromate was not given. KCl (c.p. grade) was recrystallized twice from water.

ESTIMATED ERROR:

Soly: nothing specified

Temp: ± 0.02°C (authors)

REFERENCES:

COMPONENTS:

(1) Barium bromate; $Ba(BrO_3)_2$; [13967-90-3]

(2) Potassium bromate; $KBrO_3$; [7758-01-2]

(3) Water; H_2O; [7732-18-5]

ORIGINAL MEASUREMENTS:

Harkins, W. D.

J. Am. Chem. Soc. 1911, *33*, 1807-27.

VARIABLES:

$T/K = 298$

$10^3c_2/\text{mol dm}^{-3} = 0$ to 99.85

PREPARED BY:

Hiroshi Miyamoto

EXPERIMENTAL VALUES:

t/°C	Potassium Bromate	Barium Bromate	
	$10^3c_2/\text{mol dm}^{-3}$	g dm^{-3}	$10^3c_1/\text{mol dm}^{-3}$ [b]
25	0	7.96[a]	20.25
	24.988	5.216	13.208
	49.971	3.415	8.687
	99.85	1.72	4.375

[a] The author reports that the solubility of barium bromate in pure water is 0.793 grams to 100 grams solution and the density of the saturated solution is 1.0038. The value (7.96 g/dm^3) was calculated by the compiler.

[b] The value was calculated by the compiler.

COMMENTS AND/OR ADDITIONAL DATA:

On the table of the original paper the word "solubility Ag_2SO_4" appeared. The compiler presumes that this is a misprint of $Ba(BrO_3)_2$.

AUXILIARY INFORMATION

METHOD/APPARATUS/PROCEDURE:

Barium bromate with H_2O and aqueous $KBrO_3$ solutions were placed in glass-stoppered bottles. The bottles were sealed with paraffin, and rotated in a thermostat for about 24 hours. Saturation was approached both from undersaturation and supersaturation. The solutions were filtered, by means of air pressure, through a thin inverted submerged asbestos filter which was held in a glass tube between two perforated Pt disks. The first portion of solution was rejected, and the remainder of the filtrate was used for analyses. The concentrations of barium bromate and $KBrO_3$ solutions were determined by precipitation as AgBr after a reduction by hydrazine. The results of this method were checked by analyzing solutions of $KBrO_3$ made by weighing out the dry salt. The concentration of the $KBrO_3$ solutions were also determined by titration against a thiosulfate solution which was standardized against pure iodine, copper sulfate and KI.

SOURCE AND PURITY OF MATERIALS:

Barium bromate was prepared by mixing dilute solutions of barium chloride (Kahlbaum) and potassium bromate (Kahlbaum). The product was recrystallized from water until it was entirely free from chloride. The number of hydrated waters was not given.
Potassium bromate (Kahlbaum) was recrystallized twice from water. The product was dried at 130°C.

ESTIMATED ERROR:

Soly: the deviations from the mean value were within ± 0.7 %.
Temp: ± 0.01°C (author)

REFERENCES:

COMPONENTS:

(1) Barium bromate; $Ba(BrO_3)_2$; [13967-90-3]

(2) Potassium nitrate; KNO_3; [7757-79-1]

(3) Water; H_2O; [7732-18-5]

ORIGINAL MEASUREMENTS:

Harkins, W. D.

J. Am. Chem. Soc. 1911, *33*, 1807-27.

VARIABLES:

$T/K = 298$

$10^3 c_2/mol\ dm^{-3} = 0$ to 99.97

PREPARED BY:

Hiroshi Miyamoto

EXPERIMENTAL VALUES:

t/°C	Potassium Nitrate $10^3 c_2/mol\ dm^{-3}$	Barium Bromate $g\ dm^{-3}$	Barium Bromate $10^3 c_1/mol\ dm^{-3}$ [b]
25	0	7.96[a]	20.25
	25.018	8.62	21.93
	50.032	9.91	25.21
	99.97	10.25	26.07

[a] The author reports that the solubility of barium bromate in pure water is 0.793 grams to 100 grams solution and the density of the saturated solution is 1.0038. The value (7.96 $g\ dm^{-3}$) was calculated by the compiler.

[b] The value was calculated by the compiler.

COMMENTS AND/OR ADDITIONAL DATA:

On the table of the original paper the word "solubility Ag_2SO_4" appeared. The compiler presumes that this is a misprint of $Ba(BrO_3)_2$.

AUXILIARY INFORMATION

METHOD/APPARATUS/PROCEDURE:

Barium bromate with H_2O and aqueous KNO_3 solutions were placed in glass-stoppered bottles. The bottles were sealed with paraffin, and rotated in a thermostat for about 24 hours. Saturation was approached both from under-saturation and supersaturation. The solutions were filtered, by means of air pressure, through a thin inverted submerged asbestos filter which was held in a glass tube between two perforated Pt disks. The first portion of solution was rejected, and the remainder of the filtrate was used for analyses.

The concentration of barium bromate was determined by precipitation as AgBr after a reduction by hydrazine.

SOURCE AND PURITY OF MATERIALS:

Barium bromate was prepared by mixing dilute solutions of barium chloride (Kahlbaum) and potassium bromate (Kahlbaum). The product was recrystallized from water until it was entirely free from chloride. The number of hydrated waters was not given.

KNO_3 was purified by 7 crystallizations from water and was dried at 160°C in a current of dry air.

ESTIMATED ERROR:

Soly: the deviations from the mean value were within ± 0.1 %

Temp: ± 0.01°C (author)

REFERENCES:

COMPONENTS:

(1) Barium bromate; $Ba(BrO_3)_2$; [13967-90-3]

(2) Magnesium nitrate; $Mg(NO_3)_2$; [10377-60-3]

(3) Water; H_2O; [7732-18-5]

ORIGINAL MEASUREMENTS:

Harkins, W. D.

J. Am. Chem. Soc. 1911, *33*, 1807-27.

VARIABLES:

T/K = 298

c_2/eq dm^{-3} = 0.1

PREPARED BY:

Hiroshi Miyamoto

EXPERIMENTAL VALUES:

The solubility of $Ba(BrO_3)_2$ in 0.1 equivalent per liter aqueous $Mg(NO_3)_2$ solution is 8.196 grams per liter (author) and 0.2085 mol dm^{-3} (compiler).

COMMENTS AND/OR ADDITIONAL DATA:

On the table of the original paper the word "solubility Ag_2SO_4" appeared. The compiler presumes that this is a misprint of $Ba(BrO_3)_2$.

AUXILIARY INFORMATION

METHOD/APPARATUS/PROCEDURE:

Barium bromate with water and aqueous $Mg(NO_3)_2$ solutions were placed in glass-stoppered bottles. The bottles were sealed with paraffin, and rotated in a thermostat for about 24 hours. Saturation was approached both from undersaturation and supersaturation. The solutions were filtered, by means of air pressure, through a thin inverted submerged asbestos filter which was held in a glass tube between two perforated Pt disks, in a thermostat. The first portion of solution was objected, and the remainder of the filtrate was used for analyses. The concentration of barium bromate was determined by precipitation as AgBr after a reduction by hydrazine.

SOURCE AND PURITY OF MATERIALS:

Barium bromate was prepared by mixing dilute solutions of $BaCl_2$ (Kahlbaum) and $KBrO_3$ (Kahlbaum). The product was recrystallized from H_2O until it was entirely free from chloride. The number of hydrated waters was not given. Magnesium nitrate was purified by two methods as follows: (1) Kahlbaum's magnesium nitrate was dissolved in H_2O, and boiled with magnesium carbonate solution, the solution was filtered, the filtrate made acid with HNO_3 and partly evaporated, and the salt separated by cooling and filtration. (2) Merck's magnesium nitrate was dissolved in concentrated HNO_3, the residue filtered off, and the salt crystallized by evaporation and cooling. The product was recrystallized from water.

ESTIMATED ERROR:

Soly: the deviations from the mean value were within ± 0.1 %

Temp: ± 0.01°C (author)

COMPONENTS:

(1) Barium bromate; $Ba(BrO_3)_2$; [13967-90-3]

(2) Barium nitrate; $Ba(NO_3)_2$; [10022-31-8]

(3) Water; H_2O; [7732-18-5]

ORIGINAL MEASUREMENTS:

Harkins, W. D.

J. Am. Chem. Soc. 1911, *33*, 1807-27.

VARIABLES:

$T/K = 298$

$10^3 c_2/\text{eq dm}^{-3} = 0$ to 199.95

PREPARED BY:

Hiroshi Miyamoto

EXPERIMENTAL VALUES:

t/°C	Barium Nitrate	Barium Bromate	
	$10^3 c_2/\text{eq dm}^{-3}$	$g\ dm^{-3}$	$10^3 c_1/\text{mol dm}^{-3}$ [b]
25	0	7.96[a]	20.25
	25.018	7.221	18.368
	50.039	6.83	17.373
	99.97	6.415	16.318
	199.95	7.085	18.022

[a] The author reports that the solubility of barium bromate in pure water is 0.793 grams to 100 grams solution and the density of the saturated solution is 1.0038. The value (7.96 g/dm^3) was calculated by the compiler.

[b] The value was calculated by the compiler.

COMMENTS AND/OR ADDITIONAL DATA:

On the table of the original paper the word "solubility Ag_2SO_4 in millimols." appeared. The compiler presumes that this is a misprint of $Ba(BrO_3)_2$.

AUXILIARY INFORMATION

METHOD/APPARATUS/PROCEDURE:

Barium bromate crystals with H_2O and aqueous $Ba(NO_3)_2$ solutions were placed in glass-stoppered bottles. The bottles were sealed with paraffin, and rotated in a thermostat for about 24 hours. Saturation was approached both from undersaturation and supersaturation. The solutions were filtered, by means of air pressure, through a thin inverted submerged asbestos filter which was held in a glass tube between two perforated Pt disks, in a thermostat. The first portion of solution was objected, and the remainder of the filtrate was used for analyses. The concentration of barium bromate was determined by precipitation as AgBr after a reduction by hydrazine.

SOURCE AND PURITY OF MATERIALS:

Barium bromate was prepared by mixing dilute solutions of barium chloride (Kahlbaum) and potassium bromate (Kahlbaum). The product was recrystallized from water until it was entirely free from chloride. The number of hydrated waters of the salt was not given. Barium nitrate (Kahlbaum) was recrystallized twice from water. The product was dried at 130°C.

ESTIMATED ERROR:

Soly: the deviations from the mean value were within ± 0.2 %.

Temp: ± 0.01°C (author)

REFERENCES:

COMPONENTS:

(1) Barium bromate; $Ba(BrO_3)_2$; [13967-90-3]

(2) Glycine; $C_2H_5NO_2$; [56-40-6]

(3) Water; H_2O; [7732-18-5]

ORIGINAL MEASUREMENTS:

Keefer, R. M.; Reiber, H. G.; Bisson, C. S.

J. Am. Chem. Soc. 1940, *62*, 2951-5.

VARIABLES:

$T/K = 298$

$m_2/\text{mol kg}^{-1} = 0$ to 0.1008

PREPARED BY:

Hiroshi Miyamoto

EXPERIMENTAL VALUES:

$t/^{\circ}C$	Glycine $m_2/\text{mol kg}^{-1}$	Barium Bromate $10^2 m_1/\text{mol kg}^{-1}$
25	0	2.008
	0.0251	2.045
	0.0503	2.081
	0.0755	2.113
	0.1008	2.150

AUXILIARY INFORMATION

METHOD/APPARATUS/PROCEDURE:

An excess of $Ba(BrO_3)_2$ crystals and aqueous glycine solution were placed in glass-stoppered Pyrex flasks. The flasks were rotated in a thermostat at 25°C for at least 12 hours. The solutions saturated with barium bromate were analyzed for barium by iodometric analysis, and by precipitating the barium as $BaSO_4$ after the bromate ion was removed by evaporating the saturated barium bromate solution to dryness in the presence of HCl and KBr.

SOURCE AND PURITY OF MATERIALS:

Barium bromate was prepared by slowly adding with stirring an equivalent amount of 0.20 M barium chloride soln to 1000 ml of 0.16 M potassium bromate. The precipitate was filtered, washed, and then air dried at room temp. The number of hydrated waters for the barium bromate was not given. Glycine (c.p. grade) was recrystallized twice from water by addition of EtOH. The product was dried in a vacuum oven at about 35°C.

ESTIMATED ERROR:

Soly: nothing specified
Temp: ± 0.02°C (authors)

COMPONENTS:

(1) Barium chloride; $BaCl_2$; [10361-37-2]

(2) Barium bromate; $Ba(BrO_3)_2$; [13967-90-3]

(3) Water; H_2O; [7732-18-5]

ORIGINAL MEASUREMENTS:

Ricci, J. E.; Freedman, A. J.

J. Am. Chem. Soc. 1952, *74*, 1769-73.

EXPERIMENTAL VALUES:

t/°C	Composition of the Saturated Solutions				Density	Nature of the Solid Phase[a]
	Barium Bromate		Barium Chloride		ρ/g cm^{-3}	
	mass %	mol % (compiler)	mass %	mol % (compiler)		
10	0.456[b]	0.0210	0.00	0.000	1.001	A.W
	0.232	0.0113	6.57	0.606	--	A.W
	0.201	0.0105	12.98	1.277	--	A.W
	0.190	0.0106	18.99	1.992	--	A.W
	0.187	0.0108	22.34	2.434	--	A.W
	0.185	0.0107	22.52	2.458	1.237	A.W + A.B.2W
	0.161	0.00944	23.78	2.633	1.253	A.B.2W
	0.159	0.00934	23.92	2.653	1.264	A.B.2W
	0.144	0.00856	24.91	2.795	1.267	A.B.2W + B.2W
	0.000	0.00000	24.93	2.793	1.265	B.2W
25	0.788[b]	0.0364	0.00	0.000	1.003	A.W
	0.446	0.0217	5.91	0.543	--	A.W
	0.373	0.0197	14.21	1.419	--	A.W
	0.338	0.0191	20.16	2.146	--	A.W
	0.323	0.0190	23.64	2.619	--	A.W
	0.321	0.0189	23.93	2.660	1.254	A.W. + A.B.2W
	0.296	0.0176	24.64	2.761	1.265	A.B.2W
	0.266	0.0159	25.52	2.889	1.274	A.B.2W
	0.249	0.0150	26.18	2.986	1.283	A.B.2W
	0.226	0.0138	27.02	3.113	1.294	A.B.2W + B.2W
	0.00	0.0000	27.06	3.110	1.292	B.2W
45	1.529[b]	0.07110	0.00	0.000	--	A.W
	0.825	0.0428	11.94	1.170	--	A.W
	0.658	0.0382	22.40	2.456	--	A.W
	0.623	0.0375	25.53	2.903	1.274	A.W. + A.B.2W
	0.560	0.0339	26.04	2.977	1.275	A.B.2W
	0.493	0.0302	27.20	3.151	1.290	A.B.2W
	0.473	0.0291	27.46	3.190	1.291	A.B.2W
	0.392	0.0247	29.60	3.528	1.318	A.B.2W + B.2W
	0.000	0.0000	29.78	3.539	1.321	B.2W

[a] A = $Ba(BrO_3)_2$; B = $BaCl_2$; W = H_2O; 2W = $2H_2O$

[b] For binary systems the compiler computes the following

Soly of $Ba(BrO_3)_2$ = 0.01165 mol kg^{-1} at 10°C

= 0.02007 mol kg^{-1} at 25°C

= 0.03950 mol kg^{-1} at 45°C

COMPONENTS:

(1) Barium chloride; $BaCl_2$; [10361-37-2]

(2) Barium bromate; $Ba(BrO_3)_2$; [13967-90-3]

(3) Water; H_2O; [7732-18-5]

ORIGINAL MEASUREMENTS:

Ricci, J. E.; Freedman, A. J.

J. Am. Chem. Soc. 1952, *74*, 1769-73.

VARIABLES: T/K = 283, 298, and 313

Concentration of $BaCl_2$

PREPARED BY:

Hiroshi Miyamoto

COMMENTS AND/OR ADDITIONAL DATA:

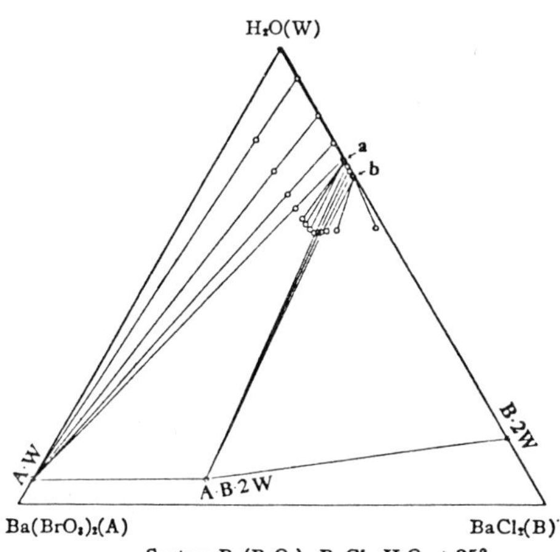

System $Ba(BrO_3)_2$–$BaCl_2$–H_2O at 25°.

AUXILIARY INFORMATION

METHOD/APPARATUS/PROCEDURE:

Complexes were made up from water, $Ba(BrO_3)_2 \cdot H_2O$ and $BaCl \cdot 2H_2O$. Equilibrium was checked by repeated analysis, the minimum time of stirring being 8 days.
The filtered saturated solution was analyzed for the iodate with standard sodium thiosulfate solution.

SOURCE AND PURITY OF MATERIALS:

C.p. grade $Ba(BrO_3)_2 . H_2O$ was used as received. The purity of which was checked with the following results; by iodometry 95.73 mass % $Ba(BrO_3)_2$ and by dehydration to constant weight at 110°C, 95.53 mass %. The theoretical value is 95.60 mass %.
C.p. grade barium chloride was used without further purification.

ESTIMATED ERROR:

Soly: nothing specified
Temp: not given

ACKNOWLEDGEMENT:

The figure reprinted from the *J. Am. Chem. Soc.* by permission of the copyright owners, The American Chemical Society.

COMPONENTS:

(1) Barium chlorate; $Ba(ClO_3)_2$; [13477-00-4]

(2) Barium bromate; $Ba(BrO_3)_2$; [13967-90-3]

(3) Water; H_2O; [7732-18-5]

ORIGINAL MEASUREMENTS:

Ricci, J. E.; Smiley, S. H.

J. Am. Chem. Soc. 1944, *66*, 1011-5.

VARIABLES:

T/K = 298.10

$Ba(ClO_3)_2$/mass % = 0 - 27.54

PREPARED BY:

Hiroshi Miyamoto

EXPERIMENTAL VALUES:

t/°C	Compositions of Saturated Solutions				Solution Density[b]
	Barium Bromate		Barium Chlorate		
	mass %	mol % (compiler)	mass %	mol % (compiler)	ρ/g cm^{-3}
24.95	0.791[a]	0.0365	--	--	1.001
	0.609	0.0284	1.292	0.0779	1.011
	0.553	0.0260	2.304	0.1402	1.016
	0.491	0.0237	4.850	0.3024	(1.043)
	0.446	0.0220	7.019	0.4471	1.060
	0.423	0.0214	9.370	0.6112	(1.085)
	0.402	0.0205	10.50	0.6929	(1.095)
	0.347	0.0185	14.41	0.9909	(1.131)
	0.347	0.0185	14.56	1.003	(1.132)
	0.310	0.0169	16.83	1.188	1.151
	0.282	0.0156	18.07	1.293	1.165
	0.249	0.0140	19.53	1.421	(1.178)
	0.235	0.0134	20.50	1.508	1.186
	0.207	0.0120	21.72	1.620	1.202
	0.176	0.0103	22.91	1.733	(1.209)
	0.145	0.00858	23.85	1.824	(1.217)
	0.112	0.00671	24.87	1.925	(1.227)
	0.078	0.00473	25.87	2.027	1.242
	0.057	0.00348	26.51	2.093	(1.241)
	--	--	27.54	2.201	1.249

[a] For binary system the compiler computes the following Soly of $Ba(BrO_3)_2$ = 0.02028 mol kg^{-1} at 24.95°C.

[b] Densities in parentheses are interpolated by the authors.

AUXILIARY INFORMATION

METHOD/APPARATUS/PROCEDURE:

Barium bromate, barium chlorate and water were shaken in a thermostat. Equilibrium was approached from two directions, in about half of the experiments from undersaturation, and in the others from supersaturation. The period of stirring was from one to six, averaging three and one-half months.
The liquid solution was analyzed after sampling by means of calibrated pipets fitted with filtering tips. Water was determined by drying at 110°C and the bromate was determined iodometrically (1) in the presence of chlorate. These two determinations allowed the calculation of percentage of each salt in various solutions. The nature of solid phases was not given in the paper. It can be inferred from the phase diagram.

SOURCE AND PURITY OF MATERIALS:

C.p. grade $Ba(BrO_3)_2 \cdot H_2O$ was used as received. The purity was checked with the following results: by iodometry, 95.73 % $Ba(BrO_3)_2$, and by dehydration to constant weight at 110°C, 95.53 %. The theoretical value is 95.60 %.
C.p. grade $Ba(ClO_3)_2 \cdot H_2O$ was used as received and found to contain 5.69 % of water by drying at 110°C as compared with the theoretical 5.60 %.

ESTIMATED ERROR:

Soly: nothing specified
Temp: ± 0.02°C (authors)

REFERENCES:

1. Swensen, T.; Ricci, J. E. *J. Am. Chem. Soc.* 1939, *61*, 1974.

COMPONENTS:

(1) Barium chlorate; $Ba(ClO_3)_2$; [13477-00-4]

(2) Barium bromate; $Ba(BrO_3)_2$; [13967-90-3]

(3) Water; H_2O; [7732-18-5]

ORIGINAL MEASUREMENTS:

Ricci, J. E.; Smiley, S. H.

J. Am. Chem. Soc. 1944, *66*, 1011-5.

COMMENTS AND/OR ADDITIONAL DATA:

The phase diagrams are given below (based on mass %).

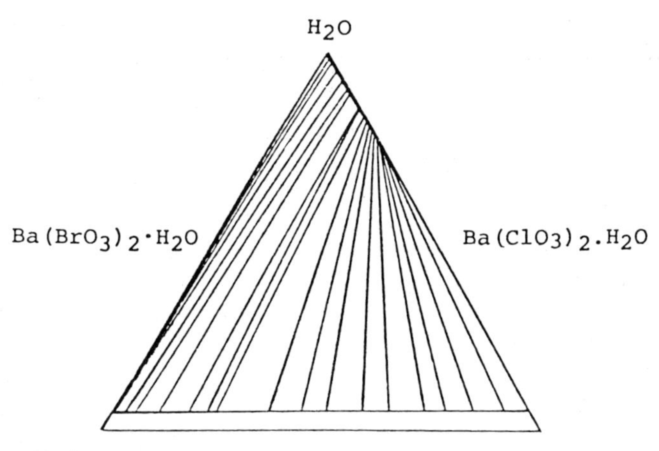

System $Ba(BrO_3)_2$-$Ba(ClO_3)_2$-H_2O at 25°

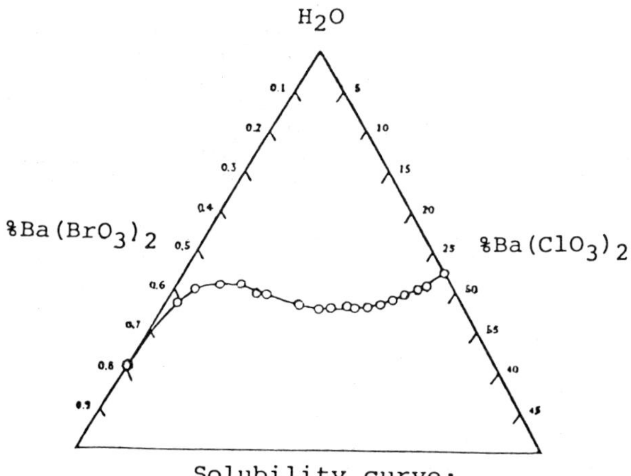

Solubility curve:

$Ba(BrO_3)_2$-$Ba(ClO_2)_2$-H_2O

at 25°

ACKNOWLEDGEMENT:

The figures reprinted from the *J. Am. Chem. Soc.* by permission of the copyright owners, The American Chemical Society.

COMPONENTS:

(1) Barium bromate; $Ba(BrO_3)_2$; [13967-90-3]

(2) Barium bromide; $BaBr_2$; [10553-31-8]

(3) Water; H_2O; [7732-18-5]

ORIGINAL MEASUREMENTS:

Ricci, J. E.; Freedman, A. J.

J. Am. Chem. Soc. 1952, *74*, 1769-73.

VARIABLES:

T/K = 298
composition

PREPARED BY:

Hiroshi Miyamoto

EXPERIMENTAL VALUES:

t/°C	Composition of Saturated Solutions				Nature of the Solid Phase[a]
	Barium Bromate		Barium Bromide		
	mass %	mol % (compiler)	mass %	mol % (compiler)	
25	0.395	0.0206	12.71	0.8788	A + B
	0.320	0.0190	24.15	1.901	"
	0.257	0.0179	36.19	3.337	"
	0.229	0.0180	44.21	4.601	"
	0.220	0.0191	49.92[b]	5.722	"

[a] A = $Ba(BrO_3)_2 \cdot H_2O$; B = $BaBr_2 \cdot 2H_2O$

[b] Saturated solution

AUXILIARY INFORMATION

METHOD/APPARATUS/PROCEDURE:

Complexes were made up from water, $Ba(BrO_3)_2 \cdot H_2O$ and $BaBr_2 \cdot 2H_2O$. The analysis for the system involved the determination of total solid and iodometric titration of the bromate. No other information is given.

SOURCE AND PURITY OF MATERIALS:

C.p. grade $Ba(BrO_3)_2 \cdot H_2O$ was used as received. The purity of which was checked with the following results; by iodometry 95.73 mass % $Ba(BrO_3)_2$ and by dehydration to constant weight at 110°C, 95.53 mass %. The theoretical value is 95.60 mass %. The source of barium bromide was not given.

ESTIMATED ERROR:

Nothing specified

REFERENCES:

COMPONENTS:

(1) Barium nitrate; $Ba(NO_3)_2$; [10022-31-8]

(2) Barium bromate; $Ba(BrO_3)_2$; [13967-90-3]

(3) Water; H_2O; [7732-18-5]

ORIGINAL MEASUREMENTS:

Ricci, J. E.; Freedman, A. J.

J. Am. Chem. Soc. 1952, *74*, 1769-73.

VARIABLES:

T/K = 298
Composition

PREPARED BY:

Hiroshi Miyamoto

EXPERIMENTAL VALUES:

t/°C	Composition of Saturated Solutions				Nature of the Solid Phase[a]
	Barium Bromate		Barium Nitrate		
	mass %	mol % (compiler)	mass %	mol % (compiler)	
25	0.593	0.0280	2.44	0.173	A + B
	0.570	0.0275	4.74	0.344	"
	0.568	0.0280	7.07	0.525	"
	0.571	0.0286	8.60	0.648	"
	0.573	0.0289	9.23	0.700	"

[a] A = $Ba(BrO_3)_2 \cdot H_2O$; B = $Ba(NO_3)_2$

AUXILIARY INFORMATION

METHOD/APPARATUS/PROCEDURE:

Complexes were made up from water, $Ba(BrO_3)_2 \cdot H_2O$ and $Ba(NO_3)_2$. The analysis for this system involved the determination of total solid and iodometric titration of the bromate. No other information is given.

SOURCE AND PURITY OF MATERIALS:

C.p. grade $Ba(BrO_3)_2 \cdot H_2O$ was used as received. The purity of which was checked with the following results; by iodometry 95.73 mass % $Ba(BrO_3)_2$ and by dehydration to constant weight at 110°C, 95.53 mass %. The theoretical value is 95.60 mass %. C.p. grade barium nitrate was used without further purification.

ESTIMATED ERROR:

Nothing specified

REFERENCES:

COMPONENTS:

(1) Barium iodate; $Ba(IO_3)_2$; [10567-69-8]

(2) Water; H_2O; [7732-18-5]

EVALUATOR:
Hiroshi Miyamoto
Department of Chemistry
Niigata University
Niigata, Japan

March, 1982

CRITICAL EVALUATION:

1. The binary system; $Ba(IO_3)_2$-H_2O

Solubilities in the binary $Ba(IO_3)_2$-H_2O have been reported in 19 publications (1-19) which are summarized in Table 1.

Table 1 Solubility studies of barium iodate in water

Reference	T/K	Solid Phase	Soly and/or Soly Product	Method of Analysis
Trautz; Anschütz (1)	273-373	$Ba(IO_3)_2 \cdot H_2O$	Soly	gravimetric
Hill; Zink (2)	room temp	*	Soly	iodometric
Harkins; Winninghoff (3)	298	*	Soly	iodometric
Polessitskij (4)	298-373	*	Soly	gravimetric ($Ba(IO_3)_2$) gravimetric (Ba^{2+}) iodometric (IO_3^-)
Macdougall; Davies (5)	298	$Ba(IO_3)_2 \cdot H_2O$	Soly, K_{s0}	iodometric
Naidich; Ricci (6)	298	$Ba(IO_3)_2 \cdot H_2O$	Soly	iodometric
Davis; Ricci; Sauter (7)	298	$Ba(IO_3)_2 \cdot H_2O$	Soly	iodometric
Keefer; Reiber; Bisson (8)	298	*	Soly	iodometric
Pedersen (9)	291	$Ba(IO_3)_2 \cdot H_2O$	Soly	iodometric
Derr; Vosburgh (10)	298	*	Soly, $K°_{s0}$	iodometric
Davies; Wyatt (11)	298	$Ba(IO_3)_2 \cdot H_2O$	Soly	iodometric
Ricci (12)	298	$Ba(IO_3)_2 \cdot H_2O$	Soly	iodometric
Monk (13)	298	*	Soly	iodometric
Monk (14)	298	*	Soly, K_{s0}	iodometric
Ricci; Freedman (15)	298	$Ba(IO_3)_2 \cdot H_2O$	Soly	iodometric
Bousquet; Mathurin; Vermande (16)	273-303 313-359	$Ba(IO_3)_2 \cdot H_2O$ $Ba(IO_3)_2$	$K°_{s0}$	iodometric "
Miyamoto (17)	298	$Ba(IO_3)_2 \cdot H_2O$	Soly	iodometric
Jones; Madigan; Wilson (18)	278.2, 283.2, 298.2	$Ba(IO_3)_2 \cdot H_2O$	Soly	conductometric
Miyamoto; Suzuki; Yanai (19)	293, 298, 303	$Ba(IO_3)_2 \cdot H_2O$	Soly	iodometric

* The degree of hydration is not given in the original paper.

COMPONENTS:	EVALUATOR:
(1) Barium iodate; $Ba(IO_3)_2$; [10567-69-8] (2) Water; H_2O; [7732-18-5]	Hiroshi Miyamoto Department of Chemistry Niigata University Niigata, Japan March, 1982

CRITICAL EVALUATION:

Many of these studies also deal with the aqueous ternary system, and two studies (1,4) are concerned solely with the binary system.

Milad, Morsi, Soliman and Seleem (20) measured solubilities of barium iodate in aqueous and aqueous ethanol media containing $LiNO_3$ and $LiClO_4$ Fedorov, Robov, Shmyd'ko, Vorontsova and Mironov (21) determined solubilities of barium iodate in aqueous $LiNO_3$ solution containing $LiClO_4$. Neither invertigator's groups (20,21) reported either the solubility of barium iodate in pure water, or the solubility product at zero ionic strength.

Depending upon temperature and composition, equilibrated solid phases of varying degrees of hydration have been reported. The following solid phases have been identified:

$Ba(IO_3)_2 \cdot H_2O$	[7787-34-0]
$Ba(IO_3)_2$	[10567-69-8]

Bousquet, Mathurin and Vermande (16) report that below 303K the stable phase is the monohydrate, and above 313K the anhydrous salt is the stable phase. The transition temperature between the monohydrate and the anhydrous salt is 303.2K. However, the thermal analyses of the salt was not reported. Trautz and Anschütz (1), and Polessitskij (4) measured solubilities of barium iodate in pure water over the temperature range from 273-373K. In these publications, the solubility of barium iodate was found to increase monotonically. Trautz and Anschütz reported the solid phase as the monohydrate whereas Polessitskij did not specify the nature of the solid phase, but the evaluator presumes it to be the monohydrate.

Many investigators (1,5-7,9,11,12,15-19) used the monohydrate in solubility determinations. The degree of hydration of the salt is not given in some articles. The evaluator assumes that the investigators who did not report the degree of hydration also used the monohydrate because the monohydrate is the stable phase at 298K. Only the studies of Bousquet (16) report the use of the monohydrate and anhydrous salt in the solubility determinations.

The data to be considered in this critical evaluation are summarized in Table 2.

COMPONENTS:

(1) Barium iodate; $Ba(IO_3)_2$; [10567-69-8]

(2) Water; H_2O; [7732-18-5]

EVALUATOR:

Hiroshi Miyamoto
Department of Chemistry
Niigata University
Niigata, Japan

March, 1982

CRITICAL EVALUATION:

Table 2 Summary of solubility data in the binary $Ba(IO_3)_2$-H_2O system

T/K	$10^3 c_1$/mol dm^{-3}	ref
273.2	0.164[a]	(1)
"	0.374	(4)
275.2	0.400	(18)
283.2	0.287	(1)
"	0.538	(18)
288.2	0.626	(4)
291.1	0.6694	(9)
293.2	0.452	(1)
"	0.702	(19)
303.2	0.637	(1)
"	0.923	(19)
313.2	0.842	(1)
"	1.19	(4)
323.2	1.15	(1)
"	1.49	(4)
333.2	1.52	(1)
"	1.79	(4)

T/K	$10^3 c_1$/mol dm^{-3}	ref
298.1	0.810	(5)
298.2	0.575	(1)
"	0.789	(3)
"	0.809	(10)
"	0.812	(11)
"	0.812	(13)
"	0.812	(14)
"	0.8145	(7)
"	0.8177	(6)
"	0.818	(17)
"	0.818	(19)
"	0.820	(18)
"	0.833	(4)
343.2	1.91	(1)
"	2.196	(4)
351.2	2.669	(4)
353.2	2.363	(1)
363.2	2.899	(1)
373.2	3.777	(4)
"	4.052[b]	(1)

a: The value at eutectic point (273.04K).

b: The value at 372.4K/735 mmHg.

Solubility at 273.2K. The value reported in (1) obtained by Trautz and Anschütz is considerably lower than that of Polessitskij (4) as shown in Table 2. This difference is similar to differences reported at other temperatures, that is, at 298K the result of Trautz and Anschütz is certainly too small about 25%. The difference is greater than expected from the estimated precision suggesting the presence of a systematic error in the data of ref (1). Therefore, all the results of Trautz and Anschütz are rejected. The value, 0.374 mmol dm^{-3}, reported by Polessitskij is designated as a tentative value.

Solubility at 275.2K. Only one result obtained by Jenes, Madizan and Wilson has been reported. The value, 0.400 mmol dm^{-3}, is designated as a tentative value.

Solubility at 283.2K. The result reported in (1) obtained by Trautz and Anschütz is considerably lower than that of Jenes, Madigan and Wilson (18), and as discussed in the above analysis, the data from ref (1) are rejected. The tentative value of the solubility at 283K is therefore taken from ref (18) and is 0.538 mmol dm^{-3}.

Solubility at 288.2K. Only one result has been reported by Polessitskij (4). The value, 0.626 mmol dm^{-3}, is designated as a tentative value.

COMPONENTS:	EVALUATOR:
(1) Barium iodate; $Ba(IO_3)_2$; [10567-69-8] (2) Water; H_2O; [7732-18-5]	Hiroshi Miyamoto Department of Chemistry Niigata University Niigata, Japan March, 1982

CRITICAL EVALUATION:

Solubility at 291.1K. Only one result has been reported by Pedersen (9). The value, 0.6694 mmol dm^{-3}, is designated as a tentative value.

Solubility at 293.2K. The rejected result reported in (1) obtained by Trautz and Anschütz is considerably lower than that reported by Miyamoto, Suzuki and Yanai (19), and it is felt that the value from (19) is more accurate. The tentative value of the solubility at 293 K is 0.702 mmol dm^{-3}.

Solubility at 298.2K. This value has been reported in 13 publications. The stable solid phase at this temperature is the monohydrate form. The result reported in (1) obtained by Trautz and Anschütz is lower than that of other investigators, and the result reported by Polessitskij (4) is higher than that of others. The result of Trautz and Anschütz differs by about 30% from the mean of 13 values. Therefore, the result is rejected. The results of Harkins and Winninghoff (3) and Polessitskij (4) are also rejected. The arithmetic mean of the remaining 10 values is 0.814 mmol dm^{-3}, and the standard deviation is 0.004 mmol dm^{-3}. The recommended value of the solubility at 298 K is 0.814 mmol dm^{-3}.

Solubility at 303.2K. The result reported in (1) obtained by Trautz and Anschütz is considerably lower than that reported by Miyamoto, Suzuki, and Yanai (19), and as in the above analysis it is felt that the data in (19) are more accurate. The tentative value of the solubility at 303K is 0.923 mmol dm^{-3}.

Solubility at 313.2, 323.2, 333.2, and 343.2K. The solubility of $Ba(IO_3)_2$ in pure water at 313.2, 323.2, 333.2, and 343.2K has been reported by Trautz and Anschütz (1), and Polessitskij (4). The results reported in (1) obtained by Trautz and Anschütz are again much lower than those of Polessitskij and as in the above analysis it is felt that the data in (4) are more accurate. The tentative values of the solubilities at 313, 323, 333, and 343K are 1.19, 1.49, 1.79, and 2.196 mmol dm^{-3}, respectively.

Solubility at 351.2K. The solubility at 351.2K has been reported by only Polessitskij (4). The tentative value of the solubility at 351.2K is 2.669 mmol dm^{-3}.

The experimental solubility data obtained by Polessitskij (4), Pedersen (9), Jones, Madigan, and Wilson (18), and Miyamoto, Suzuki, and Yanai (19) are considered to be sufficiently reliable for use in a smoothing equation. All data reported in the 3 publications were used except the solubility at 298K, and in which case the mean of the 12 determinations was used.

The fitting equation used was as follows:

$$\ln S = A + B/(T/100\text{K}) + C \ln (T/100\text{K})$$

By using $T/100$K as the variable rather than $T/$K the coefficients in the smoothed equation are of roughly equal magnitude.

The best fit for the 14 data points was found to be:

$$\ln (S/\text{mmol dm}^{-3}) = 15.72672 - 35.26709/(T/100\text{K}) - 3.763181 \ln (T/100\text{K}) \; : \; \sigma = 0.032$$

where S is the solubility of barium iodate in water.

Recommended and tentative values for the solubility of barium iodate in water with the values calculated from the smoothing equation are given in Table 3.

COMPONENTS:

(1) Barium iodate; $Ba(IO_3)_2$; [10567-69-8]

(2) Water; H_2O; [7732-18-5]

EVALUATOR:

Hiroshi Miyamoto
Department of Chemistry
Niigata University
Niigata, Japan

March, 1982

CRITICAL EVALUATION:

Table 3 Recommended and tentative values for the solubility of $Ba(IO_3)_2$ in water

T/K	10^3c_1(exptl)/mol dm^{-3}	$10^3\sigma$/mol dm^{-3}	10^3c_1(calcd)/mol dm^{-3}	Solid Phase
273.2	0.374	--	0.381	$Ba(IO_3)_2 \cdot H_2O$
275.2	0.400	--	0.407	"
283.2	0.538	--	0.525	"
288.2	0.626	--	0.610	"
291.1	0.6694	--	0.664	"
293.2	0.702	--	0.705	"
298.2	0.814[a]	0.004	0.809	"
303.2	0.923	--	0.924	"
313.2	1.19	--	1.19	$Ba(IO_3)_2$
323.2	1.49	--	1.49	"
333.2	1.79	--	1.85	"
343.2	2.196	--	2.25	"
351.2	2.669	--	2.61	"
373.2	3.777	--	3.75	"

a: recommended value

σ: standard deviation

The solubility (based on mol kg^{-1}) of barium iodate monohydrate in water at 298.2K has been reported by Ricci (12), and Ricci and Freedman (15). Keefer, Reiber, and Bisson (8) also reported the solubility of barium iodate in water at 298.2K, but the degree of hydration of the salt was not given in the original article. The evaluator assumes that the monohydrate was used in the solubility determination because the monohydrate is the stable solid phase at 298.2K.

The arithmetic mean of three results is 8.32 x 10^{-4} mol kg^{-1}, and the standard deviation is 0.26 x 10^{-4} mol kg^{-1}. The mean obtained is a tentative value.

2. Solubility of barium iodate in acidic solutions

Naidlich and Ricci (6) measured solubilities of barium iodate monohydrate in diluted hydrochloric acid and nitric acid. The solubility of barium iodate increases with increasing acid concentration, and the effect of nitric acid on the solubility of barium iodate is slightly larger than that of hydrochloric acid.

Naidlich and Ricci (6) calculated the dissociation constant, K_D, for iodic acid from the solubility data, either with or without the correction for the ion-pair $BaIO_3^+$, and the value was found to be 0.163 mol dm^{-3} in agreement with the best value (0.1686) derived from conductivity data (22).

Naidlich and Ricci's results (6) for the solubility of barium iodate monohydrate in dilute hydrochloric and nitric acid are designated as tentative values.

COMPONENTS:	EVALUATOR:
(1) Barium iodate; $Ba(IO_3)_2$; [10567-69-8] (2) Water; H_2O; [7732-18-5]	Hiroshi Miyamoto Department of Chemistry Niigata University Niigata, Japan March, 1982

CRITICAL EVALUATION:

3. Solubility of barium iodate in aqueous ammonia

Hill and Zink (2) report only one result for the solubility of barium iodate in a concentrated ammonia solution. They did not report the temperature, and the concentration of ammonia was roughly given in the article.

Derr and Vousburgh(10) measured solubilities in ammonia solutions, and the data are given in the compilation. They reported that barium iodate is considerably less soluble in concentrated ammonia solution than in water, and also that the decrease in solubility product is roughly proportional to the ammonia concentration. If the decrease in the solubility product is also proportional to the NH_3 concentration for NH_3 concentrations less than 0.168 mol dm^{-3}, a 1% decrease in K_{s0} would result in 0.03-0.04 mol $dm^{-3}NH_3$ solutions. If a complex ion is formed, the observed K_{s0} would decrease still further.

The results of Derr and Vousburgh are designated as tentative values.

4. Solubility of barium iodate in aqueous KCl solutions

Naidlich and Ricci (6) and Keefer, Reiber and Bisson (8) measured solubilities of barium iodate monohydrate in aqueous KCl solutions at 298.2K, and Macdougall and Davies (5) studied this system at 298.05K. The degree of hydration for barium iodate used is not given in the publication by Keefer, Reiber, and Bisson, and the evaluator again assumes that the salt was the monohydrate.

Due to the absence of reliable density data, the evaluator could not convert the results reported by Keefer, Reiber, and Bisson into molal units. Nevertheless, their results appear to be in excellent agreement with those reported in (5,6).

Solubilities of barium iodate monohydrate in aqueous KCl solutions increase with increasing KCl concentration.

The arithmetic mean of the two results (5,6) is designated as a tentative value. The tentative values are given in Table 4 with the standard deviation, and also with the corresponding result of Keefer, Reiber, and Bisson.

COMPONENTS:

(1) Barium iodate; $Ba(IO_3)_2$; [10567-69-8]

(2) Water; H_2O; [7732-18-5]

EVALUATOR:
Hiroshi Miyamoto
Department of Chemistry
Niigata University
Niigata, Japan

March, 1982

CRITICAL EVALUATION:

Table 4 Tentative values of solubility of $Ba(IO_3)_2$ in aqueous KCl solutions at 298.2K (solid phase is the monohydrate)

c_2/mol dm^{-3}	$10^3 c_1$/mol dm^{-3}	$10^3 \sigma$/mol dm^{-3}	$10^3 m_1$/mol kg^{-1}
0	0.814	0.005	0.811
0.001	0.831	0.006	--
0.002	0.840	--	--
0.0035	0.859	--	--
0.005	0.886	0.02	0.877
0.0075	0.899	--	--
0.01	0.924	0.008	0.922
0.02	0.985	--	--
0.05	1.123	0.007	1.121
0.1	1.272	0.004	1.272
0.2454	1.566	--	--
0.4908	1.895	--	--
0.9817	2.378	--	--

c_2: concentration of KCl

c_1: tentative value of solubility

m_1: result of Keefer, Reiber and Bisson

5. Solubility of barium iodate in aqueous KNO_3 solutions

Solubilities of barium iodate monohydrate in aqueous KNO_3 solutions at 298.2K have been reported by Harkins and Winninghoff (3), Polessitskij (4), Naidlich and Ricci (6), and Davies, Ricci and Sauter (7), and at 298.05K by Macdougall and Davies (5). The degree of hydration for barium iodate used was not described in the papers reported by Harkins and Winninghoff, and Polessitskij, but the evaluator assumes that the salt was the monohydrate since the monohydrate is the stable phase at 298K.

The solubility of barium iodate in aqueous KNO_3 solutions increases with increasing concentration of KNO_3, and the data are given in the compilations of publications (3,4,5,6,7).

The evaluation for solubilities of barium iodate in aqueous KNO_3 solutions varying the concentration of KNO_3 is given below.

<u>Solubility in 0.001 mol dm^{-3} KNO_3 solution</u>. This value has been reported in 2 publications (3,6). The result reported in (3) by Harkins and Winninghoff is slightly lower than that of Naidlich and Ricci (6). The arithmetic mean of two results is 0.832 mmol dm^{-3}, and the standard deviation is 0.009 mmol dm^{-3}.

<u>Solubility in 0.002 mol dm^{-3} KNO_3 solution</u>. This value has been reported in 3 publications (3,5,7). The result of Harkins and Winninghoff (3) is considerably lower than that of others (5,7), and is rejected. The arithmetic mean of the remaining two results is 0.846 mmol dm^{-3}, and the standard deviation is 0.007 mmol dm^{-3}.

<u>Solubility in 0.0035 mol dm^{-3} KNO_3 solution</u>. Only one result has been reported by Macdougall and Davies (5). The result obtained is 0.863 mmol dm^{-3}.

COMPONENTS:

(1) Barium iodate; $Ba(IO_3)_2$; [10567-69-8]

(2) Water; H_2O; [7732-18-5]

EVALUATOR:
Hiroshi Miyamoto
Department of Chemistry
Niigata University
Niigata, Japan

March, 1982

CRITICAL EVALUATION:

Solubility in 0.005 mol dm^{-3} KNO_3 solution. This value has been reported in 2 publications. The result of Macdougall and Davies (5) is slightly lower than that of Naidlich and Ricci (6). The arithmetic mean of the two results is 0.887 mmol dm^{-3}, and the standard deviation is 0.010 mmol dm^{-3}.

Solubility of 0.0075 mol dm^{-3} KNO_3 solution. Only one result has been reported by Macdougall and Davies (5). The result obtained is 0.906 mmol dm^{-3}.

Solubility in 0.010 mol dm^{-3} KNO_3 solution. This value has been reported in 4 publications (3,5,6,7). The result of Harkins and Winninghoff (3) is slightly lower than that of others (5,6,7). The arithmetic mean of the four results is 0.932 mmol dm^{-3}, and the standard deviation is 0.020 mmol dm^{-3}.

Solubility in 0.020 mol dm^{-3} KNO_3 solution. Only one result has been reported by Macdougall and Davies (5). The result reported is 1.006 mmol dm^{-3}.

Solubility in 0.050 mol dm^{-3} KNO_3 solution. The solubility of $Ba(IO_3)_2$ in 0.050 mol KNO_3 solution has been reported in 3 publications (3,5,7), and that in 0.05012 mol dm^{-3} KNO_3 solution has been reported by Naidlich and Ricci (6). The result of Naidlich and Ricci is in excellent agreement with that of Davies, Ricci and Sauter (7), but the result of Harkins and Winninghoff (3) is considerably higher than that of others (5,6,7). The arithmetic mean of three results (5,6,7) is 1.169 mmol dm^{-3}, and the standard deviation is 0.012 mol dm^{-3}.

Solubility in 0.1 mol dm^{-3} KNO_3 solution. The solubility of $Ba(IO_3)_2$ in 0.1 mol dm^{-3} KNO_3 solution has been reported by Macdougall and Davies (5), and that in 0.1002 mol dm^{-3} KNO_3 solution by Naidlich and Ricci (6). The result of Macdougall and Davies is in good agreement with that of Naidlich and Ricci. The arithmetic mean of two results is 1.364 mmol dm^{-3}, and the standard deviation is 0.004 mmol dm^{-3}.

Solubility in 0.2 mol dm^{-3} KNO_3 solution. This value has been reported in 3 publications (3,4,7). The arithmetic mean of all results is 1.652 mmol dm^{-3}, and the standard deviation is 0.055 mmol dm^{-3}.

Solubility in 0.2454, 0.4908, 0.9817, 1.4 and 2.25 mol dm^{-3} KNO_3 solution. The solubility of $Ba(IO_3)_2$ in 0.2454, 0.4908 and 0.9817 mol dm^{-3} KNO_3 solution has been reported by Naidlich and Ricci (6), and that in 1.4 and 2.25 mol dm^{-3} KNO_3 solution was reported by Polessitskij (4).

The tentative solubility values in aqueous KNO_3 solutions are given in Table 5.

COMPONENTS:

(1) Barium iodate; $Ba(IO_3)_2$; [10567-69-8]

(2) Water; H_2O; [7732-18-5]

EVALUATOR:
Hiroshi Miyamoto
Department of Chemistry
Niigata University
Niigata, Japan

March, 1982

CRITICAL EVALUATION:

Table 5 Tentative values of solubility of barium iodate in aqueous KNO_3 solution at 298.2K (solid phase is the monohydrate)

c_2/mol dm^{-3}	$10^3 c_1$/mol dm^{-3}	$10^3 \sigma$/mol dm^{-3}
0	0.814	0.004
0.001	0.832	0.009
0.002	0.846	0.007
0.0035	0.863	--
0.0050	0.887	0.010
0.0075	0.906	--
0.010	0.932	0.020
0.020	1.006	--
0.050	1.169	0.012
0.1	1.364	0.004
0.2	1.652	0.055
0.2454	1.760	--
0.4908	2.290	--
0.9817	3.237	--
1.4	4.208	--
2.25	5.686	--

c_1: tentative value of solubility

c_2: concentration of KNO_3

σ : standard deviation

6. Solubility of barium iodate in aqueous $Ba(NO_3)_2$ solutions

Solubilities of barium iodate monohydrate in aqueous $Ba(NO_3)_2$ solutions at 298.2K have been reported in 3 publications (3,7,15). The degree of hydration for the iodate was not given in the paper by Harkins and Winninghoff. The evaluator assumes that the monohydrate was used for the determination of the solubility. The absence of reliable density data prevented the evaluator from converting the results reported in mol dm^{-3} units by Ricci and Freedman (15) into molal units. Therefore, it is not possible to compare the results of Ricci and Freedman to those of others (3,7).

The solubility of barium iodate monohydrate in aqueous $Ba(NO_3)_2$ solution decreases with increasing $Ba(NO_3)_2$ concentration. In aqueous solutions of lower $Ba(NO_3)_2$ concentration, the results of Harkins and Winninghoff are slightly lower than those of Davis, Ricci and Sauter (7).

The arithmetic mean of two results reported in (3) and (7) are designated as tentative values, and the results in 0.100 and 0.200 mol dm^{-3} $Ba(NO_3)_2$ solution reported by Harkins and Winninghoff are also designated as tentative values. The tentative values with the standard deviations are given in Table 6.

COMPONENTS:

(1) Barium iodate; $Ba(IO_3)_2$; [10567-69-8]

(2) Water; H_2O; [7732-18-5]

EVALUATOR:
Hiroshi Miyamoto
Department of Chemistry
Niigata University
Niigata, Japan

March, 1982

CRITICAL EVALUATION:

Table 6 Tentative values of solubility in aqueous $Ba(NO_3)_2$ solutions at 298.2K

c_2/mol dm^{-3}	$10^3 c_1$/mol dm^{-3}	$10^3 \sigma$/mol dm^{-3}
0.0005	0.688	0.10
0.001	0.613	0.10
0.0025	0.492	0.06
0.01	0.340	0.04
0.025	0.296	0.15
0.100	0.283	--
0.200	0.279	--

c_1: tentative value of solubility

c_2: concentration of $Ba(NO_3)_2$

σ : standard deviation

7. Solubility of barium iodate in ethanol-water mixed solvents

Solubilities of barium iodate monohydrate in the mixtures of ethanol and water have been reported by Hill and Zink (2), and Monk (14). The units of the concentration of solvent and the temperature are not given in the article reported by Hill and Zink. The degree of hydration of the salt used is not given in either publication, but the evaluator assumes that the solid phase is the monohydrate.

Solubilities of barium iodate in ethanol-water mixed solvents were determined by Monk for the purpose of testing the applicability of the Born equation (23) to solubility phenomena effect in different media. The solubilities obtained in these systems decrease with increasing the ethanol concentration, and also decrease with decreasing solvent dielectric constant.

The results obtained by Monk (14) are designated as tentative values.

8. Solubility of barium iodate in mixtures of various organic solvents and water

Pedersen (9) iodometrically determined the solubility of barium iodate monohydrate in the mixtures of 1,4-dioxane and water at 291.2K.

Monk (14) measured solubilities of barium iodate in mixtures of various organic solvents and water at 298K. He used methanol, ethanol, 1-propanol, 1,2-ethandiol (ethylene glycol), 1,2,3-propanetriol (glycerol), 2-propanone (acetone), 1,4-dioxane and ethyl acetate.

Miyamoto (17) determined solubilities in tetrahydrofuran-water mixtures at 298K iodometrically, and Miyamoto, Suzuki and Yanai (19) reported results in N,N-dimethylformamide-water mixtures at 293, 298 and 303K.

Miyamoto (17) and Miyamoto, Suzuki and Yanai (19) report that the monohydrate was used for the determination of solubilities, and although Monk did not report the degree of hydration, the evaluator again assumes that the monohydrate was used in the solubility experiments.

COMPONENTS:

(1) Barium iodate; $Ba(IO_3)_2$; [10567-69-8]

(2) Water; H_2O; [7732-18-5]

EVALUATOR:

Hiroshi Miyamoto
Department of Chemistry
Niigata University
Niigata, Japan

March, 1982

CRITICAL EVALUATION:

Several authors (9,14,17,19) studied the solubilities in various mixed solvents with the purpose of testing the applicability of the Born equation (23) to solubility phenomena. The solubility in the various mixed solvents studied decreases with increasing concentration of organic solvent, and decreases with decreasing dielectric constant. Monk(14) concludes that these results indicate that the chemical character of the solvent is of major importance in influencing the decrease in solubility with decreasing dielectric constant of the solvent.

The results reported in (9) by Pederson, in (14) by Monk, in (17) by Miyamoto, and in (19) by Miyamoto, Suzuki, and Yanai are designated as tentative values.

9. Solubility of barium iodate in aqueous glycine solutions

At 298.2K Keefer, Reiber, and Bisson (8) reported solubilities based on mol kg^{-1}units, and Monk (13) reported solubilities based on mol dm^{-3} units. The degree of hydration of barium iodate was not given in either publication, but the evaluator assumes that the monohydrate was used. The absence of reliable density data prevented the evaluator from converting results reported in mol dm^{-3} units into molal units and vice versa. Therefore, it is not possible to compare the results of Keefer, Reiber, and Bisson to those of Monk.

In both publications the authors report that the solubility of barium iodate in aqueous glycine solution increases with increasing concentration of the acid.

Monk (13) calculated the dissociation constant (0.17 mol dm^{-3}) of barium glycinate cation (BaG^+) from data of barium iodate in aqueous glycine solutions containing NaOH. The magnitude of the dissociation constant for BaG^+ is such that the concentration of this ion is negligible. However, Keefer, Reiber, and Bisson (8) did not consider this complex.

10. Solubility of barium iodate in aqueous alanine solutions

At 298.2K Keefer, Reiber, and Bisson (8) report solubilities based on mol kg^{-1} units, and Monk (13) reportes solubilities based on mol dm^{-3} units. The degree of hydration of barium iodate is not given in either publication, but the evaluator again assumes that the monohydrate was used. The absence of reliable density data prevented the evaluator from converting results reported in molar units into molal units and vice versa. Therefore it is not possible to compare the results of Keefer, Reiber, and Bisson to those of Monk.

In both publications the authors report that the solubility of barium iodate in aqueous alanine solution increases with increasing concentration of alanine.

Monk (13) calculated the dissociation constant (0.17 mol dm^{-3}) of barium alaninate cation (BaA^+) from data of barium iodate in aqueous alanine solutions containing NaOH. The magnitude of the dissociation constant BaA^+ is such that the concentration of this ion is negligible. However, Keefer, Reiber, and Bisson (8) did not consider this complex.

The values reported in (8) obtained by Keefer, Reiber, and Bisson and in (13) obtained by Monk are designated as tentative values.

COMPONENTS:

(1) Barium iodate; $Ba(IO_3)_2$; [10567-69-8]

(2) Water; H_2O; [7732-18-5]

EVALUATOR:

Hiroshi Miyamoto
Department of Chemistry
Niigata University
Niigata, Japan

March, 1982

CRITICAL EVALUATION:

11. Solubility product of barium iodate in aqueous solution

Macdougall and Davies (5) measured solubilities of barium iodate monohydrate in a number of aqueous salt solutions at 298.2K but did not calculate the thermodynamic solubility product of barium iodate: only the concentration solubility products were obtained from the solubility data. The value is not considered further because other investigators have reported the thermodynamic solubility product.

Derr and Vosburgh (10) have reported the solubility of barium iodate in ammonia solution at 298.2K, but did not give the degree of hydration of the salt used. However, the evaluator assumes that the authors used the monohydrate for the determination of the solubility because the solubility obtained is in excellent agreement with that found by Macdougall and Davies (5) in which the monohydrate is the stable solid phase. They calculated the activity solubility product of barium iodate in water from the solubility data observed at 298.2K. The value obtained is 1.53×10^{-9} mol^3 dm^{-9}.

Monk (14) measured solubilities of barium iodate in various aqueous-organic solvent mixtures, and obtained the activity solubility product at 298.2K. The value reported by Monk is 1.552×10^{-9} mol^3 dm^{-9}.

Bousquet, Mathurin and Vermande (16) studied the solubility products of barium iodate monohydrate over the temperature range 273 to 303K, and those of the anhydrous salt over the range from 313 to 359K. The solubility products reported in (16) except the value at 298.2K, are designated as tentative values. The value at 298.2K is 1.60×10^{-9} mol dm^{-9}, and the value agrees with that of other investigators.

Jones, Madigan and Wilson (18) measured the solubility of barium iodate in water at 298K, and again the evaluator assumes that the monohydrate was used. The solubility obtained was transformed to activity solubility product with values of the mean ionic activity coefficients calculated from Davies' modification (24) of the Guntelberg equation. The value obtained is 1.586×10^{-9} mol^3 dm^{-9}.

<u>Solubility product at 298.2K. The arithmetic mean of four results based on mol dm^{-3} concentration units</u> (10,14,16,18) is 1.567×10^{-9} mol^3 dm^{-9}, and the standard deviation, σ, is 0.026×10^{-9} mol^3 dm^{-9}. The mean is designated as a recommended value.

The recommended and tentative values given in Table 7 were fitted to the following equations:

$$\ln K^{\circ}_{s0}(1) = -32.66882 - 20.21218/(T/100\text{K}) + 17.57018 \ln (T/100\text{K}) \quad : \quad \sigma = 0.47 \times 10^{-10}$$

$$\ln K^{\circ}_{s0}(2) = -78.27746 + 59.02588/(T/100\text{K}) + 35.23994 \ln (T/100\text{K}) \quad : \quad \sigma = 0.22 \times 10^{-8}$$

where $K^{\circ}_{s0}(1)$ and $K^{\circ}_{s0}(2)$ are the thermodynamic solubility products for the monohydrate and the anhydrous salt, respectively.

COMPONENTS:	EVALUATOR:
(1) Barium iodate; $Ba(IO_3)_2$; [10567-69-8] (2) Water; H_2O; [7732-18-5]	Hiroshi Miyamoto Department of Chemistry Niigata University Niigata, Japan March, 1982

CRITICAL EVALUATION:

Table 7 Recommended and tentative values for the solubility product of barium iodate in aqueous solution

T/K	$10^9 K^\circ_{s0}$(exptl)/mol^3 dm^{-9}	$10^9 K^\circ_{s0}$(calcd)/mol^3 dm^{-9}
	Solid phase: $Ba(IO_3)_2 \cdot H_2O$	
273.2	0.1826	0.1853
281.2	0.3936	0.3800
291.2	0.884	0.898
298.2	1.567[a]	1.606
303.2	2.455	2.405
	Solid phase: $Ba(IO_3)_2$	
313.2	4.57	4.59
323.2	7.77	7.76
333.2	13.3	13.1
343.2	22.5	22.2
352.2	32.9	35.6
356.2	46.5	43.9

a: recommended value

REFERENCES:

1. Trautz, M.; Anschütz, A. *Z. Phys. Chem.* 1906, *56*, 236.
2. Hill, A. E.; Zink, W. A. H. *J. Am. Chem. Soc.* 1909, *31*, 43.
3. Harkins, W. D.; Winninghoff, W. J. *J. Am. Chem. Soc.* 1911, *33*, 1827.
4. Polessitskij, A. *C. R. (Dokl.) Acad. Sci. URSS* 1935, *4*, 193.
5. Macdougall, G.; Davies, C. W. *J. Chem. Soc.* 1935, 1416.
6. Naidich, S.; Ricci, J. E. *J. Am. Chem. Soc.* 1939, *61*, 3268.
7. Davis, T. W.; Ricci, J. E.; Sauter, C. G. *J. Am. Chem. Soc.* 1939, *61*, 3274.
8. Keefer, R. M.; Reiber, H. G.; Bisson, C. S. *J. Am. Chem. Soc.* 1940, *62*, 2951.
9. Pedersen, K. J. *K. Dan. Vidensk. Selsk. Nat.-Fis. Medd.* 1941, *18*, 1.
10. Derr, P. F.; Vosburgh, W. C. *J. Am. Chem. Soc.* 1943, *65*, 2408.
11. Davies, C. W.; Wyatt, P. A. H. *Trans. Faraday Soc.* 1949, *45*, 770.
12. Ricci, J. E. *J. Am. Chem. Soc.* 1951, *73*, 1375.
13. Monk, C. B. *Trans. Faraday Soc.* 1951, *47*, 1233.
14. Monk, C. B. *J. Chem. Soc.* 1951, 2723.
15. Ricci, J. E.; Freedman, A. J. *J. Am. Chem. Soc.* 1952, *74*, 1769.
16. Bousquet, J.; Mathurin, D.; Vermande, P. *Bull. Soc. Chim. Fr.* 1969, 1111.

COMPONENTS:

(1) Barium iodate; $Ba(IO_3)_2$; [10567-69-8]

(2) Water; H_2O; [7732-18-5]

EVALUATOR:

Hiroshi Miyamoto
Department of Chemistry
Niigata University
Niigata, Japan

March, 1982

CRITICAL EVALUATION:

17. Miyamoto, H. *Nippon Kagaku Kaishi* 1972, 659.

18. Jones, A. L.; Madigan, G. A.; Wilson, I. R. *J. Cryst. Growth* 1973, *20*, 99.

19. Miyamoto, H.; Suzuki, K.; Yanai, K. *Nippon Kagaku Kaishi* 1978, 1150.

20. Milad, N. E.; Morsi, S. E.; Soliman, S. T.; Seleem, L. M. N. *Egypt. J. Chem.* 1973, *16*, 395.

21. Fedorov, V. A.; Robov, A. M.; Shmyd'ko, I. I.; Vorontsova, N. A.; Mironov, V. E. *Zh. Neorg. Khim.* 1974, *19*, 1746; *Russ. J. Inorg. Chem. (Engl. Transl.)* 1974, *19*, 950.

22. Fuoss, R. M.; Kraus, C. A. *J. Am. Chem. Soc.* 1933, *55*, 476.

23. Born, M. *Z. Phys.* 1920, *1*, 45.

24. Davies, C. W. *Ion Association*. Butterworths. London. 1962.

COMPONENTS:

(1) Barium iodate; $Ba(IO_3)_2$; [10567-69-8]

(2) Water; H_2O; [7732-18-5]

ORIGINAL MEASUREMENTS:

Trautz, M.; Anschütz, A.

Z. Physik. Chem. 1906, *56*, 236-42.

VARIABLES:

T/K = 273.10 - 372.4

PREPARED BY:

Hiroshi Miyamoto

EXPERIMENTAL VALUES:

t/°C	Barium Iodate	
	mass %	$10^4 c_1$/mol dm^{-3} (compiler)
-0.046 ± 0.002 (eutectic point)	0.008	1.64
+10	0.014	2.87
20	0.022	4.52
25	0.028	5.75
30	0.031	6.37
40	0.041	8.42
50	0.056	11.5
60	0.074	15.2
70	0.093	19.1
80	0.115	23.63
90	0.141	28.99
99.2/735mm (= ca 100/760mm)	0.197	40.52

AUXILIARY INFORMATION

METHOD/APPARATUS/PROCEDURE:

$Ba(IO_3)_2$ crystals and water were shaken in a thermostat at 10-90°C for 14 hours.
Equilibrium at 100°C was established in a vapor of boiling water for 6-7 hours, (the temperature was checked against the boiling point of pure water). Aliquots of saturated solutions were removed by means of a pipette fitted with cotton wool.
The solution was placed in a stoppered tube and the sample was weighed. $Ba(IO_3)_2$ was determined gravimetrically by evaporation of the solvent. After the solution saturated with the barium iodate was frozen at near 0°C, the melted part of the solution was analyzed for the iodate content, and the melting point of the frozen part was measured by using a Beckmann thermometer.

SOURCE AND PURITY OF MATERIALS:

Barium iodate was recrystallized from water.
Other information was not given.

ESTIMATED ERROR:

Soly: the deviations from the mean were about ± 5 %
Temp: ± 0.04°C (authors)

REFERENCES:

COMPONENTS:

(1) Barium iodate; $Ba(IO_3)_2$; [10567-69-8]

(2) Water; H_2O; [7732-18-5]

ORIGINAL MEASUREMENTS:

Polessitskij, A.

C. R. Dokl. Acad. Sci. USSR 1935, *4*, 193-6.

VARIABLES:

T/K = 273 to 373

PREPARED BY:

Hiroshi Miyamoto

EXPERIMENTAL VALUES:

t/°C	Barium Iodate	
	s_1/mg dm^{-3}	10^3c_1/mol dm^{-3}
0	182	0.374
15	305	0.626
25	406.3	0.833
40	580.0	1.19
50	727	1.49
60	873	1.79
70	1070	2.196
78	1300	2.669
100	1840	3.777

AUXILIARY INFORMATION

METHOD/APPARATUS/PROCEDURE:

$Ba(IO_3)_2$ crystals were stirred with water in a thermostat for 24 hours. After settling the solutions for one hour, samples were withdrawn with a pipet with cotton-wool.
The solubilities at high temperatures were determined in a special apparatus which facilitates to keep the saturated solutions at a constant temperature by the vapor of a boiling liquid (H_2O-100°C. C_2H_5OH-78°C). Three analytical methods were used: (1) Evaporation of 200 cm^3 of the solution and drying at 90°C, (2) Determination of Ba as $BaSO_4$ from 200 cm^3 of the solution, (3) Iodometric titration of IO_3^-.

SOURCE AND PURITY OF MATERIALS:

$Ba(IO_3)_2$ was prepared by adding recrystallized $BaCl_2 \cdot 2H_2O$ to an equivalent amount of KIO_3. The precipitate was filtered off and washed with hot water, or washed by decantation with a large amount of cold water, under strong stirring.

ESTIMATED ERROR:

Soly: three analytical methods gave the same results within ± 3 %.
Temp: not given.

REFERENCES:

COMPONENTS:

(1) Barium iodate; $Ba(IO_3)_2$; [10567-69-8]

(2) Water; H_2O; [7732-18-5]

ORIGINAL MEASUREMENTS:

Bousquet, J.; Mathurin, D.; Vermande, P.

Bull. Soc. Chim. Fr. 1969, 1111-5.

VARIABLES:

T/K = 273 to 359

PREPARED BY:

Hiroshi Miyamoto

EXPERIMENTAL VALUES:

t/°C	Barium Iodate monohydrate $10^{10} K^\circ_{s0}$/mol^{3}dm^{-9}	Barium Iodate anhydrate $10^9 K^\circ_{s0}$/mol^{3}dm^{-9}
0	1.828	
8	3.936	
17	8.84	
25	16.0	
30	24.55	
40		4.57
50		7.77
60		13.3
70		22.5
79		32.9
86		46.5

The solubility product, K°_{s0}, of $Ba(IO_3)_2 \cdot xH_2O$ was given in the following:

$$K^\circ_{s0} = (c_{Ba^{2+}} \times c^2_{IO_3^-})(y_{Ba^{2+}} \times y^2_{IO_3^-}) = 4S^3 y^3_\pm \qquad (1)$$

where S represents solubility of iodate, $y_\pm$ is an activity coefficient, and is given by modified Debye-Hückel equation

$$-\log y_\pm = Z_+ Z_- A \sqrt{I} - BI \qquad (2)$$

From (1) and (2)

$$Y = -BI + 1/3 \log K^\circ_{s0} \qquad (3)$$

where $Y = 1/3 \log (4S^3) - Z_+Z_-A\sqrt{I}$, and A = 0.5115 at 25°C.
Solubility product (K°_{s0}) and unknown constant (B) are evaluated from the intercept and the slope of Y vs I plots. The solubilities of $Ba(IO_3)_2$ in aqueous NaCl solutions were determined in order to obtain Y vs I plots, but the data of solubilities were not given in the paper.

METHOD/APPARATUS/PROCEDURE:

Aqueous NaCl solutions and the specified hydrated crystals were placed into glass-stoppered Erlenmeyer flasks. The flasks were stirred in a thermostat for 1-15 hours. The iodate content was determined iodometrically.

SOURCE AND PURITY OF MATERIALS:

BDH labeled $Ba(IO_3)_2 \cdot H_2O$ was used. The anhydrate was prepared from the monohydrate by dehydration at 200°C.

ESTIMATED ERROR:

Soly: nothing specified
Temp: ± 0.05°C (authors)

REFERENCES:

COMPONENTS:

(1) Barium iodate; $Ba(IO_3)_2$; [10567-69-8]

(2) Water; H_2O; [7732-18-5]

ORIGINAL MEASUREMENTS:

Jones, A. L.; Madigan, G. A.; Wilson, I. R.

J. Cryst. Growth 1973, *20*, 99-102.

VARIABLES:

Four crystal types
T/K = 275.2, 283.2 and 298.2

PREPARED BY:

Hiroshi Miyamoto

EXPERIMENTAL VALUES:

t/°C	Crystal type[a]	Measurement method[b]		Barium Iodate $10^4 c_1$/mol dm^{-3}	Barium Iodate $10^{10} K^{\circ}_{s0}$/mol^3 dm^{-9} [c]
2.0	P	I		3.99	
	P	II		4.02	
			(Av)	4.00	2.04
10.0	P	I		5.38	
	P	II		5.38	
			(Av)	5.38	4.79
25.0	A	I		8.22	
	B	I		8.23	
	C	I		8.18	
	C	II		8.18	
	P	I		8.18	
	P	II		8.19	
			(Av)	8.20	15.86

[a] The preparations of the crystal types A, B and C are given in (A), (B) and (C) in "Source and purity of materials." The primary precipitate (5-50 μm size) in preparation of barium iodate is named "P".

[b] Two series of the conductivity measurements were carried out in this study. The details of the method are described in "Method: I and II."

[c] The mean solubilities were transformed to solubility products with values of the mean ionic activity coefficient calculated from Davies modification of the Guntelberg equation (1).

AUXILIARY INFORMATION

METHOD/APPARATUS/PROCEDURE:

A conductivity method was used. Conductances were measured at 1592 Hz with a Wayne-Kerr conductance bridge accurate to 0.1% full scale. Two series of measurements were made. In those of series I, the conductivity was monitored until a small, constant, rate of increase was found, similar to that expected for glass dissolution from separate experiments. Extrapolation of this constant rate to zero time gave the value taken as the solubility. In series II, the conductivity was followed similarly until within ca. 0.25% of the expected equilibrium value. The temperature was then raised by ca. 2 K until the conductivity rose by ca. 8%, due to the increases in molar conductivity and in solubility. The temperature was then restored to its initial value,

continued . . .

SOURCE AND PURITY OF MATERIALS:

Finely divided barium iodate (5-50μm) was prepared by pouring barium chloride solution (200 cm^3; 0.05 mol dm^{-3}) and potassium iodate solution (200 cm^3; 0.1 mol dm^{-3}) simultaneously into 600 cm^3 distilled water, with continuous stirring. The precipitate was washed many times by decantation, using conductivity water. Large crystals were prepared by the following three methods:

(A) The precipitate was dissolved in conductivity water at 80-85°C to produce an approximately saturated solution. After filtration, the solution was allowed to cool slowly. The crystals formed appeared to be cubes truncated corners, of linear dimension 50-150 μm.

(B) The method was similar to A but used nitric acid (1 mol dm^{-3}) as solvent. The product contained

continued . . .

COMPONENTS:

(1) Barium iodate; $Ba(IO_3)_2$; [10567-69-8]

(2) Water; H_2O; [7732-18-5]

ORIGINAL MEASUREMENTS:

Jones, A. L.; Madigan, G. A.; Wilson, I. R.

J. Cryst. Growth 1973, *20*, 99-102.

AUXILIARY INFORMATION

METHOD/APPARATUS/PROCEDURE: continued . . and the conductivity began to fall. Extrapolation of measurements after this, to zero rate, again gave an estimate of solubility. In measurements of series II, it was difficult to make an accurate estimate of conductivity changes due to glass dissolution. Blank experiments and comparison of series I and II agree in suggesting that they are almost negligible.
The concentration of barium iodate was calculated from conductivity measurements using the method of Righellato and Davies (2).

SOURCE AND PURITY OF MATERIALS: continued crystals up to 0.5 mm in size.
(C) $Ba(IO_3)_2 \cdot H_2O$ in granular form is obtained by the reaction in solution of barium ion with iodate ion formed by the slow reduction of periodic acid by lactic acid at room temperature. In separate preparations this gave truncated cubes of side 0.3 to 0.8 mm, and flattened cubes, 0.8 to 3 mm long and 0.4 to 0.8 mm thick. Before use all crystals were aged for at least 2 weeks under conductivity water, with several changes of water.

ESTIMATED ERROR: Soly: Standard deviation 0.02 at 25 ^{0}C.
Temp.: ± 0.03 K (authors)

REFERENCES:

1. Robinson, R. W.; Stokes, R. H. *Electrolyte Solutions* Butterworths. London. 1959, 231.
2. Righellato, E. C.; Davies, C. W. *Trans. Faraday Soc.* 1930, *26*, 592.

COMPONENTS:

(1) Barium iodate; $Ba(IO_3)_2$; [10567-69-8]
(2) Lithium nitrate; $LiNO_3$; [7790-69-4]
(3) Lithium perchlorate; $LiClO_4$; [7791-03-9]
(4) Water; H_2O; [7732-18-5]

ORIGINAL MEASUREMENTS:

Federov, V. A.; Robov, A. M.; Shmyd'ko, I. I.; Vorontsova, N. A.; Mironov, V. E.

Zh. Neorg. Khim. 1974, *19*, 1746-50; *Russ. J. Inorg. Chem.(Engl. Transl.)* 1974, *19*, 950-3.

VARIABLES:

T/K = 298
Ionic Strength

PREPARED BY:

Hiroshi Miyamoto

EXPERIMENTAL VALUES:

t/°C	Nitrate Ion	Barium Iodate				
	c_2/g-ion dm^{-3}	10^3c_1/mol dm^{-3}				
		I = 0.5	1.0	2.0	3.0	4.0
25	0	1.63	1.84	2.09	2.28	2.16
	0.1	1.68	--	--	--	--
	0.2	1.76	2.04	2.30	--	--
	0.3	1.88	--	--	--	--
	0.4	2.04	2.22	--	2.68	2.68
	0.5	2.22	--	2.62	--	--
	0.6		2.37	--	--	--
	0.8		2.59	2.94	3.10	3.20
	1.0		2.77	3.16	--	--
	1.2			--	3.56	3.74
	1.3			3.48	--	--
	1.5			3.70	--	--
	1.6			--	4.06	4.30
	1.8			4.00	--	--
	2.0			4.20	4.47	4.90
	2.4				5.17	5.50
	2.8				5.76	6.15
	3.0				6.07	--
	3.2					6.78
	3.6					7.47
	4.0					8.14

AUXILIARY INFORMATION

METHOD/APPARATUS/PROCEDURE:

Equilibrium between the solid phase and the solution was reached by vigorous agitation with a magnetic stirrer in stoppered vessels in a thermostat. Equilibrium was established after stirring for 4-6 hours and was checked by removing specimens after equal intervals of time. The concentrations for $Ba(IO_3)_2$ in the saturated solutions were determined iodometrically.

SOURCE AND PURITY OF MATERIALS:

The author stated that $Ba(IO_3)_2$ was made by well-known method, but the details of the method were not given.
Chemically pure grade $LiClO_4$ and $LiNO_3$ used were recrystallized from twice-distilled water. Before recrystallization, the solutions were boiled with active carbon.

ESTIMATED ERROR:

Soly: the reproducibility of the results averages ± 1.5 - 2%
Temp: not given

REFERENCES:

COMPONENTS:

(1) Barium iodate; $Ba(IO_3)_2$; [10567-69-8]

(2) Sodium thiosulfate; $Na_2S_2O_3$; [7772-98-7]

(3) Water; H_2O; [7732-18-5]

ORIGINAL MEASUREMENTS:

Davies, C. W.; Wyatt, P. A. H.

Trans. Faraday Soc. 1949, *45*, 770-3.

VARIABLES:

T/K = 298 and 308

$10^3 c_2$/mol dm^{-3} = 0 to 20.090

PREPARED BY:

Hiroshi Miyamoto

EXPERIMENTAL VALUES:

t/°C	Sodium Thiosulfate $10^3 c_2$/mol dm^{-3}	Barium Iodate $10^3 c_1$/mol dm^{-3}
25	0	0.812
	5.025	1.022
	8.030	1.097
	10.050	1.137
	15.080	1.223
	20.090	1.295
35	0	1.049
	5.010	1.322
	8.016	1.415
	10.020	1.467
	15.03	1.579
	20.06	1.680

COMMENTS AND/OR ADDITIONAL DATA:

The concentrations of the individual ionic species and ion pairs were calculated by successive approximation from the relation

$$\log [M^{2+}][X^-]/[MX^+] = \log K_D^\circ + 2I^{1/2}/(1 + I^{1/2}) - 0.40I$$

where I = ionic strength and K_D° is the dissociation constant of the ion pairs. For the ion pairs $BaIO_3^+$ and BaS_2O_3, the values taken for K_D° were 0.08 and 0.0061, respectively.

AUXILIARY INFORMATION

METHOD/APPARATUS/PROCEDURE:

The saturating column method was used as described by Money and Davies (1). The saturator was immersed in a thermostat regulated 25 ± 0.01 and 35 ± 0.02°C. Samples of saturated solution were withdrawn in warmed pipets and analyzed by iodometric titration for the iodate.

SOURCE AND PURITY OF MATERIALS:

$Ba(IO_3)_2 \cdot H_2O$ crystals were prepared by slow dropwise addition of solutions of A.R. grade $BaCl_2$ and KIO_3 to a large volume of water. $Na_2S_2O_3$ was an A.R. grade sample that had been recrystallized and dried over a saturated $CaCl_2$ solution.

ESTIMATED ERROR:

Soly: duplicate determinations of the solubilities agreed to almost within 0.1%, but occasionally slightly larger.

Temp: 25°C: ± 0.01°C,
35°C: ± 0.02°C (authors)

REFERENCES:

1. Money, R. W.; Davies, C. W. *J. Chem. Soc.* 1934, 400.

COMPONENTS:

(1) Barium iodate; $Ba(IO_3)_2$; [10567-69-8]

(2) Sodium bromoacetate; $C_2H_2O_2BrNa$; [1068-52-6]

(3) Water; H_2O; [7732-18-5]

ORIGINAL MEASUREMENTS:

Davies, C. W.; Wyatt, P. A. H.

Trans. Faraday Soc. 1949, *45*, 770-3.

VARIABLES:

$T/K = 298$

$10^3 c_2/\text{mol dm}^{-3} = 0$ to 25.00

PREPARED BY:

Hiroshi Miyamoto

EXPERIMENTAL VALUES:

t/°C	Sodium Bromoacetate $10^3 c_2/\text{mol dm}^{-3}$	Barium Iodate $10^3 c_1/\text{mol dm}^{-3}$
25	0	0.812
	10.00	0.920
	15.00	0.954
	20.00	0.986
	25.00	1.012

COMMENTS AND/OR ADDITIONAL DATA:

The concentrations of the individual ionic species and ion pairs were calculated by successive approximations from the relation

$$\log [M^{2+}][X^-]/[MX^+] = \log K_D^\circ + 2I^{1/2}/(1 + I^{1/2}) - 0.40I$$

where I = ionic strength and K_D° is the dissociation constant of the ion pair. For the ion pairs $BaIO_3^+$ and $BaBrAc^+$, the values taken for K_D° were 0.08 and 0.6, respectively.

AUXILIARY INFORMATION

METHOD/APPARATUS/PROCEDURE:

The saturating column method was used as described by Money and Davies (1). The saturator was immersed in a thermostat regulated 25 ± 0.01°C. Samples of saturated solution were withdrawn in warmed pipets and analyzed by iodometric titration for the iodate.

SOURCE AND PURITY OF MATERIALS:

$Ba(IO_3)_2 \cdot H_2O$ crystals were prepared by slow dropwise addition of solution of A.R. grade $BaCl_2$ and KIO_3 to a large volume of water.
Sodium bromoacetate solutions were made up by neutralizing with standard NaOH solution a sample of acid which had been redistilled at 15mm pressure and stored in the dark over conc H_2SO_4.

ESTIMATED ERROR:

Soly: duplicate determinations of the solubilities agreed to almost within 0.1%, but occasionally slightly larger.
Temp: ± 0.01°C (authors)

REFERENCES:

1. Money, R. W.; Davies, C. W. *J. Chem. Soc.* 1934, 400.

COMPONENTS:

(1) Barium iodate; $Ba(IO_3)_2$; [10567-69-8]

(2) Potassium chloride; KCl; [7447-40-7]

(3) Water; H_2O; [7732-18-5]

ORIGINAL MEASUREMENTS:

Macdougall, G.; Davies, C. W.

J. Chem. Soc. 1935, 1416-9.

VARIABLES:

$T/K = 298.08$

$c_2/\text{mol dm}^{-3} = 0$ to 0.1

PREPARED BY:

Hiroshi Miyamoto
Mark Salomon

EXPERIMENTAL VALUES:

t/°C	Potassium Chloride $c_2/\text{mol dm}^{-3}$	Barium Iodate $10^3c_1/\text{mol dm}^{-3}$	Ionic Strength $I^{1/2}/(\text{mol dm}^{-3})^{1/2}$	$(1/3)\log[Ba^{2+}][IO_3^-]^2$
24.93	0	0.810	0.0491	$\bar{3}.1048$
	0.001	0.827	0.0588	$\bar{3}.1132$
	0.002	0.840	0.0670	$\bar{3}.1202$
	0.0035	0.859	0.0777	$\bar{3}.1297$
	0.005	0.874	0.0871	$\bar{3}.1372$
	0.0075	0.899	0.1008	$\bar{3}.1489$
	0.01	0.918	0.1128	$\bar{3}.1579$
	0.02	0.985	0.1513	$\bar{3}.1868$
	0.05	1.117	0.2308	$\bar{3}.2383$
	0.1	1.269	0.3220	$\bar{3}.2887$

COMMENTS AND/OR ADDITIONAL DATA:

The concentrations of the individual ionic species and ion pairs were calculated by successive approximations from the relation

$$\log[M^{2+}][M^-]/[MX^+] = \log K_D^\circ + 2I^{1/2} - 2I$$

where I = ionic strength and K_D° is the dissociation constant of the ion pair. For the ion pairs $BaIO_3^+$ and $BaCl^+$, the values taken for K_D° were 0.08 and 1.35, respectively.

AUXILIARY INFORMATION

METHOD/APPARATUS/PROCEDURE:

Saturating column method as in (1) and modified as in (2). A bulb containing the solvent soln is attached to a column containing the slightly soluble salt, and the solvent is allowed to flow through the column at a rate sufficient to insure satn (1). The modification (2) consisted of connecting the column by capillary tubing to a second parallel arm in which the satd soln collected. The entire apparatus was placed in a thermostat. A portion of the satd soln was run through the satg column a second time. 100 cm^3 samples were taken for analysis using a calibrated pipet and running it into acid KI soln. The liberated I_2 was titrd by weight against approx 0.15N thiosulfate soln, 0.01N I_2 soln being used for the back titrn. Each data point is the mean of two detn. Conductivity measurements on binary $Ba(IO_3)_2$-H_2O solns from very dilute to satn were made at 24.93°C. Only the value for the satd soln was reported which was used to calc the ion pair disscn const at infinite diln.

SOURCE AND PURITY OF MATERIALS:

$Ba(IO_3)_2 \cdot H_2O$ was prepared by dripwise addition of solns of A.R. grade $Ba(OH)_2$ and HIO_3 (in slight excess) into conductivity water in an apparatus "protected from the atmosphere." The crystalline precipitate was washed until its solubility was constant.

ESTIMATED ERROR:

Soly: the mean of each point agreed to within ± 0.3%.

Temp: ± 0.01°K

REFERENCES:

1. Brönsted, J. N.; La Mer, V. K. *J. Am. Chem. Soc.* 1924, *46*, 555.
2. Money, R. W.; Davies, C. W. *J. Chem. Soc.* 1934, 400.

COMPONENTS:

(1) Barium iodate; $Ba(IO_3)_2$; [10567-69-8]

(2) Potassium chloride; KCl; [7447-40-7]

(3) Water; H_2O; [7732-18-5]

ORIGINAL MEASUREMENTS:

Naidich, S.; Ricci, J. E.

J. Am. Chem. Soc. 1939, *61*, 3268-73.

VARIABLES:

$T/K = 298$

$c_2/\text{mol dm}^{-3} = 0$ to 0.9817

PREPARED BY:

Hiroshi Miyamoto

EXPERIMENTAL VALUES:

t/°C	Potassium Chloride $c_2/\text{mol dm}^{-3}$	Barium Iodate $10^4 c_1/\text{mol dm}^{-3}$
25	0	8.177
	0.001	8.357
	0.005	8.977
	0.01	9.297
	0.05012	11.28
	0.1002	12.75
	0.2454	15.66
	0.4908	18.95
	0.9817	23.78

AUXILIARY INFORMATION

METHOD/APPARATUS/PROCEDURE:

The KCl solution used as the solvent was prepared by weight directly from recrystallized KCl, and followed by dilution.
$Ba(IO_3)_2$ crystals and the KCl solutions were immersed in Pyrex glass-stoppered bottles. The bottles were rotated in a waterbath for several days. The determinations were made both from under- and super-sautration, the analyses being repeated after a day of two in almost every case.
The saturated solutions were withdrawn by means of suction a 50 or 100 ml sample into a calibrated pipet, fitted with quantitative filter paper at the tip, and was delivered into an Erlenmeyer flask. The iodate content in the saturated solutions was determined iodometrically.

SOURCE AND PURITY OF MATERIALS:

C.p. grade $Ba(IO_3)_2 \cdot H_2O$ was washed 8-10 times with distilled water, and dried at about 100°C. Analysis of the product gave 96.5% $Ba(IO_3)_2$ as compared with the theoretical figure of 96.43% for the monohydrate.

ESTIMATED ERROR:

Soly: nothing specified
Temp: ± 0.01°C (authors)

REFERENCES:

COMPONENTS:

(1) Barium iodate; $Ba(IO_3)_2$; [10567-69-8]

(2) Potassium chloride; KCl; [7447-40-7]

(3) Water; H_2O; [7732-18-5]

ORIGINAL MEASUREMENTS:

Keefer, R. M.; Reiber, H. G.; Bisson, C. S.

J. Am. Chem. Soc. 1940, *62*, 2951-5.

VARIABLES:

T/K = 298

$10^2 m_2$/mol kg^{-1} = 0 to 10.07

PREPARED BY:

Hiroshi Miyamoto

EXPERIMENTAL VALUES:

t/°C	Potassium Chloride $10^2 m_2$/mol kg^{-1}	Barium Iodate $10^4 m_1$/mol kg^{-1}
25	0	8.11
	0.1267	8.31
	0.2533	8.47
	0.5064	8.77
	1.003	9.22
	1.254	9.40
	2.514	10.16
	5.025	11.21
	7.543	12.06
	10.07	12.72

AUXILIARY INFORMATION

METHOD/APPARATUS/PROCEDURE:

Excess of $Ba(IO_3)_2$ and aqueous KCl solution were placed in a glass-stoppered Pyrex flask. The flasks were rotated in a thermostat at 25°C for at least 12 hours. Equilibrium was obtained in 4-5 hours. The saturated solutions were analyzed iodometrically. Analyses and solubility measurements were performed in duplicate. Densities of all solutions were determined, but the data were not given in the original paper.

SOURCE AND PURITY OF MATERIALS:

$Ba(IO_3)_2$ was prepared from 0.2 mol kg^{-1} $BaCl_2$ solution and 0.2 mol kg^{-1} KIO_3 solution. The precipitate was filtered, washed and dried at room temperature. The number of hydrated water was not given.
C.p. grade KCl was recrystallized from water.

ESTIMATED ERROR:

Soly: nothing specified
Temp: ± 0.02°C (authors)

REFERENCES:

COMPONENTS:

(1) Barium iodate; $Ba(IO_3)_2$; [10567-69-8]

(2) Potassium chlorate; $KClO_3$; [3811-04-9]

(3) Water; H_2O; [7732-18-5]

ORIGINAL MEASUREMENTS:

Macdougall, G.; Davies, C. W.

J. Chem. Soc. 1935, 1416-9.

VARIABLES:

T/K = 298.08

c_2/mol dm^{-3} = 0 to 0.075

PREPARED BY:

Hiroshi Miyamoto
Mark Salomon

EXPERIMENTAL VALUES:

t/°C	Potassium Chlorate c_2/mol dm^{-3}	Barium Iodate $10^3 c_1$/mol dm^{-3}
24.93	0	0.810
	0.005	0.880
	0.01	0.924
	0.02854	1.031
	0.075	1.184

COMMENTS AND/OR ADDITIONAL DATA:

The concentrations of the individual ionic species and ion pairs were calculated by successive approximations from the relation

$$\log[M^{2+}][X^-]/[MX^+] = \log K_D^\circ + 2I^{1/2} - 2I$$

where I = ionic strength and K_D° is the dissociation constant of the ion pair. For the ion pairs, $BaIO_3^+$ and $BaClO_3^+$, the values taken for K_D° were 0.08 and 0.2, respectively.

AUXILIARY INFORMATION

METHOD/APPARATUS/PROCEDURE:

Saturating column method as in (1) and modified as in (2). A bulb containing the solvent soln is attached to a column containing the slightly soluble salt, and the solvent is allowed to flow through the column at a rate sufficient to insure satn (1). The modification (2) consisted of connecting the column by capillary tubing to a second parallel arm in which the satd soln collected. The entire apparatus was placed in a thermostat. A portion of the satd soln was run through the satg column a second time. 100 cm³ samples were taken for analysis using a calibrated pipet and running it into acid KI soln. The liberated I_2 was titrd by weight against approx 0.15 N thiosulfate soln, 0.01N I_2 soln being used for the back titrn. Each data point is the mean of two detn. Conductivity measurements on binary $Ba(IO_3)_2$-H_2O soln from very dilute to satn were made at 24.93°C. Only the value for the satd soln was reported which was used to calc the ion pair dissocn const at infinite dilution.

SOURCE AND PURITY OF MATERIALS:

$Ba(IO_3)_2 \cdot H_2O$ was prepared by dripwise addn of solns of A.R. grade $Ba(OH)_2$ and HIO_3 (in slight excess) into conductivity water in an apparatus "protected from the atmosphere." The crystalline precipitate was washed until its solubility was constant.

ESTIMATED ERROR:

Soly: the mean of each point agreed to within ± 0.3%

Temp: ± 0.01°K

REFERENCES:

1. Brönsted, J. N.; La Mer, V. K. *J. Am. Chem. Soc.* 1924, *46*, 555.
2. Money, R. W.; Davies, C. W. *J. Chem. Soc.* 1934, 400.

COMPONENTS:

(1) Barium iodate; $Ba(IO_3)_2$; [10567-69-8]

(2) Potassium perchlorate; $KClO_4$; [7778-74-7]

(3) Water; H_2O; [7732-18-5]

ORIGINAL MEASUREMENTS:

Macdougall, G.; Davies, C. W.

J. Chem. Soc. 1935, 1416-9.

VARIABLES:

T/K = 298.08

c_2/mol dm^{-3} = 0 to 0.075

PREPARED BY:

Hiroshi Miyamoto
Mark Salomon

EXPERIMENTAL VALUES:

t/°C	Potassium Perchlorate c_2/mol dm^{-3}	Barium Iodate 10^3c_1/mol dm^{-3}	Ionic Strength $I^{1/2}$/(mol dm^{-3})$^{1/2}$	(1/3) log$[Ba^{2+}][IO_3^-]^2$
24.93	0	0.810	0.0491	$\bar{3}$.1048
	0.004	0.866	0.0811	$\bar{3}$.1333
	0.008845	0.905	0.1074	$\bar{3}$.1523
	0.035	1.045	0.1951	$\bar{3}$.2128
	0.075	1.164	0.2800	$\bar{3}$.2568

AUXILIARY INFORMATION

METHOD/APPARATUS/PROCEDURE:

Saturating column method as in (1) and modified as in (2). A bulb containing the solvent soln is attached to a column containing the slightly soluble salt, and the solvent is allowed to flow through the column at a rate sufficient to insure satn (1). The modification (2) consisted of connecting the column by capillary tubing to a second parallel arm in which the satd soln collected. The entire apparatus was placed in a thermostat. A portion of the satd soln was run through the satg column a second time. 100 cm^3 samples were taken for analysis using a calibrated pipet and running it into acid KI soln. The liberated I_2 was titrd by weight against approx 0.15N thiosulfate soln, 0.01N I_2 soln being used for the back titrn. Each data point is the mean of two detn.

SOURCE AND PURITY OF MATERIALS:

$Ba(IO_3)_2 \cdot H_2O$ was prepared by dripwise addn of solns of A.R. grade $Ba(OH)_2$ and HIO_3 (in slight excess) into conductivity water in an apparatus "protected from the atmosphere." The crystalline precipitate was washed until its solubility was constant.

ESTIMATED ERROR:

Soly: the mean of each point agreed to within ± 0.3%

Temp: ± 0.01°K

REFERENCES:

1. Brönsted, J. N.; La Mer, V. K. *J. Am. Chem. Soc.* 1924, *46*, 555.
2. Money, R. W.; Davies, C. W. *J. Chem. Soc.* 1934, 400.

COMPONENTS:

(1) Barium iodate; $Ba(IO_3)_2$; [10567-69-8]

(2) Potassium bromate; $KBrO_3$; [7758-01-2]

(3) Water; H_2O; [7732-18-5]

ORIGINAL MEASUREMENTS:

Davis, T. W.; Ricci, J. E.; Sauter, C. G.

J. Am. Chem. Soc. 1939, *61*, 3274-84.

VARIABLES:

$T/K = 298$

$10^3 c_2/mol\ dm^{-3} = 0$ to 5.045

PREPARED BY:

Hiroshi Miyamoto

EXPERIMENTAL VALUES:

t/°C	Potassium Bromate $10^3 c_2/mol\ dm^{-3}$	Barium Iodate $10^4 c_1/mol\ dm^{-3}$
25	0	8.145
	0.997	8.3
	2.996	8.5
	5.045	9.0

AUXILIARY INFORMATION

METHOD/APPARATUS/PROCEDURE:

The equilibrium procedure and the analytical method of barium iodate in the presence of varying concentrations of the electrolyte are not given. As the solubilities of barium iodate in dioxane-water mixtures are described in the paper, the compiler assumed that the method concerning to the iodate in the electrolyte solution was similar to those in dioxane-water systems. Barium iodate crystals and aqueous $KBrO_3$ solution were placed in glass-stoppered Pyrex bottles. The bottles were rotated at 25°C for one or more days. Samples for analysis were withdrawn by suction through quantitative filters into calibrated 100 ml pipets after allowing some time for the undissolved salt to settle. The iodate content was determined iodometrically.

SOURCE AND PURITY OF MATERIALS:

C.p. grade barium iodate monohydrate was washed 8 to 10 times with distilled water, sedimented in tall cylinders to remove smaller size particles, and dried at 100°C before use. Iodometric titration gave 96.5% $Ba(IO_3)_2$ compared with 96.43% theoretical for the monohydrate. C.p. grade $KBrO_3$ was used.

ESTIMATED ERROR:

Soly: nothing specified

Temp: ± 0.02°C (authors)

REFERENCES:

COMPONENTS:

(1) Barium iodate; $Ba(IO_3)_2$; [10567-69-8]

(2) Potassium iodate; KIO_3; [7758-05-6]

(3) Water; H_2O; [7732-18-5]

ORIGINAL MEASUREMENTS:

Harkins, W. D.; Winninghoff, W. J.

J. Am. Chem. Soc. 1911, *33*, 1827-36.

VARIABLES:

$T/K = 298$

$10^3 s_2/\text{eq dm}^{-3} = 0$ to 1.0608

PREPARED BY:

Hiroshi Miyamoto

EXPERIMENTAL VALUES:

t/°C	Potassium Iodate		Barium Iodate	
	$10^3 s_2/\text{eq dm}^{-3}$	$10^3 c_2/\text{mol dm}^{-3}$	$10^3 s_1/\text{eq dm}^{-3}$	$10^4 c_1/\text{mol dm}^{-3}$
25	0	0	1.579[a]	7.89
	0.10608	0.10608	1.510[b]	7.55
	0.5304	0.5304	1.242[b]	6.21
	1.0608	1.0608	0.9418[b]	4.709

[a] mean of 7 detns with a standard deviation $\sigma = 0.0021$

[b] mean of 2 detns

AUXILIARY INFORMATION

METHOD/APPARATUS/PROCEDURE:

Though the details of equilibration procedure were not given, some of the data were obtained by approaching equilibrium from the side of supersaturation.
The concentration of the saturating salt was determined by adding KI to the solution of the iodate, and liberating iodine by adding HCl. The iodine was determined by titration with $Na_2S_2O_3$ solution.

SOURCE AND PURITY OF MATERIALS:

$Ba(IO_3)_2$ was made by precipitating $Ba(NO_3)_2$ with KIO_3 in a very dilute solution in which the nitrate was kept constantly in excess. The number of hydrated water is not given. KIO_3 was purified by recrystallization.

ESTIMATED ERROR:

Soly: above described
Temp: not given

REFERENCES:

COMPONENTS:

(1) Barium iodate; $Ba(IO_3)_2$; [10567-69-8]

(2) Potassium nitrate; KNO_3; [7757-79-1]

(3) Water; H_2O; [7732-18-5]

ORIGINAL MEASUREMENTS:

Harkins, W. D.; Winninghoff, W. J.

J. Am. Chem. Soc. 1911, *33*, 1827-36.

VARIABLES:

T/K = 298

KNO_3/eq dm^{-3} = 0 to 0.200

PREPARED BY:

Hiroshi Miyamoto

EXPERIMENTAL VALUES:

t/°C	Potassium Nitrate		Barium Iodate	
	s_2/eq dm^{-3}	c_2/mol dm^{-3}	$10^3 s_1$/eq dm^{-3}	$10^4 c_1$/mol dm^{-3}
25	0	0	1.579[a]	7.89
	0.002	0.002	1.624[b]	8.12
	0.010	0.010	1.826[b]	9.13
	0.050	0.050	2.640[b]	13.2
	0.200	0.200	3.190[b]	16.0

[a] mean of 7 detns with a standard deviation σ = 0.0021

[b] mean of 2 detns

AUXILIARY INFORMATION

METHOD/APPARATUS/PROCEDURE:

Though the details of equilibration procedure were not given, some of the data were obtained by approaching equilibrium from the side of supersaturation.
The concentration of the saturating salt was determined by adding KI to the solution of the iodate, and liberating iodine by adding HCl. The iodine was determined by titration with $Na_2S_2O_3$ solution.

SOURCE AND PURITY OF MATERIALS:

$Ba(IO_3)_2$ was made by precipitating $Ba(NO_3)_2$ with KIO_3 in a very dilute solution in which the nitrate was kept constantly in excess. The number of hydrated water is not given.
KNO_3 was purified by recrystallization.

ESTIMATED ERROR:

Soly: above described
Temp: not given

REFERENCES:

COMPONENTS:

(1) Barium iodate; $Ba(IO_3)_2$; [10567-69-8]

(2) Potassium nitrate; KNO_3; [7757-79-1]

(3) Water; H_2O; [7732-18-5]

ORIGINAL MEASUREMENTS:

Macdougall, G.; Davies, C. W.

J. Chem. Soc. 1935, 1416-9.

VARIABLES:

$T/K = 298.08$

$c_2/\text{mol dm}^{-3} = 0$ to 0.1

PREPARED BY:

Hiroshi Miyamoto
Mark Salomon

EXPERIMENTAL VALUES:

t/°C	Potassium Nitrate $c_2/\text{mol dm}^{-3}$	Barium Iodate $10^3c_1/\text{mol dm}^{-3}$	Ionic Strength $I^{1/2}/(\text{mol dm}^{-3})^{1/2}$	1/3 $\log[Ba^{2+}][IO_3^-]^2$
24.93	0	0.810	0.0491	$\bar{3}.1048$
	0.001	0.826	0.0587	$\bar{3}.1124$
	0.002	0.841	0.0669	$\bar{3}.1194$
	0.0035	0.863	0.0776	$\bar{3}.1291$
	0.005	0.880	0.0869	$\bar{3}.1367$
	0.0075	0.906	0.1004	$\bar{3}.1474$
	0.01	0.932	0.1123	$\bar{3}.1582$
	0.02	1.006	0.1506	$\bar{3}.1852$
	0.05	1.156	0.2279	$\bar{3}.2321$
	0.1	1.361	0.3164	$\bar{3}.2859$

COMMENTS AND/OR ADDITIONAL DATA:

The concentrations of the individual ionic species and ion pairs were calculated by successive approximations from the relation

$$\log[M^{2+}][X^-]/[MX^+] = \log K_D^\circ + 2I^{1/2} - 2I$$

where I = ionic strength and K_D° is the dissociation constant of the ion pair. For the ion pairs $BaIO_3^+$ and $BaNO_3^+$, the values taken for K_D° were 0.08 and 0.12, respectively.

AUXILIARY INFORMATION

METHOD/APPARATUS/PROCEDURE:

Saturating column method as in (1) and modified as in (2). A bulb containing the solvent soln is attached to a column containing the slightly soluble salt, and the solvent is allowed to flow through the column at a rate sufficient insure satn (1). The modification (2) consisted of connecting the column by capillary tubing to a second parallel arm in which the satd soln collected. The entire apparatus was placed in a thermostat. A portion of the satd soln was run through the satg column a second time. 100 cm^3 samples were taken for analysis using a calibrated pipet and running it into acid KI soln. The liberated I_2 was titrd by weight against approx 0.15N thiosulfate soln, 0.01N I_2 soln being used for the back titrn. Each data point is the mean of two detns. Conductivity measurements on binary $Ba(IO_3)_2$-H_2O solns from very dilute to satn were made at 24.93°C. Only the value for the satd soln was reported which was used to calc the ion pair dissocn const at infinite diln.

SOURCE AND PURITY OF MATERIALS:

$Ba(IO_3)_2 \cdot H_2O$ was prepared by dripwise addn of solns of A.R. grade $Ba(OH)_2$ and HIO_3 (in slight excess) into conductivity water in an apparatus "protected from the atmosphere." The crystalline precipitate was washed until its solubility was constant.

ESTIMATED ERROR:

Soly: the mean of each point agreed to within ± 0.3%

Temp: ± 0.01K

REFERENCES:

1. Brönsted, J. N.; La Mer, V. K. *J. Am. Chem. Soc.* 1924, *46*, 555.
2. Money, R. W.; Davies, C. W. *J. Chem. Soc.* 1934, 400.

COMPONENTS:

(1) Barium iodate; $Ba(IO_3)_2$; [10567-69-8]

(2) Potassium nitrate; KNO_3; [7757-79-1]

(3) Water; H_2O; [7732-18-5]

ORIGINAL MEASUREMENTS:

Polessitskij, A.

C. R. Dokl. Acad. Sci. USSR 1935, *4*, 193-6.

VARIABLES:

T/K = 298

c_2/mol dm^{-3} = 0.2 to 2.25

PREPARED BY:

Hiroshi Miyamoto

EXPERIMENTAL VALUES:

t/°C	Potassium Nitrate	Barium Iodate	
	c_2/mol dm^{-3}	s_1/mg dm^{-3}	$10^3 c_1$/mol dm^{-3} [a]
25	0.2	833	1.71
	1.4	2050	4.208
	2.25	2770	5.686

[a] Compiler calculations using 1977 IUPAC recommended atomic weights.

AUXILIARY INFORMATION

METHOD/APPARATUS/PROCEDURE:

$Ba(IO_3)_2$ crystals were stirred with KNO_3 aqueous solutions in a thermostat for 24 hours. After settling the solution for one hour, samples were withdrawn with a pipet with cotton-wool. Three analytical methods were used: (1) Evaporation of 200 cm^3 of the solution and drying at 90°C, (2) Determination of Ba as $BaSO_4$, from 200 cm^3 of the solution, (3) Iodometric titration of IO_3^-.

SOURCE AND PURITY OF MATERIALS:

$Ba(IO_3)_2$ was prepared by adding recrystallized $BaCl_2 \cdot 2H_2O$ to an equivalent amount of KIO_3. The precipitate was filtered off and washed with hot water, or washed by decantation with a large amount of cold water, under strong stirring.

ESTIMATED ERROR:

Soly: three analytical methods gave the same results within ± 3 %.
Temp: not given

REFERENCES:

COMPONENTS:

(1) Barium iodate; $Ba(IO_3)_2$; [10567-69-8]

(2) Potassium nitrate; KNO_3; [7757-79-1]

(3) Water; H_2O; [7732-18-5]

ORIGINAL MEASUREMENTS:

Davis, T. W.; Ricci, J. E.; Sauter, C. G.

J. Am. Chem. Soc. 1939, *61*, 3274-84.

VARIABLES:

T/K = 298

c_2/mol dm^{-3} = 0 to 0.2

PREPARED BY:

Hiroshi Miyamoto

EXPERIMENTAL VALUES:

t/°C	Potassium Nitrate c_2/mol dm^{-3}	Barium Iodate $10^4 c_1$/mol dm^{-3}
25	0	8.145
	0.002	8.513
	0.01	9.383
	0.05	11.74
	0.2	16.45

AUXILIARY INFORMATION

METHOD/APPARATUS/PROCEDURE:

The equilibrium procedure and the analytical method of barium iodate in the presence of varying concentrations of the electrolyte are not given. As the solubilities of barium iodate in dioxane-water mixtures are described in the paper, the compiler assumed that the method concerning to the iodate in the electrolyte solution was similar to those in dioxane-water systems. Barium iodate crystals and aqueous KNO_3 solution were placed in glass-stoppered Pyrex bottles. The bottles were rotated at 25°C for one or more days. Samples for analysis were withdrawn by suction through quantitative filters into calibrated 100 ml pipets after allowing some time for the undissolved salt to settle. The iodate content was determined iodometrically.

SOURCE AND PURITY OF MATERIALS:

C.p. grade barium iodate monohydrate was washed 8 to 10 times with distilled water, sedimented in tall cylinders to remove smaller size particles, and dried at 100°C before use. Iodometric titration gave 96.5% $Ba(IO_3)_2$ compared with 96.43% theoretical for the monohydrate. C.p. grade KNO_3 was used.

ESTIMATED ERROR:

Soly: nothing specified

Temp: ± 0.02°C (authors)

REFERENCES:

COMPONENTS:

(1) Barium iodate; $Ba(IO_3)_2$; [10567-69-8]

(2) Potassium nitrate; KNO_3; [7757-79-1]

(3) Water; H_2O; [7732-18-5]

ORIGINAL MEASUREMENTS:

Naidich, S.; Ricci, J. E.

J. Am. Chem. Soc. 1939, *61*, 3268-73.

VARIABLES:

T/K = 298

c_2/mol dm^{-3} = 0 to 0.9817

PREPARED BY:

Hiroshi Miyamoto

EXPERIMENTAL VALUES:

t/°C	Potassium Nitrate c_2/mol dm^{-3}	Barium Iodate 10^4c_1/mol dm^{-3}
25	0	8.177
	0.001	8.387
	0.005	8.937
	0.01	9.427
	0.05012	11.78
	0.1002	13.66
	0.2454	17.60
	0.4908	22.90
	0.9817	32.37

AUXILIARY INFORMATION

METHOD/APPARATUS/PROCEDURE:

The KNO_3 solution used as the solvent was prepared by weight directly from recrystallized KNO_3, and followed by proper dilution.
$Ba(IO_3)_2$ crystals and the KNO_3 solutions were immersed in Pyrex glass-stoppered bottles. The bottles were rotated in a waterbath for several days. The determination were made both from under- and super-saturation, the analyses being repeated after a day or two in almost every case.
The saturated solutions were withdrawn by means of suction a 50 or 100 ml sample into a calibrated pipet, fitted with quantitative filter paper at the tip, and was delivered into an Erlenmeyer flask.
The iodate content in the saturated solutions was determined iodometrically.

SOURCE AND PURITY OF MATERIALS:

C.p. grade $Ba(IO_3)_2 \cdot H_2O$ was washed 8-10 times with distilled water, and dried at about 100°C. Analysis of the product gave 96.5% $Ba(IO_3)_2$ as compared with the theoretical figure of 96.43% for the monohydrate.

ESTIMATED ERROR:

Soly: nothing specified
Temp: ± 0.01°C (authors)

REFERENCES:

COMPONENTS:

(1) Barium iodate; $Ba(IO_3)_2$; [10567-69-8]

(2) Magnesium chloride; $MgCl_2$; [7786-30-3]

(3) Water; H_2O; [7732-18-5]

ORIGINAL MEASUREMENTS:

Davis, T. W.; Ricci, J. E.; Sauter, C. G.

J. Am. Chem. Soc. 1939, *61*, 3274-84.

VARIABLES:

T/K = 298

$10^2 c_2$/mol dm^{-3} = 0 to 10.14

PREPARED BY:

Hiroshi Miyamoto

EXPERIMENTAL VALUES:

t/°C	Magnesium Chloride $10^2 c_2$/mol dm^{-3}	Barium Iodate $10^4 c_1$/mol dm^{-3}
25	0	8.145
	0.1023	8.596
	0.5069	9.591
	2.532	11.99
	10.14	16.11

AUXILIARY INFORMATION

METHOD/APPARATUS/PROCEDURE:

The equilibrium procedure and the analytical method of barium iodate in the presence of varying concentrations of the electrolyte are not given. As the solubilities of barium iodate in dioxane-water mixtures are described in the paper, the compiler assumed that the method concerning to the iodate in the electrolyte solution was similar to those in dioxane-water systems. Barium iodate crystals and aqueous $MgCl_2$ solution were placed in glass-stoppered Pyrex bottles. The bottles were rotated at 25°C for one or more days. Samples for analysis were withdrawn by suction through quantitative filters into calibrated 100 ml pipets after allowing some time for the undissolved salt to settle. The iodate content was determined iodometrically.

SOURCE AND PURITY OF MATERIALS:

C.p. grade barium iodate monohydrate was washed eight to ten times with distilled water, sedimented in tall cylinders to remove smaller size particles, and dried at 100°C before use. Iodometric titration gave 96.5 % $Ba(IO_3)_2$ compared with 96.43 % theoretical for the monohydrate. C.p. grade $MgCl_2$ was used.

ESTIMATED ERROR:

Soly: nothing specified
Temp: ± 0.02°C (authors)

REFERENCES:

COMPONENTS:

(1) Barium iodate; $Ba(IO_3)_2$; [10567-69-8]

(2) Calcium chloride; $CaCl_2$; [10043-52-4]

(3) Water; H_2O; [7732-18-5]

ORIGINAL MEASUREMENTS:

Macdougall, G.; Davies, C. W.

J. Chem. Soc. 1935, 1416-9.

VARIABLES:

T/K = 298.08

c_2/mol dm^{-3} = 0 to 0.009928

PREPARED BY:

Hiroshi Miyamoto
Mark Salomon

EXPERIMENTAL VALUES:

t/°C	Calcium Chloride c_2/mol dm^{-3}	Barium Iodate 10^3c_1/mol dm^{-3}	Ionic Strength $I^{1/2}$/(mol dm^{-3})$^{1/2}$	(1/3)log$[Ba^{2+}][IO_3^-]^2$
24.93	0.0	0.810	0.0491	$\bar{3}$.1048
	0.002061	0.884	0.0937	$\bar{3}$.1398
	0.004961	0.955	0.1328	$\bar{3}$.1698
	0.009928	1.046	0.1809	$\bar{3}$.2046

COMMENTS AND/OR ADDITIONAL DATA:

The concentrations of the individual ionic species and ion pairs were calculated by successive approximations from the relation

$$\log[M^{2+}][X^-]/[MX^+] = \log K_D^\circ + 2I^{1/2} - 2I$$

where I = ionic strength and K_D° is the dissociation constant of the ion pair. For the ion pairs $BaIO_3^+$, $BaCl^+$, $CaIO_3^+$, the values taken for K_D° were 0.08, 1.35, and 0.13, respectively.

AUXILIARY INFORMATION

METHOD/APPARATUS/PROCEDURE:

Saturating column method as in (1) and modified as in (2). A bulb containing the solvent solution is attached to a column containing the slightly soluble salt, and the solvent is allowed to flow through the column at a rate sufficient to insure saturation (1). The modification (2) consisted of connecting the column by capillary tubing to a second parallel arm in which the saturated solution collected. The entire apparatus was placed in a thermostat. A portion of the satd sln was run through the saturating column a second time. 100cm^3 samples were taken for analysis using a calibrated pipet and running it into acid KI sln. The liberated I_2 was titrd by weight against approx 0.15N thiosulfate sln, 0.01N I_2 sln being used for the back titrn. Each data point is the mean of two determinations.
Conductivity measurements on binary $Ba(IO_3)_2$-H_2O slns from very dilute to satn were made at 24.93°C. Only the value for the satd sln was reported which was used to calc the ion pair dissociation constant at infinite diln: K_D°/mol dm^{-3} = K_D° = 0.083 at 24.93°C.

SOURCE AND PURITY OF MATERIALS:

$Ba(IO_3)_2 \cdot H_2O$ was prepared by dropwise addn of solutions of A.R. grade $Ba(OH)_2$ and HIO_3 (in slight excess) into conductivity water in an apparatus "protected from the atmosphere." The crystalline precipitate was washed until its solubility was constant.

ESTIMATED ERROR:

Soly: the mean of each point agreed to within ± 0.3%

Temp: ± 0.01 K

REFERENCES:

1. Brönsted, J. N.; La Mer, V. K. *J. Am. Chem. Soc.* 1924, *46*, 555.
2. Money, R. W.; Davies, C. W. *J. Chem. Soc.* 1934, 400.

COMPONENTS:

(1) Barium iodate; $Ba(IO_3)_2$; [10567-69-8]

(2) Calcium nitrate; $Ca(NO_3)_2$; [10124-37-5]

(3) Water; H_2O; [7732-18-5]

ORIGINAL MEASUREMENTS:

Polessitskij, A.

C. R. Dokl. Acad. Sci. USSR 1935, *4*, 193-6.

VARIABLES:

T/K = 298

c_2/mol dm^{-3} = 1.36 to 5.18

PREPARED BY:

Hiroshi Miyamoto

EXPERIMENTAL VALUES:

t/°C	Calcium Nitrate	Barium Iodate	
	c_2/mol dm^{-3}	s_1/mg dm^{-3}	$10^2 c_1$/mol dm^{-3}
25	1.36	4266	0.874
	2.27	5196	1.064
	3.07	6457	1.324
	3.60	6500	1.333
	4.22	5519	1.133
	5.18	4022	0.825

AUXILIARY INFORMATION

METHOD/APPARATUS/PROCEDURE:

$Ba(IO_3)_2$ crystals were stirred with aqueous $Ca(NO_3)_2$ solutions in a thermostat for 24 hours. After settling the solutions for one hour, samples were withdrawn with a pipet with cotton-wool.
Three analytical methods were used:
(1) Evaporation of 200 cm^3 of the solution and drying at 90°C,
(2) Determination of Ba as $BaSO_4$ from 200 cm^2 of the solution,
(3) Iodometric titration of IO_3^-.

SOURCE AND PURITY OF MATERIALS:

$Ba(IO_3)_2$ was prepared by adding recrystallized $BaCl_2 \cdot 2H_2O$ to an equivalent amount of KIO_3. The precipitate was filtered off and washed with hot water, or washed by decantation with a large amount of cold water, under strong stirring.

ESTIMATED ERROR:

Soly: three analytical methods gave the same results within ± 3 %
Temp: not given

REFERENCES:

COMPONENTS:

(1) Barium iodate; $Ba(IO_3)_2$; [10567-69-8]

(2) Barium nitrate; $Ba(NO_3)_2$; [10022-31-8]

(3) Water; H_2O; [7732-18-5]

ORIGINAL MEASUREMENTS:

Harkins, W. D.; Winninghoff, W. J.

J. Am. Chem. Soc. 1911, *33*, 1827-36.

VARIABLES:

T/K = 298

$Ba(NO_3)_2$/eq dm^{-3} = 0 to 0.200

PREPARED BY:

Hiroshi Miyamoto

EXPERIMENTAL VALUES:

t/°C	Barium Nitrate		Barium Iodate	
	s_2/eq dm^{-3}	c_2/mol dm^{-3}	$10^3 s_1$/eq dm^{-3}	$10^4 c_1$/mol dm^{-3}
25	0	0	1.579[a]	7.89
	0.001	0.0005	1.362[b]	6.81
	0.002	0.0010	1.212[c]	6.06
	0.005	0.0025	0.9753[d]	4.88
	0.020	0.010	0.6744[e]	3.37
	0.050	0.025	0.6131[f]	3.07
	0.100	0.050	0.5659[g]	2.83
	0.200	0.100	0.5580[g]	2.79

[a] mean of 7 detns with a standard deviation σ = 0.0021

[b] mean of 4 detns with a standard deviation σ = 0.0016

[c] mean of 4 detns with a standard deviation σ = 0.0013

[d] mean of 4 detns with a standard deviation σ = 0.0011

[e] mean of 4 detns with a standard deviation σ = 0.0006

[f] mean of 4 detns with a standard deviation σ = 0.0004

[g] mean of 2 detns

AUXILIARY INFORMATION

METHOD/APPARATUS/PROCEDURE:

Though the details of equilibration procedure were not given, some of the data were obtained by approaching equilibrium from the side of supersaturation.
The concentration of the saturating salt was determined by adding KI to the solution of the iodate, and liberating iodine by adding HCl. The iodine was determined by titration with $Na_2S_2O_3$ solution.

SOURCE AND PURITY OF MATERIALS:

$Ba(IO_3)_2$ was made by precipitating $Ba(NO_3)_2$ with KIO_3 in a very dilute solution in which the nitrate was kept constantly in excess. The salt was washed with conductivity water. The number of hydrated water is not given.
$Ba(NO_3)_2$ was purified by recrystallization.

ESTIMATED ERROR:

Soly: above described
Temp: not given

REFERENCES:

COMPONENTS:

(1) Barium iodate; $Ba(IO_3)_2$; [10567-69-8]

(2) Barium nitrate; $Ba(NO_3)_2$; [10022-31-8]

(3) Water; H_2O; [7732-18-5]

ORIGINAL MEASUREMENTS:

Davis, T. W.; Ricci, J. E.; Sauter, C. G.

J. Am. Chem. Soc. 1939, *61*, 3274-84.

VARIABLES:

$T/K = 298$
$10^2 c_2/\text{mol dm}^{-3} = 0$ to 2.5

PREPARED BY:

Hiroshi Miyamoto

EXPERIMENTAL VALUES:

t/°C	Barium Nitrate $10^2 c_2/\text{mol dm}^{-3}$	Barium Iodate $10^4 c_1/\text{mol dm}^{-3}$
25	0	8.145
	0.05	6.955
	0.1	6.206
	0.25	4.963
	1	3.426
	2.5	2.854

AUXILIARY INFORMATION

METHOD/APPARATUS/PROCEDURE:

The equilibrium procedure and the analytical method of barium iodate in the presence of varying concentrations of the electrolyte are not given. As the solubilities of barium iodate in dioxane-water mixtures are described in the paper, the compiler assumed that the method concerning to the iodate in the electrolyte solution was similar to those in dioxane-water systems. Barium iodate crystals and aqueous $Ba(NO_3)_2$ solution were placed in glass-stoppered Pyrex bottles. The bottles were rotated at 25°C for one or more days. Samples for analysis were withdrawn by suction through quantitative filters into calibrated 100 ml pipets after allowing some time for the undissolved salt to settle. The iodate content was determined iodometrically.

SOURCE AND PURITY OF MATERIALS:

C.p. grade barium iodate monohydrate was washed eight to ten times with distilled water, sedimented in tall cylinders to remove smaller size particles, and dried at 100°C before use. Iodometric titration gave 96.5 % $Ba(IO_3)_2$ compared with 96.43 % theoretical for the monohydrate.
C.p. grade $Ba(NO_3)_2$ was used.

ESTIMATED ERROR:

Soly: nothing specified
Temp: ± 0.02°C (authors)

REFERENCES:

COMPONENTS:

(1) Barium iodate; $Ba(IO_3)_2$; [10567-69-8]

(2) Glycine; $C_2H_5NO_2$; [56-40-6]

(3) Water; H_2O; [7732-18-5]

ORIGINAL MEASUREMENTS:

Keefer, R. M.; Reiber, H. G.; Bisson, C. B.

J. Am. Chem. Soc. 1940, *62*, 2951-3.

VARIABLES:

T/K = 298

m_2/mol kg^{-1} = 0.0251 to 0.8175

PREPARED BY:

Hiroshi Miyamoto

EXPERIMENTAL VALUES:

t/°C	Glycine m_2/mol kg^{-1}	Barium Iodate $10^4 m_1$/mol kg^{-1}
25	0.0251	8.31
	0.0503	8.51
	0.0755	8.71
	0.1008	8.95
	0.1990	9.77
	0.8175	15.52

AUXILIARY INFORMATION

METHOD/APPARATUS/PROCEDURE:

Glycine solutions were prepared from distilled water using a calibrated volumetric equipment. An excess of air-dried barium iodate was placed in a glass-stoppered Pyrex flask and 200 ml of glycine was added. The flasks were rotated in a thermostat for at least 12 hours. Equilibrium was obtained in 4-5 hours. The saturated solutions were analyzed iodometrically. Analyses and solubility measurements were performed in duplicate. Densities of all solutions were determined, but the data were not given in the original paper.

SOURCE AND PURITY OF MATERIALS:

Barium iodate was prepared from 0.2 mol kg^{-1} $BaCl_2$ solution and 0.2 mol kg^{-1} KIO_3 solution. The precipitate was filtered, washed and dried at room temperature. The number of hydrated water was not given.

C.p. grade glycine was recrystallized twice from water by the addition of EtOH. The product was dried in a vacuum oven at about 35°C.

ESTIMATED ERROR:

Soly: nothing specified

Temp: ± 0.02°C (authors)

REFERENCES:

COMPONENTS:

(1) Barium iodate; $Ba(IO_3)_2$; [10567-69-8]
(2) Sodium hydroxide; NaOH; [1310-73-2]
(3) Glycine; $C_2H_5NO_2$; [56-40-6]
(4) Water; H_2O; [7732-18-5]

ORIGINAL MEASUREMENTS:

Monk, C. B.

Trans. Faraday Soc. 1951, *47*, 1233-40.

VARIABLES:

T/K = 298
Concentration of NaOH and glycine

PREPARED BY:

Hiroshi Miyamoto

EXPERIMENTAL VALUES:

t/°C	Glycine 10^4c_3/mol dm^{-3}	Sodium Hydroxide 10^4c_2/mol dm^{-3}	Barium Iodate 10^4c_1/mol dm^{-3}	Dissociation Constant K_D/mol dm^{-3}
25	303.6	303.6	10.56	0.18
	404.8	404.8	11.05	0.17
	744.8	744.8	12.39	0.16

The dissociation constant K_D of barium glycinate was calculated from the solubility data. Allowance was made for the presence of $Ba(OH)^+$, $Ba(IO_3)^+$ and for undissociated $NaIO_3$(l). The result derived is of such an order that negligible concentration of those ions are formed in the presence of aqueous glycine solution.

AUXILIARY INFORMATION

METHOD/APPARATUS/PROCEDURE:

The saturating column method was used (1). The saturation was ensured by passing a portion of the solution through the saturating column a second time. The analyses were effected by withdrawing 100 cm^3 of the saturated solution in a calibrated pipet, and running the solution into an acidic KI solution. The liberated iodine was titrated by weight against 0.15 mol dm^{-3} $Na_2S_2O_3$ solution, 0.005 mol dm^{-3} iodine solution being used for the back titration.

SOURCE AND PURITY OF MATERIALS:

$Ba(IO_3)_2$ was made by allowing KIO_3 and $BaCl_2$ solutions to drip slowly into hot water.
A.R. grade glycine was used, the acid was dried to constant weight in a vacuum oven at 80°C.
The source of NaOH was not given in the original paper.

ESTIMATED ERROR:

Soly: nothing specified
Temp: ± 0.03°C (author)

REFERENCES:

1. Macdougall, G.; Davies, C. W. *J. Chem. Soc.* 1935, 1416.

COMPONENTS:

(1) Barium iodate; $Ba(IO_3)_2$; [10567-69-8]

(2) Potassium chloride; KCl; [7447-40-7]

(3) Glycine; $C_2H_5NO_2$; [54-40-6]

(4) Water; H_2O; [7732-18-5]

ORIGINAL MEASUREMENTS:

Keefer, R. M.; Reiber, H. G.; Bisson, C. B.

J. Am. Chem. Soc. 1940, *62*, 2951-5.

VARIABLES:

T/K = 298

Concentration of KCl and glycine

PREPARED BY:

Hiroshi Miyamoto

EXPERIMENTAL VALUES:

t/°C	Potassium Chloride m_2/mol kg^{-1}	Glycine m_3/mol kg^{-1}	Barium Iodate $10^4 m_1$/mol kg^{-1}
25	0.02516	0.02511	10.35
	0.02519	0.05030	10.56
	0.02522	0.07552	10.78
	0.02525	0.1008	11.00
	0.05036	0.02513	11.42
	0.05041	0.05032	11.66
	0.05047	0.07558	11.90
	0.05053	0.1009	12.13
	0.1009	0.02517	12.97
	0.1010	0.05041	13.21
	0.1011	0.07570	13.46
	0.1012	0.1010	13.71

AUXILIARY INFORMATION

METHOD/APPARATUS/PROCEDURE:

Glycine and KCl solutions were prepared from boiled distilled water using a volumetric equipment. An excess of air-dried barium iodate was placed in a glass-stoppered Pyrex flask and 200 ml of the solution was added. The flasks were rotated in a thermostat at 25°C for at least 12 hours. Equilibrium was obtained in 4-5 hours. The saturated solutions were analyzed iodometrically. Analyses and solubility measurements were performed in duplicate. Densities of all solutions were determined, but the data were not given in the original paper.

SOURCE AND PURITY OF MATERIALS:

Barium iodate was prepared from 0.2 mol kg^{-1} $BaCl_2$ solution and 0.2 mol kg^{-1} KIO_3 solution. The precipitate was filtered, washed, and dried at room temperature. The number of hydrated water was not given. C.p. grade KCl was recrystallized from water. C.p. grade glycine was recrystallized twice from water by the addition of EtOH. The product was dried in vacuum oven at about 35°C.

ESTIMATED ERROR:

Soly: nothing specified

Temp: ± 0.02°C (authors)

REFERENCES:

COMPONENTS:

(1) Barium iodate; $Ba(IO_3)_2$; [10567-69-8]
(2) Potassium iodate; KIO_3; [7758-05-6]
(3) Glycine; $C_2H_5NO_2$; [56-40-6]
(4) Water; H_2O; [7732-18-5]

ORIGINAL MEASUREMENTS:

Monk, C. B.

Trans. Faraday Soc. 1951, *47*, 1233-40.

VARIABLES:

T/K = 298
Concentration of glycine and KIO_3

PREPARED BY:

Hiroshi Miyamoto

EXPERIMENTAL VALUES:

t/°C	Glycine 10^3c_3/mol dm^{-3}	Potassium Iodate 10^3c_2/mol dm^{-3}	Barium Iodate 10^4c_1/mol dm^{-3}
25	3572.0	40.49	2.78
	0	0	8.12
	1537.0	0	9.31
	3572.0	0	11.18

AUXILIARY INFORMATION

METHOD/APPARATUS/PROCEDURE:

The saturating column method was used (1). The saturation was ensured by passing a portion of the solution through the saturating column a second time. The analyses were effected by withdrawing 100 cm^3 of the saturated solution in a calibrated pipet, and running the solution into an acidic KI solution. The liberated iodine was titrated by weight against 0.15 mol dm^{-3} $Na_2S_2O_3$ solution, 0.005 mol dm^{-3} iodine solution being used for the back titration.

SOURCE AND PURITY OF MATERIALS:

$Ba(IO_3)_2$ was made by allowing KIO_3 and $BaCl_2$ solutions to drip slowly into hot water.
A.R. grade glycine was used, the acid was dried to constant weight in a vacuum oven at 80°C.
The source of KIO_3 was not given in the original paper.

ESTIMATED ERROR:

Soly: nothing specified
Temp: ± 0.03°C (author)

REFERENCES:

1. Macdougall, G.; Davies, C. W. *J. Chem. Soc.* 1935, 1416.

COMPONENTS:

(1) Barium iodate; $Ba(IO_3)_2$; [10567-69-8]

(2) Alanine; $C_3H_7NO_2$; [302-72-7]

(3) Water; H_2O; [7732-18-5]

ORIGINAL MEASUREMENTS:

Keefer, R. M.; Reiber, H. G.; Bisson, C. S.

J. Am. Chem. Soc. 1940, *62*, 2951-5.

VARIABLES:

T/K = 298

m_2/mol kg^{-1} = 0.0251 to 0.1008

PREPARED BY:

Hiroshi Miyamoto

EXPERIMENTAL VALUES:

t/°C	Alanine m_2/mol kg^{-1}	Barium Iodate $10^4 m_1$/mol kg^{-1}
25	0.0251	8.29
	0.0503	8.43
	0.0755	8.58
	0.1008	8.76

AUXILIARY INFORMATION

METHOD/APPARATUS/PROCEDURE:

Alanine solutions were prepared from boiled distilled water using a calibrated volumetric equipment. An excess of air-dried barium iodate was placed in a glass-stoppered Pyrex flask and 200 ml of alanine was added. The flasks were rotated in a thermostat at 25°C for at least 12 hours. Equilibrium was obtained in 4-5 hours.
The saturated solutions were analyzed iodometrically. Analyses and solubility measurements were done in duplicate. Densities of all solutions were determined, but the data were not given in the original paper.

SOURCE AND PURITY OF MATERIALS:

Barium iodate was prepared from 0.2 mol kg^{-1} $BaCl_2$ solution and 0.2 mol kg^{-1} KIO_3 solution. The precipitate was filtered, washed and dried at room temperature. The number of hydrated water was not given.
Alanine (c.p. grade) was recrystallized twice from water by the addition of EtOH. The product was dried in vacuum oven at about 35°C.

ESTIMATED ERROR:

Soly: nothing specified
Temp: ± 0.02°C (authors)

REFERENCES:

COMPONENTS:

(1) Barium iodate; $Ba(IO_3)_2$; [10567-69-8]

(2) Sodium hydroxide; NaOH; [1310-73-2]

(3) Alanine; $C_3H_7NO_2$; [302-72-7]

(4) Water; H_2O; [7732-18-5]

ORIGINAL MEASUREMENTS:

Monk, C. B.

Trans. Faraday Soc. 1951, *47*, 1233-40.

VARIABLES:

T/K = 298

Concentration of NaOH and alanine

PREPARED BY:

Hiroshi Miyamoto

EXPERIMENTAL VALUES:

t/°C	Alanine $10^4 c_3$/mol dm^{-3}	Sodium Hydroxide $10^4 c_2$/mol dm^{-3}	Barium Iodate $10^4 c_1$/mol dm^{-3}	Dissociation Constant K_D/mol dm^{-3}
25	274.2	274.2	10.43	0.16
	358.1	358.1	10.85	0.17
	731.9	731.9	12.28	0.17

The dissociation constant of barium alaninate was calculated from the solubility data. Allowance was made for the presence of $Ba(OH)^+$, $Ba(IO_3)^+$ and for undissociated $NaIO_3$(l). The result derived is of such an order that negligible concentrations of those ions are formed in presence of aqueous alanine solution.

AUXILIARY INFORMATION

METHOD/APPARATUS/PROCEDURE:

The saturating column method was used (1). The saturation was ensured by passing a portion of the solution through the saturating column a second time. The analyses were effected by withdrawing 100 cm^3 of the saturated solution in a calibrated pipet, and running the solution into an acidic KI solution. The liberated iodine was titrated by weight against 0.15 mol dm^{-3} $Na_2S_2O_3$ solution, 0.005 mol dm^{-3} iodine solution being used for the back titration.

SOURCE AND PURITY OF MATERIALS:

$Ba(IO_3)_2$ was made by allowing KIO_3 and $BaCl_2$ solutions to drip slowly into hot water.

Laboratory grade alanine was recrystallized from aqueous alcohol. The acid was dried to constant weight in a vacuum oven at 80°C.

The source of NaOH was not given in the original paper.

ESTIMATED ERROR:

Soly: nothing specified

Temp: ± 0.03°C (author)

REFERENCES:

1. Macdougall, G.; Davies, C. W. *J. Chem. Soc.* 1935, 1416.

COMPONENTS:

(1) Barium iodate; $Ba(IO_3)_2$; [10567-69-8]
(2) Potassium iodate; KIO_3; [7758-05-6]
(3) Alanine; $C_3H_7NO_2$; [302-72-7]
(4) Water; H_2O; [7732-18-5]

ORIGINAL MEASUREMENTS:

Monk, C. B.

Trans. Faraday Soc. 1951, *47*, 1233-40.

VARIABLES:

T/K = 298
Concentration of alanine and KIO_3

PREPARED BY:

Hiroshi Miyamoto

EXPERIMENTAL VALUES:

t/°C	Alanine 10^3c_3/mol dm^{-3}	Potassium Iodate 10^3c_2/mol dm^{-3}	Barium Iodate 10^4c_1/mol dm^{-3}
25	0	0	8.12
	1500	2.78	8.07
	1500	0	9.13

AUXILIARY INFORMATION

METHOD/APPARATUS/PROCEDURE:

The saturating column method was used (1). The saturation was ensured by passing a portion of the solution through the saturating column a second time. The analyses were effected by withdrawing 100 cm^3 of the saturated solution in a calibrated pipet, and running the solution into an acidic KI solution. The liberated iodine was titrated by weight against 0.15 mol dm^{-3} $Na_2S_2O_3$ solution, 0.005 mol dm^{-3} iodine solution being used for the back titration.

SOURCE AND PURITY OF MATERIALS:

$Ba(IO_3)_2$ was made by allowing KIO_3 and $BaCI_2$ solutions to drip slowly into hot water.
Laboratory grade alanine was recrystallized from aqueous alcohol. The acid was dried to constant weight in a vacuum oven at 80°C.
The source of KIO_3 was not given in the original paper.

ESTIMATED ERROR:

Soly: nothing specified
Temp: ± 0.03°C (author)

REFERENCES:

1. Macdougall, G.; Davies, C. W. *J. Chem. Soc.* 1935, 1416.

COMPONENTS:

(1) Barium iodate; $Ba(IO_3)_2$; [10567-69-8]

(2) Glycyl glycine; $C_4H_8N_2O_3$; [556-50-3]

(3) Water; H_2O; [7732-18-5]

ORIGINAL MEASUREMENTS:

Monk, C. B.

Trans. Faraday Soc. 1951, *47*, 1233-40.

VARIABLES:

T/K = 298

$10^3 c_2$/mol dm^{-3} = 0 to 632.2

PREPARED BY:

Hiroshi Miyamoto

EXPERIMENTAL VALUES:

t/°C	Glycyl Glycine $10^3 c_2$/mol dm^{-3}	Barium Iodate $10^4 c_1$/mol dm^{-3}
25	0	8.12
	344.0	8.53
	578.0	8.95
	632.2	9.12

AUXILIARY INFORMATION

METHOD/APPARATUS/PROCEDURE:

The saturating column method was used (1). The saturation was ensured by passing a portion of the solution through the saturating column a second time. The analyses were effected by withdrawing 100 cm^3 of the saturated solution in a calibrated pipet, and running the solution into an acidic KI solution. The liberated iodine was titrated by weight against 0.15 mol dm^{-3} $Na_2S_2O_3$ solution, 0.005 mol dm^{-3} iodine solution being used for the back titration.

SOURCE AND PURITY OF MATERIALS:

$Ba(IO_3)_2$ was made by allowing KIO_3 and $BaCl_2$ solutions to drip slowly into hot water.
Glycyl glycine used was of Roche Products chemicals.

ESTIMATED ERROR:

Soly: nothing specified
Temp: ± 0.03°C (author)

REFERENCES:

1. Macdougall, G.; Davies, C. W. *J. Chem. Soc.* 1935, 1416.

COMPONENTS:

(1) Barium iodate; $Ba(IO_3)_2$; [10567-69-8]

(2) Barium chloride; $BaCl_2$; [13477-00-4]

(3) Water; H_2O; [7732-18-5]

ORIGINAL MEASUREMENTS:

Ricci, J. E.

J. Am. Chem. Soc. 1951, *73*, 1375-6.

VARIABLES:

$T/K = 298$

$BaCl_2$/mass % = 0.00 to 27.11

PREPARED BY:

Hiroshi Miyamoto

EXPERIMENTAL VALUES:

t/°C	Composition of Saturated Solutions				Nature of the Solid Phase[a]
	Barium Chloride		Barium Iodate		
	mass %	mol % (compiler)	mass %	mol % (compiler)	
25	0.00	0.00	0.040[c]	1.480	A
	19.63	2.070	0.0189	0.852	A
	20.50	2.183	0.0187	0.851	A
	21.56	2.323	0.0198	0.912	A
	25.29	2.846	0.0231	1.11	A(m)
	25.59	2.890	0.0240	1.16	A(m)[b]
	27.09	3.118	0.0267	1.31	(A + C)(m)
	22.06	2.390	--	--	A + B
	22.05	2.389	--	--	A + B
	(22.38)?	2.434	0.0201	0.935	A + B
	22.06	2.391	0.0201	0.931	A + B
	22.72	2.481	0.0189	0.882	B
	23.76	2.626	0.0186	0.879	B
	24.25	2.696	(0.020)	0.950	B
	24.73	2.764	0.0179	0.855	B
	25.45	2.869	0.0171	0.824	B
	25.51	2.878	(0.019)	0.916	B
	26.01	2.952	0.0165	0.801	B
	26.82	3.074	0.0152	0.745	B
	27.09	3.115	0.0153	0.752	B + C
	27.11	3.112	0.0153	0.752	B + C
	27.11	3.117	0.00	0.000	C

[a] A = $Ba(IO_3)_2 \cdot H_2O$; B = $Ba(IO_3)_2 \cdot BaCl_2 \cdot 2H_2O$; C = $BaCl_2 \cdot 2H_2O$; m = metastable

[b] Probably in course of change, not completely at equilibrium.

[c] For binary system the compiler computes the following

Soly of $Ba(IO_3)_2 = 8.21 \times 10^{-4}$ mol kg^{-1}

AUXILIARY INFORMATION

METHOD/APPARATUS/PROCEDURE:

Complexes were made up from water, $Ba(IO_3)_2 \cdot H_2O$ and $BaCl_2 \cdot 2H_2O$. After 1 or 2 weeks of stirring the filtered saturated solution was analyzed for the iodate with standard thiosulfate and for total solid by evaporation. Reanalysis after a similar additional period of stirring was used to confirm equilibrium.

SOURCE AND PURITY OF MATERIALS:

C.p. grade $Ba(IO_3)_2 \cdot 6H_2O$ and $BaCl_2 \cdot 2H_2O$ were used and checked for purity by direct analysis.

ESTIMATED ERROR:

Nothing specified

REFERENCES:

COMPONENTS:

(1) Barium iodate; $Ba(IO_3)_2$; [10567-69-8]

(2) Barium chloride; $BaCl_2$; [13477-00-4]

(3) Water; H_2O; [7732-18-5]

ORIGINAL MEASUREMENTS:

Ricci, J. E.

J. Am. Chem. Soc. 1951, *73*, 1375-6.

COMMENTS AND/OR ADDITIONAL DATA:

The phase diagram is given below (based on mass %).

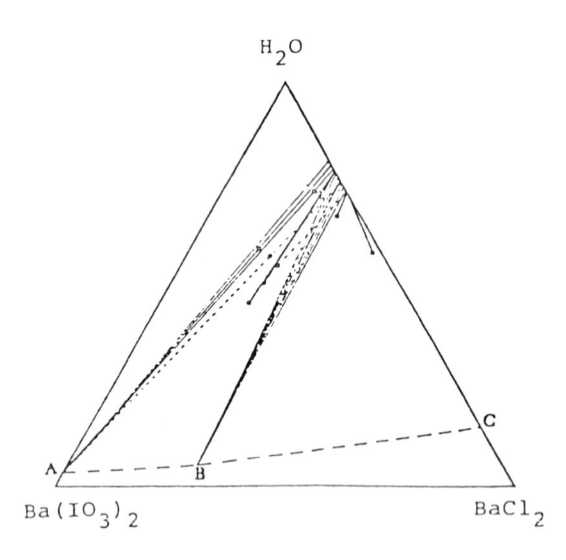

AUXILIARY INFORMATION

METHOD/APPARATUS/PROCEDURE:

SOURCE AND PURITY OF MATERIALS:

ESTIMATED ERROR:

ACKNOWLEDGEMENT:

The figure reprinted from the *J. Am. Chem. Soc.* by permission of the copyright owners, The American Chemical Society.

COMPONENTS:

(1) Barium iodate; $Ba(IO_3)_2$; [10567-69-8]

(2) Barium bromide; $BaBr_2$; [10553-31-8]

(3) Water; H_2O; [7732-18-5]

ORIGINAL MEASUREMENTS:

Ricci, J. E.; Freedman, A. J.

J. Am. Chem. Soc. 1952, *74*, 1769-73.

VARIABLES:

T/K = 298

$BaBr_2$/mass % = 0 to 50.07

PREPARED BY:

Hiroshi Miyamoto

EXPERIMENTAL VALUES:

t/°C	Barium Bromide			Barium Iodate		
	mass %	mol % (compiler)	m_2/mol kg^{-1} (compiler)	mass %	mol % (compiler)	$10^4 m_1$/mol kg^{-1}
25	0	0	0	0.042	1.55	8.63
	11.19	0.7582	0.4241	0.013	0.537	3.01
	22.34	1.715	0.9683	0.017	0.796	4.49
	33.46	2.960	1.693	0.026	1.40	8.02
	44.61	4.659	2.713	0.045	2.87	16.7
	49.02	5.515	3.240	0.068	4.67	27.4
	50.07[a]	5.739	3.380	0.075	5.24	30.9

[a] Saturated solution

AUXILIARY INFORMATION

METHOD/APPARATUS/PROCEDURE:

Mixtures of $Ba(IO_3)_2$-$BaBr_2$-H_2O of unknown composition, enclosed in Pyrex tubes, were rotated in a constant temperature water-bath. Calibrated pipets with filter paper tips were used in sampling the saturated solutions for analysis.
The total salt content was determined by evaporation to constant weight at 300°C. Bromide was determined by Volhard method with filtration of AgBr, and the iodate content was determined iodometrically.

SOURCE AND PURITY OF MATERIALS:

C.p. grade $Ba(IO_3)_2 \cdot H_2O$ used contained too little water, therefore it was leached with water until its solubility was constant, and it was then rinsed with some acetone and dried in air. The product contained 96.34% $Ba(IO_3)_2$ as compared with the theoretical 96.43%. The method of preparation of $BaBr_2 \cdot 2H_2O$ has been described in the literature (1).

ESTIMATED ERROR:

Nothing specified

REFERENCES:

1. Ricci, J. E.; Freedman, A. J. *J. Am. Chem. Soc.* 1952, *74*, 1765.

COMPONENTS:

(1) Barium iodate; $Ba(IO_3)_2$; [10567-69-8]

(2) Barium nitrate; $Ba(NO_3)_2$; [10022-31-8]

(3) Water; H_2O; [7732-18-5]

ORIGINAL MEASUREMENTS:

Ricci, J. E.; Freedman, A. J.

J. Am. Chem. Soc. 1952, *74*, 1769-73.

VARIABLES:

T/K = 298

$Ba(NO_3)_2$/mass % = 0 to 9.22

PREPARED BY:

Hiroshi Miyamoto

EXPERIMENTAL VALUES:

t/°C	Barium Nitrate			Barium Iodate		
	mass %	mol % (compiler)	m_2/mol kg^{-1} (compiler)	mass %	mol % (compiler)	$10^4 m_1$/mol kg^{-1} (compiler)
25	0	0	0	0.042	1.55	8.63
	2.50	0.176	0.098	0.019	0.720	4.00
	4.75	0.343	0.191	0.023	0.890	4.96
	7.14	0.527	0.294	0.027	1.07	5.97
	8.69	0.652	0.364	0.031	1.25	6.97
	9.22[a]	0.695	0.389	0.033	1.34	7.47

[a] Saturated solution

AUXILIARY INFORMATION

METHOD/APPARATUS/PROCEDURE:

Mixtures of $Ba(IO_3)_2$-$Ba(NO_3)_2$-H_2O of unknown composition, enclosed in Pyrex tubes, were rotated in a constant temperature water-bath. Calibrated pipets with filter paper tip were used in sampling the saturated solutions for analysis.
The total salt content was determined by evaporation to constant weight at 300°C, and the iodate content was determined iodometrically.

SOURCE AND PURITY OF MATERIALS:

C.p. grade $Ba(IO_3)_2 \cdot H_2O$ used contained too little water, therefore it was leached with water until its solubility was constant, and it was then rinsed with some acetone and dried in air. The product contained 96.34% $Ba(IO_3)_2$ as compared with the theoretical 96.43%. C.p. grade $Ba(NO_3)_2$ was used without further purification.

ESTIMATED ERROR:

Nothing specified

REFERENCES:

COMPONENTS:

(1) Barium iodate; $Ba(IO_3)_2$; [10567-69-8]

(2) Ammonia; NH_3; [7664-41-7]

(3) Water; H_2O; [7732-18-5]

ORIGINAL MEASUREMENTS:

Derr, P. F.; Vosburgh, W. C.

J. Am. Chem. Soc. 1943, *65*, 2408-11.

VARIABLES:

T/K = 298
c_2/mol dm^{-3} = 0 to 1.016

PREPARED BY:

Hiroshi Miyamoto

EXPERIMENTAL VALUES:

t/°C	Ammonia	Barium Iodate	
	c_2/mol dm^{-3}	$10^4 c_1$/mol dm^{-3}	$10^5 K^0_{s0}$/mol^{3}dm^{-9}
25	0	8.09	1.53
	0.1682	8.24	1.46
	0.3912	8.30	1.44
	0.894	7.70	1.09
	1.016	7.65	1.06

The activity solubility product was calculated from

$$K^{\circ}_{s0} = [Ba^{2+}][IO_3^-]y^3$$

where $y = I^{1/2}/(1 + I^{1/2})$ and I is the ionic strength.

AUXILIARY INFORMATION

METHOD/APPARATUS/PROCEDURE:

A roughly measured volume of ammonia solutions was isothermally saturated with $Ba(IO_3)_2$. Samples for analysis were withdrawn by forcing the solution through a filter and into a pipet by air pressure to avoid loss of NH_3.
The total ammonia was determined by titration with standard acid. The total iodate was determined iodometrically.

SOURCE AND PURITY OF MATERIALS:

Barium iodate was precipitated from solutions of reagent grade chemicals, digested at high temperature, carefully washed, and preserved under water. The name of the chemicals used and the number of hydrated water was not given.

ESTIMATED ERROR:

Nothing specified

REFERENCES:

COMPONENTS:

(1) Barium iodate; $Ba(IO_3)_2$; [10567-69-8]

(2) Ammonium hydroxide; NH_4OH; [1336-21-6]

(3) Water; H_2O; [7732-18-5]

ORIGINAL MEASUREMENTS:

Hill, A. E.; Zink, W. A. H.

J. Am. Chem. Soc. 1909, *31*, 43-9.

VARIABLES:

Room temperature

PREPARED BY:

Hiroshi Miyamoto

EXPERIMENTAL VALUES:

Solubility of barium iodate in concentrated NH_4OH (sp gr 0.90) at room temperature is:

C_{Ba} = 5.6 mg/100 cc (authors)

$C_{Ba} = 4.0_7 \times 10^{-4}$ mol dm^{-3} (compiler)

AUXILIARY INFORMATION

METHOD/APPARATUS/PROCEDURE:

Solutions containing freshly precipitated barium iodate was rotated at room temperature for 12 hours with ordinary concentrated ammonia (sp gr 0.90). Most likely the iodate content of the saturated solutions were determined iodometrically.

COMMENTS AND/OR ADDITIONAL DATA:

The purpose of this paper was to propose a new volumetric titration method for determination of Ba, in which Ba^{++} was precipitated as $Ba(IO_3)_2$ and after being washed with concentrated NH_4OH which was not to dissolve any appreciable amount of $Ba(IO_3)_2$ the precipitate was analyzed by iodometric titration. To ascertain the suitability of the solvent as washing solution the solubility data were obtained.

SOURCE AND PURITY OF MATERIALS:

Barium iodate was prepared by adding barium chloride solution to excess of potassium iodate solution. The number of hydrated water was not given.

ESTIMATED ERROR:

Nothing specified

REFERENCES:

COMPONENTS:

(1) Barium iodate; $Ba(IO_3)_2$; [10567-69-8]

(2) Hydrochloric acid; HCl; [7647-01-0]

(3) Water; H_2O; [7732-18-5]

ORIGINAL MEASUREMENTS:

Naidich, S.; Ricci, J. E.

J. Am. Chem. Soc. 1939, *61*, 3268-73.

VARIABLES:

T/K = 298

c_2/mol dm^{-3} = 0 to 0.9817

PREPARED BY:

Hiroshi Miyamoto

EXPERIMENTAL VALUES:

t/°C	Hydrochloric Acid c_2/mol dm^{-3}	Barium Iodate $10^4 c_1$/mol dm^{-3}
25	0	8.177
	0.0001	8.249
	0.0005	8.328
	0.001	8.400
	0.0025	8.649
	0.005	8.986
	0.01	9.554
	0.05012	12.74
	0.1002	15.57
	0.2454	22.28
	0.4908	32.39
	0.9817	52.43

AUXILIARY INFORMATION

METHOD/APPARATUS/PROCEDURE:

The HCl solvents were made by a series of dilution from a stock solution standardized gravimetrically. $Ba(IO_3)_2$ crystals and the HCl solvents were immersed in Pyrex glass-stoppered bottles. The bottles were rotated in a waterbath for several days. The determination were made both from under- and super-saturation, the analyses being repeated after a day or two in almost every case. The saturated solutions were withdrawn by means of suction a 50 or 100 ml sample into a calibrated pipet, fitted with quantitative filter paper at the tip, and was delivered into an Erlenmeyer flask. The iodate content in the saturated solutions was determined iodometrically.

SOURCE AND PURITY OF MATERIALS:

C.p. grade $Ba(IO_3)_2 \cdot H_2O$ was washed 8-10 times with distilled water, and dried at about 100°C. Analysis of the product gave 96.5% $Ba(IO_3)_2$ as compared with the theoretical figure of 96.43% for the monohydrate.

ESTIMATED ERROR:

Soly: nothing specified

Temp: ± 0.01°C (authors)

REFERENCES:

COMPONENTS:

(1) Barium iodate; $Ba(IO_3)_2$; [10567-69-8]

(2) Iodic acid; HIO_3; [7782-18-5]

(3) Water; H_2O; [7732-18-5]

ORIGINAL MEASUREMENTS:

Ricci, J. E.; Freedman, A. J.

J. Am. Chem. Soc. 1952, *74*, 1769-73.

VARIABLES:

T/K = 298

PREPARED BY:

Hiroshi Miyamoto

EXPERIMENTAL VALUES:

The authors state that barium could not be detected in any of the ternary systems by the sulfuric acid analyses method. That the concentration of barium iodate was negligible was confirmed by the fact that analyses for iodic acid both alklimerically and iodometrically gave identical results.

The authors measured the solubility for binary system $Ba(IO_3)_2-H_2O$, and the result is the following:

0.042 mass %	(authors)
0.863 mol kg^{-1}	(compiler)

AUXILIARY INFORMATION

METHOD/APPARATUS/PROCEDURE:

Mixtures of $Ba(IO_3)_2-HIO_3-H_2O$ enclosed in Pyrex tubes, were rotated constant temperature water-bath. Solutions containing less than 45% iodic acid were sampled with delivery pipets with filter paper tips. The more viscous solutions of higher concentration were sampled by means of specific gravity pipets fitted with ground glass caps. In the case of the solutions above 60% in concentration, the liquid-solid mixture was allowed to settle after centrifuging at high speed for a few moments; the clear liquid was then sampled without filteration. The densities of the solutions were also determined, were essentially the densities of pure solutions of I_2O_5. For the analysis, the barium content was determined by treatment with dilute sulfuric acid. Titration with sodium hydroxide, with phenolphthalein as indicator, was used for determinations of iodic acid. The representative solutions were analyzed for iodic acid both alkalimetrically and iodometrically.

SOURCE AND PURITY OF MATERIALS:

C.p. grade $Ba(IO_3)_2 \cdot H_2O$ used contained too little water, and therefore was leached with water until its solubility was constant. It was then rinsed with some acetone and dried in air. The product contained 96.34% $Ba(IO_3)_2$ as compared with the theoretical 96.43%. One sample of the iodic acid used was a commercial c.p. product containing 99.82% HIO_3 by determination of iodate and of acid. Another sample was made from c.p. grade I_2O_5 and water. The solution was evaporated under an infrared lamp until the acid began to crystallize, and the process was completed at ∿ 40°C in a stream of air. When ground and stored in vacuum over anhydrone, constant composition was reached after about two weeks, at 99.66% HIO_3. The solubility of the two samples was the same.

ESTIMATED ERROR:

Nothing specified

REFERENCES:

COMPONENTS:

(1) Barium iodate; $Ba(IO_3)_2$; [10567-69-8]

(2) Nitric acid; HNO_3; [7697-37-2]

(3) Water; H_2O; [7732-18-5]

ORIGINAL MEASUREMENTS:

Naidich, S.; Ricci, J. E.

J. Am. Chem. Soc. 1939, *61*, 3268-73.

VARIABLES:

T/K = 298

c_2/mol dm^{-3} = 0 to 0.9894

PREPARED BY:

Hiroshi Miyamoto

EXPERIMENTAL VALUES:

t/°C	Nitric Acid c_2/mol dm^{-3}	Barium Iodate 10^4c_1/mol dm^{-3}
25	0	8.177
	0.001101	8.455
	0.005505	9.178
	0.01101	9.806
	0.05505	13.75
	0.1101	17.53
	0.2474	25.83
	0.4948	39.39
	0.9895	68.78

AUXILIARY INFORMATION

METHOD/APPARATUS/PROCEDURE:

The HNO_3 solvents were made by a series of dilution from a stock solution standardized volumetrically against Na_2CO_3 using methyl organe as an indicator.
$Ba(IO_3)_2$ crystals and the HNO_3 solvents were immersed in Pyrex glass-stoppered bottles. The bottles were rotated in a waterbath for several days. The determinations were made both from under- and super-saturation, the analyses being repeated after a day or two in almost every case. The saturated solutions were withdrawn by means of suction a 50 or 100 ml sample into a calibrated pipet, fitted with quantitative filter paper at the tip, and was delivered into an Erlenmeyer flask. The iodate content in the saturated solutions was determined iodometrically.

SOURCE AND PURITY OF MATERIALS:

C.p. grade $Ba(IO_3)_2 \cdot H_2O$ was washed 8-10 times with distilled water, and dried at about 100°C. Analysis of the product gave 96.5% as compared with the theoretical figure of 96.43% for the monohydrate.

ESTIMATED ERROR:

Soly: nothing specified
Temp: ± 0.01°C (authors)

REFERENCES:

COMPONENTS:

(1) Barium iodate; $Ba(IO_3)_2$; [10567-69-8]

(2) Methanol; CH_4O; [67-56-1]

(3) Water; H_2O; [7732-18-5]

ORIGINAL MEASUREMENTS:

Monk, C. B.

J. Chem. Soc. 1951, 2723-6.

VARIABLES:

T/K = 298
methanol/mass % = 0 to 14.43

PREPARED BY:

Hiroshi Miyamoto

EXPERIMENTAL VALUES:

t/°C	Methanol		Barium Iodate	
	mass %	mol % (compiler)	$10^4 c_1$/mol dm^{-3}	$10^{10} K^\circ_{s0}$/mol^{3}dm^{-9}
25	0	0	8.12	15.52
	4.72	2.71	5.71	5.60
	9.53	5.59	4.05	2.06
	14.43	8.66	2.84	0.73

Thermodynamic solubility product constant, K°_{s0}, was calculated from

$$\log K^\circ_{s0} = \log [Ba^{2+}][IO_3^-]^2 - 3F$$

where $F = (78.54/\varepsilon)^{3/2}[I^{1/2}/(1 + I^{1/2}) - 0.2I]$, ε = dielectric constant
I = ionic strength.

AUXILIARY INFORMATION

METHOD/APPARATUS/PROCEDURE:

A saturating column method was used (1). The iodate concentrations in the saturated solutions were determined by titration with $Na_2S_2O_3$ standardized with KIO_3. Prior to the titration, excess KI was added and the solution acidified with dilute acetic acid.

SOURCE AND PURITY OF MATERIALS:

Barium iodate was prepared by allowing dilute solutions of KIO_3 and of $BaCl_2$ to drip slowly into hot water.
Methanol used was of laboratory grade.

ESTIMATED ERROR:

Soly: nothing specified
Temp: ± 0.03°C (author)

REFERENCES:

1. Davies, C. W.
J. Chem. Soc. 1938, 277.

COMPONENTS:

(1) Barium iodate; $Ba(IO_3)_2$; [10567-69-8]

(2) Ethanol; C_2H_6O; [64-17-5]

(3) Water; H_2O; [7732-18-5]

ORIGINAL MEASUREMENTS:

Hill, A. E.; Zink, W. A. H.

J. Am. Chem. Soc. 1909, *31*, 43-9.

VARIABLES:

Room temperature

PREPARED BY:

Hiroshi Miyamoto

EXPERIMENTAL VALUES:

Solubility of $Ba(IO_3)_2$ in 95%[a] ethanol[b] at room temperature is:

c_{Ba} = 3.1 mg/100cc (authors)

$c_{Ba} = 2.2_6 \times 10^{-4}$ mol dm^{-3} (compiler)

The units of the concentration of the solvent were not given.

[a] 85% in page 44 may be misprinted.

[b] The kind of alcohol was specified as C_2H_5OH in the table of page 45

AUXILIARY INFORMATION

METHOD/APPARATUS/PROCEDURE:

Solutions containing freshly precipitated barium iodate was rotated at room temperature for 12 hours with 95% alcohol. Most likely the iodate content of the saturated solutions was determined iodometrically.

COMMENTS AND/OR ADDITIONAL DATA:

The purpose of this paper was to propose a new volumetric method for determination of Ba, in which Ba^{++} was precipitated as $Ba(IO_3)_2$ and after being washed with 95% alcohol which was not to dissolve any appreciable amount of $Ba(IO_3)_2$ the precipitate was analyzed by iodometric titration. To ascertain the suitability of the solvent as washing solution the solubility data were obtained.
Effect of foreign bodies on the determination of barium was also cited.

SOURCE AND PURITY OF MATERIALS:

Barium iodate was prepared by adding barium chloride solution to excess of potassium iodate solution. The number of hydrated water was not given.

ESTIMATED ERROR:

Nothing specified

REFERENCES:

COMPONENTS:

(1) Barium iodate; $Ba(IO_3)_2$; [10567-69-8]

(2) Ethanol; C_2H_6O; [64-17-5]

(3) Water; H_2O; [7732-18-5]

ORIGINAL MEASUREMENTS:

Monk, C. B.

J. Chem. Soc. 1951, 2723-6.

VARIABLES:

T/K = 298

Ethanol/mass % = 0 to 11.59

PREPARED BY:

Hiroshi Miyamoto

EXPERIMENTAL VALUES:

t/°C	Ethanol		Barium Iodate	
	mass %	mol % (compiler)	$10^4 c_1$/mol dm^{-3}	$10^{10} K^\circ_{s0}$/mol^{3}dm^{-9}
25	0	0	8.12	15.52
	3.82	1.53	5.80	5.86
	7.67	3.15	4.18	2.27
	11.59	4.88	3.02	0.88

Thermodynamic solubility product constant, K°_{s0}, was calculated from

$$\log K^\circ_{s0} = \log [Ba^{2+}][IO_3^-]^2 - 3F$$

where $F = (78.54/\varepsilon)^{3/2}[I^{1/2}/(1 + I^{1/2}) - 0.2I]$, ε = dielectric constant, I = ionic strength.

AUXILIARY INFORMATION

METHOD/APPARATUS/PROCEDURE:

A saturating column method was used (1). The iodate concentrations in the saturated solutions were determined by titration with $Na_2S_2O_3$ standardized with KIO_3. Prior to the solution, excess KI was added and the solution acidified with dilute acetic acid.

SOURCE AND PURITY OF MATERIALS:

Barium iodate crystals was prepared by allowing dilute solutions of KIO_3 and of $BaCl_2$ to drip slowly into hot water.

Ethanol used was of laboratory grade.

ESTIMATED ERROR:

Soly: nothing specified

Temp: ± 0.03°C (author)

REFERENCES:

1. Davies, C. W. *J. Chem. Soc.* 1938, 277.

COMPONENTS:

(1) Barium iodate; $Ba(IO_3)_2$; [10567-69-8]
(2) Sodium nitrate; $NaNO_3$; [7631-99-4]
(3) Sodium hydroxide; NaOH; [1310-73-2]
(4) Ethanol; C_2H_6O; [64-17-5]
(5) Water; H_2O; [7732-18-5]

ORIGINAL MEASUREMENTS:

Milad, N. E.; Morsi, S. E.; Soliman, S. T.; Seleem, L. M. N.

Egypt. J. Chem. 1973, *16*, 395-400.

VARIABLES:

T/K = 298
EtOH/mol % = 0 - 8.72
Ionic strength and pH

PREPARED BY:

Hiroshi Miyamoto

EXPERIMENTAL VALUES:

t/°C	Ethanol mol %	Barium Iodate $10^3 c_1$/mol dm^{-3}					
		I = 0.04	0.06	0.08	0.10	0.15	0.20
25	0.00	1.130	1.248	1.292	1.381	1.546	1.630
	1.51	0.580	0.897	0.978	1.020	1.134	1.233
	4.85	0.464	0.505	0.543	0.588	0.641	0.697
	6.71	0.350	0.372	0.408	0.434	0.450	0.548
	8.72	0.261	0.290	0.315	0.323	0.377	0.412

AUXILIARY INFORMATION

METHOD/APPARATUS/PROCEDURE:

Few hundred mg portions of solid $Ba(IO_3)_2$ were transferred into 100ml flasks. To each flask different volumes of water or aqueous ethanol and $NaNO_3$ solution were added to provide a solution of a constant ionic strength. Predetermined aliquots of 0.1 mol dm^{-3} NaOH solution were added so that the final pH of the equilibrium solution be 7.0 - 7.2. The flasks were well stoppered, and mechanically shaken in an air thermostat for 4 hours. After equilibrium was attained, the solid was separated from the saturated solution by filtration.
The concentration of the iodate was determined iodometrically.

SOURCE AND PURITY OF MATERIALS:

$Ba(IO_3)_2$ was prepared by mixing 0.2 mol dm^{-3} $BaCl_2$ solution and 0.2 mol dm^{-3} KIO_3 solution. The precipitate was filtered, washed and dried at room temperature. The number of hydrated water was not given.
Pure ethanol was distilled over CaO. The distillate obtained at b.p. 78°C was used.
AnalaR $NaNO_3$ was used.

ESTIMATED ERROR:

Soly: nothing specified
Temp: ± 0.1°C (authors)

REFERENCES:

COMPONENTS:

(1) Barium iodate; $Ba(IO_3)_2$; [10567-69-8]

(2) 1-Propanol; C_3H_8O; [71-23-8]

(3) Water; H_2O; [7732-18-5]

ORIGINAL MEASUREMENTS:

Monk, C. B.

J. Chem. Soc. 1951, 2723-6.

VARIABLES:

T/K = 298

1-propanol/mass % = 0 to 12.71

PREPARED BY:

Hiroshi Miyamoto

EXPERIMENTAL VALUES:

t/°C	1-Propanol		Barium Iodate	
	mass %	mol % (compiler)	$10^4 c_1$/mol dm^{-3}	$10^{10} K^\circ_{s0}$/mol^{3}dm^{-9}
25	0	0	8.12	15.52
	4.16	1.28	5.78	5.78
	8.40	2.68	4.13	2.17
	12.71	4.18	2.94	0.80

Thermodynamic solubility product constant, K°_{s0}, was calculated from

$$\log K^\circ_{s0} = \log [Ba^{2+}][IO_3^-]^2 - 3F$$

where $F = (78.54/\varepsilon)^{3/2}[I^{1/2}/(1 + I^{1/2}) - 0.2I]$, ε = dielectric constant, I = ionic strength.

AUXILIARY INFORMATION

METHOD/APPARATUS/PROCEDURE:

A saturating column method was used (1). The iodate concentrations in the saturated solutions were determined by titration with $Na_2S_2O_3$ standardized with KIO_3. Prior to the titration, excess KI was added and the solution acidified with dilute acetic acid.

SOURCE AND PURITY OF MATERIALS:

Barium iodate crystals were prepared by allowing dilute solutions of KIO_3 and of $BaCl_2$ to drip slowly into hot water.
1-Propanol used of laboratory grade.

ESTIMATED ERROR:

Soly: nothing specified
Temp: ± 0.03°C (author)

REFERENCES:

1. Davies, C. W.
J. Chem. Soc. 1938, 277.

COMPONENTS:

(1) Barium iodate; $Ba(IO_3)_2$; [10567-69-8]

(2) 1,2-Ethanediol (ethylene glycol); $C_2H_6O_2$; [107-21-1]

(3) Water; H_2O; [7732-18-5]

ORIGINAL MEASUREMENTS:

Monk, C. B.

J. Chem. Soc. 1951, 2723-6.

VARIABLES:

T/K = 298

Ethylene glycol/mass % = 0 to 16.85

PREPARED BY:

Hiroshi Miyamoto

EXPERIMENTAL VALUES:

t/°C	Ethylene Glycol		Barium Iodate	
	mass %	mol % (compiler)	$10^4 c_1$/mol dm^{-3}	$10^{10} K^\circ_{s0}$/mol^{3}dm^{-9}
25	0	0	8.12	15.52
	5.62	1.70	7.18	10.81
	11.24	3.55	6.40	7.71
	16.85	5.56	5.73	5.58

Thermodynamic solubility product constant, K°_{s0}, was calculated from

$$\log K^\circ_{s0} = \log [Ba^{2+}][IO_3^-]^2 - 3F$$

where $F = (78.54/\varepsilon)^{3/2}[I^{1/2}/(1 + I^{1/2}) - 0.2I]$, ε = dielectric constant, I = ionic strength.

AUXILIARY INFORMATION

METHOD/APPARATUS/PROCEDURE:

A saturating column method was used (1). The iodate concentrations in the saturated solutions were determined by titration with $Na_2S_2O_3$ standardized with KIO_3. Prior to the titration, excess KI was added and the solution acidified with dilute acetic acid.

SOURCE AND PURITY OF MATERIALS:

Barium iodate crystals were prepared by allowing dilute solutions of KIO_3 and of $BaCl_2$ to drip slowly into hot water.
Ethylene glycol used was of laboratory grade.

ESTIMATED ERROR:

Soly: nothing specified
Temp: ± 0.03°C (author)

REFERENCES:

1. Davies, C. W. *J. Chem. Soc.* 1938, 277.

COMPONENTS:

(1) Barium iodate; $Ba(IO_3)_2$; [10567-69-8]

(2) 1,2,3-Propanetriol (glycerol); $C_3H_8O_3$; [56-81-5]

(3) Water; H_2O; [7732-18-5]

ORIGINAL MEASUREMENTS:

Monk, C. B.

J. Chem. Soc. 1951, 2723-6.

VARIABLES:

T/K = 298
Glycerol/mass % = 0 to 18.43

PREPARED BY:

Hiroshi Miyamoto

EXPERIMENTAL VALUES:

t/°C	Glycerol		Barium Iodate	
	mass %	mol % (compiler)	$10^4 c_1$/mol dm^{-3}	$10^9 K^\circ_{s0}$/mol^{3}dm^{-9}
25	0	0	8.12	1.55
	6.31	1.30	7.81	1.38
	12.44	2.70	7.52	1.22
	18.43	4.23	7.28	1.10

Thermodynamic solubility product constant, K°_{s0}, was calculated from

$$\log K^\circ_{s0} = \log [Ba^{2+}][IO_3^-]^2 - 3F$$

where $F = (78.54/\varepsilon)^{3/2}[I^{1/2}/(1 + I^{1/2}) - 0.2I]$, ε = dielectric constant, I = ionic strength.

AUXILIARY INFORMATION

METHOD/APPARATUS/PROCEDURE:

A saturating column method was used (1). The iodate concentrations in the saturated solutions were determined by titration with $Na_2S_2O_3$ standardized with KIO_3. Prior to the titration, excess KI was added and the solution acidified with dilute acetic acid.

SOURCE AND PURITY OF MATERIALS:

Barium iodate crystals were prepared by allowing dilute solutions of KIO_3 and of $BaCl_2$ to drip slowly into hot water.
Glycerol used was of laboratory grade.

ESTIMATED ERROR:

Soly: nothing specified
Temp: ± 0.03°C (author)

REFERENCES:

1. Davies, C. W.
J. Chem. Soc. 1938, 277.

COMPONENTS:

(1) Barium iodate; $Ba(IO_3)_2$; [10567-69-8]

(2) 2-Propanone (acetone); C_3H_6O; [67-64-1]

(3) Water; H_2O; [7732-18-5]

ORIGINAL MEASUREMENTS:

Monk, C. B.

J. Chem. Soc. 1951, 2723-6.

VARIABLES:

T/K = 298
Acetone/mass % = 0 to 12.46

PREPARED BY:

Hiroshi Miyamoto

EXPERIMENTAL VALUES:

t/°C	Acetone		Barium Iodate	
	mass %	mol % (compiler)	$10^4 c_1$/mol dm^{-3}	$10^{10} K^\circ_{s0}$/mol^{3}dm^{-9}
25	0	0	8.12	15.52
	4.09	1.31	5.99	6.38
	8.25	2.71	4.46	2.72
	12.46	4.23	3.27	1.10

Thermodynamic solubility product constant, K°_{s0}, was calculated from

$$\log K^\circ_{s0} = \log [Ba^{2+}][IO_3^-]^2 - 3F$$

where $F = (78.54/\varepsilon)^{3/2}[I^{1/2}/(1 + I^{1/2}) - 0.2I]$, ε = dielectric constant, I = ionic strength.

AUXILIARY INFORMATION

METHOD/APPARATUS/PROCEDURE:

A saturating column method was used (1). The iodate concentrations in the saturated solutions were determined by titration with $Na_2S_2O_3$ standardized with KIO_3. Prior to the titration, excess KI was added and the solution acidified with dilute acetic acid.

SOURCE AND PURITY OF MATERIALS:

Barium iodate crystals were prepared by allowing dilute solutions of KIO_3 and of $BaCl_2$ to drip slowly into hot water.
Acetone used was of AnalaR reagent.

ESTIMATED ERROR:

Soly: nothing specified
Temp: ± 0.03°C (author)

REFERENCES:

1. Davies, C. W.
J. Chem. Soc. 1938, 277.

COMPONENTS:

(1) Barium iodate; $Ba(IO_3)_2$; [10567-69-8]

(2) Tetrahydrofuran; C_4H_8O; [109-99-9]

(3) Water; H_2O; [7732-18-5]

ORIGINAL MEASUREMENTS:

Miyamoto, H.

Nippon Kagaku Kaishi 1972, 659-61.

VARIABLES:

T/K = 298

tetrahydrofuran/mass % = 0 to 40

PREPARED BY:

Hiroshi Miyamoto

EXPERIMENTAL VALUES:

t/°C	Tetrahydrofuran		Barium Iodate
	mass %	mol % (compiler)	$10^4 c_1$/mol dm^{-3}
25	0	0	8.18
	5	1.3	5.77
	10	2.7	3.98
	15	4.2	2.71
	20	5.9	1.84
	25	7.7	1.23
	30	9.7	0.78
	40	14.3	0.24

AUXILIARY INFORMATION

METHOD/APPARATUS/PROCEDURE:

Excess $Ba(IO_3)_2 \cdot H_2O$ and solvent mixtures were placed in glass-stoppered bottles. The bottles were rotated in a thermostat at 25°C for 48 hours. After the saturated solution settled, the solution was withdrawn through a siphon equipped with a glass-sintered filter.
The iodate content was determined iodometrically.

SOURCE AND PURITY OF MATERIALS:

$Ba(IO_3)_2 \cdot H_2O$ was prepared by adding solutions of $BaCl_2 \cdot 2H_2O$ (Wako Co guarantee reagent) and KIO_3 (Wako Co guarantee reagent) to a large volume of water containing KNO_3. The precipitate was filtered off, washed and dried under reduced pressure.
Tetrahydrofuran was distilled from NaOH and then redistilled from sodium metal.

ESTIMATED ERROR:

Soly: not given
Temp: ± 0.02°C (author)

REFERENCES:

COMPONENTS:

(1) Barium iodate; $Ba(IO_3)_2$; [10567-69-8]

(2) 1,4-Dioxane; $C_4H_8O_2$; [123-91-1]

(3) Water; H_2O; [7732-18-5]

ORIGINAL MEASUREMENTS:

Davis, T. W.; Ricci, J. E.; Sauter, C. G.

J. Am. Chem. Soc. 1939, *61*, 3274-84.

VARIABLES:

T/K = 298

1,4-Dioxane/mass % = 0 to 100

PREPARED BY:

Hiroshi Miyamoto

EXPERIMENTAL VALUES:

t/°C	Dioxane		Barium Iodate
	mass %	mol % (compiler)	$10^4 c_1$/mol dm^{-3}
25	0	0	8.145
	10	2.2	4.742
	20	4.9	2.526
	30	8.1	1.217
	40	12.0	0.553
	50	17.0	0.278
	60	23.5	0.140
	70	32.3	0.110
	80	45.0	0.074
	90	64.8	0.0050
	100	100	0.0000

AUXILIARY INFORMATION

METHOD/APPARATUS/PROCEDURE:

The dioxane-water mixtures were made up by weight for each experiment, in glass-stoppered Pyrex bottles. The mixture was rotated with excess $Ba(IO_3)_2 \cdot H_2O$ for one or more days at 25°C.
Samples for analysis were withdrawn by suction through quantitative filters into calibrated pipets after allowing some time for the undissolved salt to settle.
The iodate solutions were first evaporated to dryness to expel the dioxane. The iodate content was then determined iodometrically.

SOURCE AND PURITY OF MATERIALS:

C.p. grade barium iodate monohydrate was washed 8 to 10 times with distilled water, sedimented in tall cylinders to remove smaller size particles, and dried at 100°C before use. Iodometric titration gave 96.5 % $Ba(IO_3)_2$ compared with 96.43% theoretical for the monohydrate.
Dioxane came from several sources. The first step in the purification was either a refluxing with metallic sodium for several hours or treatment with sodium wire at room temperature for several days. This Na treatment was then followed by distillation through a 61 cm packed distilled column. In some cases the dioxane was first treated with alkaline permanganate, distilled, refluxed with lime and then redistilled before being treated with sodium.

ESTIMATED ERROR:

Soly: nothing specified
Temp: ± 0.02°C (authors)

COMPONENTS:

(1) Barium iodate; $Ba(IO_3)_2$; [10567-69-8]

(2) 1,4-Dioxane; $C_4H_8O_2$; [123-91-1]

(3) Water; H_2O; [7732-18-5]

ORIGINAL MEASUREMENTS:

Monk, C. B.

J. Chem. Soc. 1951, 2723-6.

VARIABLES:

T/K = 298
1,4-Dioxane/mass % = 0 to 10.0

PREPARED BY:

Hiroshi Miyamoto

EXPERIMENTAL VALUES:

t/°C	Dioxane		Barium Iodate	
	mass %	mol % (compiler)	$10^4 c_1$/mol dm^{-3}	$10^{10} K^\circ_{s0}$/mol^{3}dm^{-9}
25	0	0	8.12	15.52
	2.2	0.46	7.15	10.67
	4.7	1.00	6.04	7.16
	9.4	2.08	4.72	3.13
	10.0	2.22	4.74	3.16

Thermodynamic solubility product constant, K°_{s0}, was calculated from

$$\log K^\circ_{s0} = \log [Ba^{2+}][IO_3^-]^2 - 3F$$

where $F = (78.54/\varepsilon)^{3/2}[I^{1/2}/(1 + I^{1/2}) - 0.2I]$, ε = dielectric constant, I = ionic strength.

AUXILIARY INFORMATION

METHOD/APPARATUS/PROCEDURE:

A saturating column method was used (1). The iodate concentrations in the saturated solutions were determined by titration with $Na_2S_2O_3$ standardized with KIO_3. Prior to the titration, excess KI was added and the solution acidified with dilute acetic acid.
The saturated solutions containing dioxane were corrected for peroxides by blank titration.

SOURCE AND PURITY OF MATERIALS:

Barium iodate crystals were prepared by allowing dilute solutions of KIO_3 and of $BaCl_2$ to drip slowly into hot water.
Dioxane used was of AnalaR reagent.

ESTIMATED ERROR:

Soly: nothing specified
Temp: ± 0.03°C (author)

REFERENCES:

1. Davies, C. W.
J. Chem. Soc. 1938, 277.

COMPONENTS:

(1) Barium iodate; $Ba(IO_3)_2$; [10567-69-8]

(2) Ethyl acetate; $C_4H_8O_2$; [141-78-6]

(3) Water; H_2O; [7732-18-5]

ORIGINAL MEASUREMENTS:

Monk, C. B.

J. Chem. Soc. 1951, 2723-6.

VARIABLES:

T/K = 298

Ethyl acetate/mass % = 0 to 6.1

PREPARED BY:

Hiroshi Miyamoto

EXPERIMENTAL VALUES:

t/°C	Ethyl Acetate		Barium Iodate	
	mass %	mol % (compiler)	$10^4 c_1$/mol dm^{-3}	$10^{10} K^\circ_{s0}$/mol^{3}dm^{-9}
25	0	0	8.12	15.52
	3.8	0.80	6.42	7.80
	6.1	1.31	5.59	5.20

Thermodynamic solubility product constant, K°_{s0}, was calculated from

$$\log K^\circ_{s0} = \log [Ba^{2+}][IO_3^-]^2 - 3F$$

where $F = (78.54/\varepsilon)^{3/2}[I^{1/2}/(1 + I^{1/2}) - 0.2I]$, ε = dielectric constant, I = ionic strength.

AUXILIARY INFORMATION

METHOD/APPARATUS/PROCEDURE:

A saturating column method was used (1). The iodate concentrations in the saturated solutions were determined by titration with $Na_2S_2O_3$ standardized with KIO_3. Prior to the titration, excess KI was added and the solution acidified with dilute acetic acid.

SOURCE AND PURITY OF MATERIALS:

Barium iodate crystals were prepared by allowing dilute solutions of KIO_3 and of $BaCl_2$ to drip slowly into hot water.
Ethyl acetate used was of laboratory grade.

ESTIMATED ERROR:

Soly: nothing specified
Temp: ± 0.03°C (author)

REFERENCES:

1. Monk, C. B. *Trans. Faraday Soc.* 1951, *47*, 285; Davies, C. W.; Waind, G. M. *J. Chem. Soc.* 1950, 301;
Davies, C. W. *ibid*, 1938, 277.

COMPONENTS:

(1) Barium iodate; $Ba(IO_3)_2$; [10567-69-8]

(2) N,N-Dimethylformamide; C_3H_7NO; [68-12-2]

(3) Water; H_2O; [7732-18-5]

ORIGINAL MEASUREMENTS:

Miyamoto, H.; Suzuki, K.; Yanai, K.

Nippon Kagaku Kaishi 1978, 1150-2.

VARIABLES:

T/K = 293, 298 and 303
Dimethylformamide/mass % = 0 to 40.35

PREPARED BY:

Hiroshi Miyamoto

EXPERIMENTAL VALUES:

t/°C	Dimethylformamide mass %	Dimethylformamide mol % (compiler)	Barium Iodate $10^4 c_1$/mol dm^{-3}
20	0	0	7.02
	4.92	1.26	5.50
	9.90	2.64	4.23
	14.98	4.16	3.10
	20.45	5.96	2.16
	25.29	7.70	1.62
	29.86	9.50	1.17
	39.90	14.00	0.48
25	0	0	8.18
	4.92	1.26	6.39
	10.05	2.68	4.86
	14.90	4.14	3.66
	20.06	5.82	2.75
	24.92	7.56	2.04
	29.99	9.55	1.45
	40.35	14.29	0.69
30	0	0	9.23
	4.98	1.26	7.13
	10.76	2.89	5.35
	15.02	4.17	4.11
	20.71	6.05	2.89
	25.20	7.67	2.30
	30.36	9.70	1.65
	40.28	14.25	0.81

AUXILIARY INFORMATION

METHOD/APPARATUS/PROCEDURE:

$Ba(IO_3)_2 \cdot H_2O$ crystals and solvent mixtures were placed in glass-stoppered bottles. The bottles were placed in a thermostat at a given temperature, and rotated for 72 hours. After the saturated solutions were obtained, the solutions were separated from the solid phase using a glass-filter. After the saturated solutions were diluted with water, the concentrations of the iodate were determined iodometrically. The solubility of $Ba(IO_3)_2$ was calculated from the observed values.

SOURCE AND PURITY OF MATERIALS:

Barium iodate was prepared by adding dilute solutions of $BaCl_2$ and KIO_3 to a boiled water. The product was washed and dried at room temperature. $Ba(IO_3)_2 \cdot H_2O$ was obtained. DMF (from Mitsubishi Gas Co) was distilled under reduced pressure. After the product was dried over Na_2CO_3, the distillation of the solvent was repeated 3 times.

ESTIMATED ERROR:

Soly: the probable errors of the observed mean values were within ± 0.1 x 10^{-5} mol dm^{-3}.
Temp: ± 0.02°C (authors)

REFERENCES:

COMPONENTS:

(1) Barium iodate; $Ba(IO_3)_2$; [10567-69-8]

(2) Urea; CH_4N_2O; [57-13-6]

(3) Water; H_2O; [7732-18-5]

ORIGINAL MEASUREMENTS:

Petersen, K. J.

K. Dan. Vidensk. Selsk. Mat-Fys. Medd. 1941, *18*, 21-4.

VARIABLES:

$T/K = 291.1$

$c_2/\text{mol dm}^{-3} = 0.000$ to 1.000

PREPARED BY:

Hiroshi Miyamoto

EXPERIMENTAL VALUES:

t/°C	Urea $c_2/\text{mol dm}^{-3}$	Barium Iodate $10^3 c_1/\text{mol dm}^{-3}$
17.9	0.000	0.6694
	0.200	0.6965
	0.400	0.7240
	0.600	0.7510
	0.800	0.7787
	1.000	0.8059

AUXILIARY INFORMATION

METHOD/APPARATUS/PROCEDURE:

Excess $Ba(IO_3)_2 \cdot H_2O$ and aqueous urea solution were placed in glass stoppered-bottles. The bottles were rotated in an electrically regulated water thermostat. Samples of the saturated solutions were analyzed after different times of rotation in order to make sure that saturation was attained.
The samples were sucked from the bottle through a porous glass filter into a pipet.
The iodate contents were determined by iodometry. Analyses and solubility measurements were done in duplicate.

SOURCE AND PURITY OF MATERIALS:

$Ba(IO_3)_2 \cdot H_2O$ was prepared from barium hydroxide and iodic acid. Urea (Kahlbaum, "für wissenschaftliche Zweeke") was used without further purification. It contained traces of calcium which could not be removed by recrystallization from alcohol. 1 to 3 mg of ash and 1 to 2 x 10^{-5} moles of calcium were found per mole of urea.

ESTIMATED ERROR:

Soly: within the limit of accuracy of the analytical method (author)

Temp: nothing specified

REFERENCES:

SYSTEM INDEX

Underlined page numbers refer to evaluation text and those not underlined to compiled tables. All compounds are listed as in Chemical Abstracts. Ternary involving water and other multi-component mixtures involving water are listed as component (1) (aqueous) + other component(s) and these occur after component (1) + water. For example, Barium chloride + water comes before Barium chloride (aqueous) + iodic acid, barium salt.

A

B

D

L

M

N

REGISTRY NUMBER INDEX

Underlined numbers refer to evaluations. Other numbers refer to compiled tables.

AUTHOR INDEX

Underlined numbers refer to evaluations. Other numbers refer to compiled tables.